Production of Biobutanol from Biomass

Scrivener Publishing
100 Cummings Center, Suite 541J
Beverly, MA 01915-6106

Publishers at Scrivener
Martin Scrivener (martin@scrivenerpublishing.com)
Phillip Carmical (pcarmical@scrivenerpublishing.com)

Production of Biobutanol from Biomass

Edited by

Arindam Kuila
Department of Bioscience & Biotechnology, Banasthali Vidyapith, India

and

Mainak Mukhopadhyay
Department of Biotechnology, Swami Vivekananda University, Kolkata, West Bengal, India

Scrivener Publishing

This edition first published 2024 by John Wiley & Sons, Inc., 111 River Street, Hoboken, NJ 07030, USA and Scrivener Publishing LLC, 100 Cummings Center, Suite 541J, Beverly, MA 01915, USA
© 2024 Scrivener Publishing LLC
For more information about Scrivener publications please visit www.scrivenerpublishing.com.

All rights reserved. No part of this publication may be reproduced, stored in a retrieval system, or transmitted, in any form or by any means, electronic, mechanical, photocopying, recording, or otherwise, except as permitted by law. Advice on how to obtain permission to reuse material from this title is available at http://www.wiley.com/go/permissions.

Wiley Global Headquarters
111 River Street, Hoboken, NJ 07030, USA

For details of our global editorial offices, customer services, and more information about Wiley products visit us at www.wiley.com.

Limit of Liability/Disclaimer of Warranty
While the publisher and authors have used their best efforts in preparing this work, they make no representations or warranties with respect to the accuracy or completeness of the contents of this work and specifically disclaim all warranties, including without limitation any implied warranties of merchantability or fitness for a particular purpose. No warranty may be created or extended by sales representatives, written sales materials, or promotional statements for this work. The fact that an organization, website, or product is referred to in this work as a citation and/or potential source of further information does not mean that the publisher and authors endorse the information or services the organization, website, or product may provide or recommendations it may make. This work is sold with the understanding that the publisher is not engaged in rendering professional services. The advice and strategies contained herein may not be suitable for your situation. You should consult with a specialist where appropriate. Neither the publisher nor authors shall be liable for any loss of profit or any other commercial damages, including but not limited to special, incidental, consequential, or other damages. Further, readers should be aware that websites listed in this work may have changed or disappeared between when this work was written and when it is read.

Library of Congress Cataloging-in-Publication Data

ISBN 978-1-394-17239-9

Cover image: Pixabay.Com
Cover design by Russell Richardson

Set in size of 11pt and Minion Pro by Manila Typesetting Company, Makati, Philippines

Printed in the USA

10 9 8 7 6 5 4 3 2 1

Contents

Preface		xiii
1	**Biobutanol: An Overview**	**1**
	Bidisha Saha, Debalina Bhattacharya	
	and Mainak Mukhopadhyay	
	1.1 Introduction	2
	1.2 General Aspects of Butanol Fermentation	3
	1.2.1 Microbes That Produce Butanol, Both in Their Wild Type and After Genetic Modification	3
	1.3 *Clostridium* Species That Produce ABE and Their Respective Metabolic Characteristics	4
	1.4 Traits of the Molecularly Developed Strain and the ABE-Producing Clostridia	8
	1.5 Substrate for ABE Fermentation in Research	9
	1.6 Problem and Limitation of ABE Fermentation	9
	1.7 The Development of Butanol from Designed and Modifying Biomass	10
	1.8 Butanol Production Enhancement Using Advanced Technology	12
	1.8.1 Batch Fermentation	12
	1.8.2 Fed-Batch Fermentation	16
	1.8.3 Continuous Fermentation	17
	1.8.4 ABE Fermentation with Butanol Elimination	27
	1.9 Utilizing Pre-Treatment and Saccharification to Produce Butanol from Lignocellulosic Biomass	29
	1.10 Eliminating CCR to Produce Butanol	29
	1.11 Butanol Production from Alternative Substrate to Sugar	30
	1.12 Economics of Biobutanol	31
	1.13 Future Prospects	33
	1.14 Conclusion	36
	References	37

2 Recent Trends in the Pre-Treatment Process of Lignocellulosic Biomass for Enhanced Biofuel Production — 47
Nikita Bhati, Shreya and Arun Kumar Sharma
- 2.1 Introduction — 48
- 2.2 Composition of Lignocellulosic Biomass — 49
- 2.3 Insight on the Pre-Treatment of LCB — 51
- 2.4 Physical Pre-Treatment Method — 54
  - 2.4.1 Extrusion Method — 54
  - 2.4.2 Milling Method — 55
  - 2.4.3 Ultrasound Method — 55
  - 2.4.4 Microwave Method — 56
- 2.5 Chemical Pre-Treatment Methods — 56
  - 2.5.1 Alkali Method — 56
  - 2.5.2 Acid Method — 57
  - 2.5.3 Organosolv Method — 58
  - 2.5.4 Ionic Liquids — 58
  - 2.5.5 Supercritical Fluids — 60
  - 2.5.6 Cosolvent Enhanced Lignocellulosic Fractionation — 61
  - 2.5.7 Low Temperature Steep Delignification — 62
  - 2.5.8 Ammonia Fiber Explosion — 62
  - 2.5.9 Deep Eutectic Solvents — 63
- 2.6 Biological Pre-Treatment Methods — 64
  - 2.6.1 Combined Biological Pre-Treatment — 66
- 2.7 Future Prospects — 66
- 2.8 Conclusion — 67
- References — 67

3 Current Status of Enzymatic Hydrolysis of Cellulosic Biomass — 77
Ram Bhajan Sahu, Janki Pahlwani and Priyanka Singh
- 3.1 Introduction — 77
- 3.2 Overview on Biofuels and Its Classification — 79
  - 3.2.1 First-Generation Biofuels — 79
    - 3.2.1.1 Advantage of First-Generation Biofuel — 81
    - 3.2.1.2 Limitation of First-Generation Biofuel — 81
  - 3.2.2 Second-Generation Lignocellulosic Biofuel — 82
    - 3.2.2.1 Different Types of Feedstocks for Second-Generation Biofuels — 82
    - 3.2.2.2 Advantages — 84
    - 3.2.2.3 Disadvantages — 84
  - 3.2.3 Third-Generation Biofuels — 85
    - 3.2.3.1 Advantages — 85

		3.2.3.2 Disadvantages	86
	3.2.4	Fourth-Generation Biofuels	87
3.3	Pre-Treatment Methodologies for Hydrolysis of Lignocellulosic Biomass		87
	3.3.1	Overview	87
	3.3.2	Structural Analysis for Cellulosic Hydrolysis	90
	3.3.3	Chemical Process for Pre-Treatment of Lignocellulose	91
		3.3.3.1 Dilute Acid Pre-Treatment Process	91
	3.3.4	Ionic Liquid as Pre-Treatment Agent	93
	3.3.5	Pre-Treatment Process with Alkali Agents	94
	3.3.6	Pre-Treatment with Ultrasonic Wave	96
3.4	Conclusion		97
	References		98

4 Present Status and Future Prospect of Butanol Fermentation — 105
Rashmi Mishra, Aakansha Raj and Satyajit Saurabh

4.1	Introduction	106
4.2	Biobutanol Production	107
	4.2.1 Microbes and Biobutanol Production	110
	4.2.2 Substrate for Biobutanol Production	111
	4.2.3 ABE Fermentation Process	112
	4.2.4 Recovery of Biobutanol from Fermentation Broth	112
4.3	Perspectives	115
	4.3.1 Substrate	116
	4.3.2 Alleviate Carbon Catabolite Repression	117
	4.3.3 Fermentation Improvement	118
	4.3.4 Strain Development	119
	4.3.5 Butanol Recovery	122
4.4	Conclusion	123
	References	124

5 Strategies of Strain Improvement for Butanol Fermentation — 133
Shreya, Nikita Bhati and Arun Kumar Sharma

5.1	Introduction	134
5.2	Background	136
5.3	Microorganism	136
5.4	ABE Fermentation	137
	5.4.1 The Obstacle in ABE Fermentation from *Clostridium* sp.	138
5.5	Selection of Biomass for the Production of Butanol	138
5.6	Processes Improvement	140
5.7	Strain Improvement	141

viii   Contents

	5.7.1	Mutagenesis	142
		5.7.1.1 Spontaneous Mutations	142
		5.7.1.2 Induced Mutation	143
	5.7.2	Strain Improvement Through Genetic Engineering	144
		5.7.2.1 Recombinant DNA Technology	148
	5.7.3	Genetic Engineering in *Clostridial* sp. for Improved Butanol Tolerance and Its Production	152
5.8		Production of Butanol From Bioethanol Through Chemical Processes	153
5.9		Advances in Genetically Engineered Microbes can Produce Biobutanol	154
5.10		Economics of Biobutanol Fermentation	155
5.11		Applications of Butanol	156
5.12		Butanol Advantages	157
5.13		Conclusion	157
		References	157

6 Process Integration and Intensification of Biobutanol Production — **167**

Moumita Bishai

6.1	Introduction	167
6.2	Biobutanol	169
6.3	Biobutanol Production and Recovery	170
6.4	Process Intensification	172
	6.4.1 PI Using Bioreactors	172
	6.4.2 PI Using Membranes	173
	6.4.3 PI Using Distillation	175
	6.4.4 PI Using Liquid–Liquid Extraction	176
	6.4.5 PI Using Adsorption	177
6.5	Process Integration	178
6.6	Conclusion	184
	References	185

7 Bioprocess Development and Bioreactor Designs for Biobutanol Production — **191**

Vitor Paschoal Guanaes de Campos, Johnatt Oliveira, Eduardo Dellossso Penteado, Anthony Andrey Ramalho Diniz, Andrea Komesu and Yasmin Coelho Pio

7.1	Introduction	191
7.2	Steps in Biobutanol Production	193
7.3	Feedstock Selection	194

	7.4	Microbial Strain Selection	196
	7.5	Solvent Toxicity	196
	7.6	Fermentation Technologies	197
	7.7	Butanol Separation Techniques	200
	7.8	Current Status and Economics	203
	7.9	Concluding Remarks	204
		References	204

8 Advances in Microbial Metabolic Engineering for Increased Biobutanol Production — 209
Mansi Sharma, Pragati Chauhan, Rekha Sharma and Dinesh Kumar

8.1	Introduction		210
8.2	Metabolic Engineering		212
	8.2.1	n-Butanol	212
	8.2.2	Isobutanol	214
8.3	Microorganisms for Butanol Production		215
	8.3.1	The *Clostridium* Species	218
	8.3.2	*Escherichia coli* Species	219
	8.3.3	Other Bacteria	219
	8.3.4	Biochemistry and Physiology	220
8.4	Metabolic Engineering of Clostridia		221
	8.4.1	Genetic Tools for Clostridial Metabolic Engineering	222
	8.4.2	Optimum Selectivity Techniques for Butanol Production	222
8.5	Metabolic Engineering of *Escherichia coli*		224
8.6	Microbial Strain		226
8.7	Butanol Tolerance Improvement Through Genetic Engineering		227
8.8	Economic Viability		228
8.9	Problems and Limitations of ABE Fermentation		228
8.10	Future Outlook		229
8.11	Conclusion		230
	Acknowledgment		231
	References		231

9 Advanced CRISPR/Cas-Based Genome Editing Tools for Biobutanol Production — 239
Narendra Kumar Sharma, Mansi Srivastava and Yogesh Srivastava

9.1	Introduction	240

9.2	Microorganisms as the Primary Producer of Biobutanol		241
9.3	Acetone–Butanol–Ethanol Producing *Clostridia* and Its Limitations		243
9.4	CRISPR–Cas System for Genome Editing		244
	9.4.1	CRISPR–Cas Mediated Strategies for Genome Editing for Biobutanol Production in Microorganisms	245
		9.4.1.1 Inhibition of Contentious Pathways	245
		9.4.1.2 Redirection of the Flux of Metabolic Pathways for Better Solvent Production	247
		9.4.1.3 Enhancement of Substrate Uptake	248
	9.4.2	Improvement of the Biofuel Production	248
		9.4.2.1 Off Targets in CRISPR–Cas System	248
		9.4.2.2 Using sgRNA Design to Reduce Off Target Effects	249
		9.4.2.3 Cas9 Modifications to Reduce Off-Target Effects	249
	9.4.3	Efficient and Modified Biomass "Designed" for Biobutanol Production	250
9.5	Conclusion		251
	References		252

10 Role of Nanotechnology in Biomass-Based Biobutanol Production — 255
Pragati Chauhan, Mansi Sharma, Rekha Sharma and Dinesh Kumar

10.1	Introduction		255
10.2	Nanoparticles for Producing of Biofuel		257
	10.2.1	Magnetic Nanoparticles	257
	10.2.2	Carbon Nanotubes	258
	10.2.3	Graphene and Graphene-Derived Nanomaterial for Biofuel	260
	10.2.4	Other Nanoparticles Applied in Heterogeneous Catalysis for Biofuel Production	262
10.3	Factors Affecting the Performance of Nanoparticles in Biofuel's Manufacturing		263
	10.3.1	Synthesis Temperature	263
	10.3.2	Synthesis Pressure	263
	10.3.3	Synthesis pH	263
	10.3.4	Size of Nanoparticles	264
10.4	Role of Nanomaterials in the Synthesis of Biofuels		264
10.5	Utilization of Nanomaterials in Biofuel Production		264

		10.5.1	Production of Biodiesel Using Nanocatalysts	264
		10.5.2	Application of Nanomaterials for the Pre-Treatment of Lignocellulosic Biomass	268
		10.5.3	Application of Nanomaterials in Synthesis of Cellulase and Stability	268
		10.5.4	Application of Nanomaterials in the Hydrolysis of Lignocellulosic Biomass	269
		10.5.5	Use of Nanotechnology in Bioethanol Production	269
		10.5.6	Upgradation of Biofuel by Using Nanotechnology	272
		10.5.7	Nanoparticle Use in Biorefineries	273
	10.6	Nanotechnology in Bioethanol/Biobutanol Production		274
	10.7	Future Perspective		277
	10.8	Conclusion		278
		Acknowledgment		279
		References		279

11 Commercial Status and Future Scope of Biobutanol Production from Biomass 283

Arunima Biswas

- 11.1 Introduction — 284
- 11.2 Biobutanol—Its Brief Background Story — 286
- 11.3 Commercial Aspect of Biobutanol Production from Biomass: Strength Analysis — 287
- 11.4 Commercial Aspect of Biobutanol Production from Biomass: Weakness Analysis — 290
- 11.5 Commercial Aspect of Biobutanol Production from Biomass: Opportunities and Challenges — 293
- 11.6 Discussion: Evaluating the Future Prospects of Biobutanol — 296
- Acknowledgment — 298
- References — 298

12 Current Status and Challenges of Biobutanol Production from Biomass 301

Ram Bhajan Sahu and Priyanka Singh

- 12.1 Introduction — 301
- 12.2 Overview of Biofuel — 303
  - 12.2.1 History for Biofuel — 304
- 12.3 Classification of Bioethanol — 306
  - 12.3.1 First-Generation of Ethanol — 306
  - 12.3.2 Second-Generation Bioethanol — 308
  - 12.3.3 Third-Generation Bioethanol — 309

		12.3.4	Fourth-Generation Bioethanol	309
	12.4	Production of Biobutanol		309
		12.4.1	Pre-Treatment Stages	310
		12.4.2	Enzymatic Hydrolysis Stage	312
		12.4.3	Fermentation Stage	312
		12.4.4	Separation Stage	312
		12.4.5	Production of Butanol from Genetically Improved Strains	313
	12.5	Conclusion		317
		References		318
13	**Biobutanol: A Promising Liquid Biofuel**			**323**
	Aakansha Raj, Tasnim Arfi and Satyajit Saurabh			
	13.1	Introduction		323
		13.1.1	First-Generation Biofuels	324
		13.1.2	Second-Generation Biofuels	326
		13.1.3	Third-Generation Biofuels	326
		13.1.4	Fourth-Generation Biofuels	326
	13.2	Biobutanol		327
	13.3	Biorefinery and Biobutanol Production		329
		13.3.1	Substrates and Their Pre-Treatment for Biobutanol Production	329
			13.3.1.1 Substrate	329
			13.3.1.2 Pre-Treatment of Substrates	333
		13.3.2	Microorganisms	342
		13.3.3	Acetone–Butanol–Ethanol Fermentation	343
	13.4	Commercial Importance of Biobutanol		343
	13.5	Conclusion		346
		Abbreviations		346
		References		347

Index   **355**

Preface

N-butanol is a bulk chemical that is used as an industrial solvent and as a component in paint, coating, and adhesives, among other things. When compared to other biofuels, biobutanol has the advantages of being immiscible in water, having a higher energy content, and having a lower vapor pressure. There are various benefits to producing biobutanol from lignocellulosic biomass. However, there are challenges in producing butanol from lignocellulosic biomass, such as biomass's complex structure, low butanol yield, and high cost of production, etc. This book discusses all of these issues, as well as the current state and future prospects of lignocellulosic biobutanol production.

The 13 chapters herein discuss the current technology and future prospects of biobutanol production. The first four chapters provide an overview of the current technological status, while the next six chapters discuss different strategies for enhanced biobutanol production from lignocellulosic biomass. The last three chapters present the industrial status and techno-economic analysis of lignocellulosic biobutanol production.

This book is useful for students and researchers in the various branches of life sciences, including environmental biotechnology, bioprocess engineering, renewable energy, chemical engineering, nanotechnology, biotechnology, microbiology, etc.

We are grateful to Wiley and Scrivener Publishing, especially Linda Mohr and Martin Scrivener, for their cooperation and assistance in the timely publication of this book. We would like to express our gratitude to the writers and contributors for their efforts as well.

Dr. Arindam Kuila
Dr. Mainak Mukhopadhyay

1
Biobutanol: An Overview

Bidisha Saha[1], Debalina Bhattacharya[2] and Mainak Mukhopadhyay[3]*

[1]Department of Biotechnology, JIS University, Kolkata, West Bengal, India
[2]Department of Microbiology, Maulana Azad College, Kolkata, West Bengal, India
[3]Department of Bioscience, JIS University Kolkata, West Bengal, India

Abstract

There is hope that butanol can help offset the decline in supplies of fossil fuel-based liquid fuels. When combined with liquid fuels, butanol can be utilized as a biofuel at any concentration. The majority of the fermentative organisms employed in biobutanol synthesis are clostridia. Acetone–butanol–ethanol (ABE) fermentation is a potential niche market for these organisms, as they can convert various forms of renewable biomass into butanol. When compared with other biofuels like ethanol, butanol has various advantages. Inefficient product inhibition and heterofermentation contribute to low productivity and yield, increasing production costs for ABE fermentation. High-yield butanol synthesis has thus far relied on the application of molecular biological approaches and fermentation engineering strategies. In order to convert agricultural waste into a usable feedstock for butanol manufacturing, scientists have recently been studying methods of pre-treatment and enzymatic saccharification. This article summarizes previous studies on the topic, covering topics such as metabolic profiles and traits of clostridia that produce ABE. The study also discusses the evolution of ABE fermentation in terms of the development of extremely effective butanol production processes, including batch, continuous cultures, fed-batch with the addition of butanol removal, and the development of butanol production from biomass resources or substitute substrates to sugars.

Keywords: ABE fermentation, biobutanol production, *Clostridia* sp., fermentation process, biofuel

---

Corresponding author: m.mukhopadhyay85@gmail.com

1.1 Introduction

Increased demand for energy sources due to industrialization and motorization has been caused by the overconsumption of fuels in the form of petroleum based products as the energy source for operating a variety of engines. Fossil fuels currently account for 80% of all major energy sources used worldwide, with the transportation sector alone using about 58% of them. Although fossil fuels provide a reliable supply of energy for most of the world, they also play a significant role in the production of greenhouse gases (GHG), biodiversity loss, sea level rise, glacier retreat, and other problems. Alternative energy sources have drawn more and more attention globally as a result of anticipating the problems brought on by energy security, climatic changes, and rising raw material prices.

The shift to sustainable and environment renewable resources is essential to addressing the fuel scarcity, environmental considerations, and global warming. The usage of sustainable and renewable energy will become more prevalent in the future to meet the world's energy needs and protect the environment for coming generations. Butanol is particularly intriguing among the renewable energy sources that could be used.

Butanol has some characteristics that render it suitable for use as a conventional diesel biofuel and as a biofuel for existing automobiles, including a high energy content and low pollution output. Additionally, Butanol is nonflammable and has a low vapour pressure, which makes it safe to handle and store [1]. In the chemical and cosmetics sectors, butanol is used as a solvent [2]. By fermenting sugars from biomass, such as algal biomass [3], cane molasses [4], cassava starch [5], cheese whey [6], corn stover [7], inulin [8], palm oil mill effluent [9], sugarcane bagasse (SCB) [10], and food waste, an alcohol known as biobutanol (or butyl alcohol) can be synthesized. Acetone–butanol–ethanol (ABE) fermentation is the common name for this process [11].

Traditional fermentation processes employed on an industrial scale in the 20th century include the employment of a clostridia strain to produce butanol from biomass via ABE fermentation. This procedure can be carried out using a variety of bacteria strains, with clostridia being the most popular option. The two unique steps of acidogenesis and solventogenesis define the metabolism of clostridia, stringent anaerobe bacteria. In the initial stage, known as acidogenesis, carbon dioxide, hydrogen, organic acids are the primary end products. The subsequent process, called "solventogenesis," involves the reabsorption of the acids that were used in the formation of acetone, butanol, and ethanol (or isopropanol, which is produced by some strains of *Clostridium beijerinckii* in replacement of acetone).

A drastic alteration in the gene expression profile is what drives the transition from the acidogenic to the solventogenesis phase [12, 13]. The fundamental issue with ABE fermentation is the solvent's toxicity, which prevents clostridia from metabolizing ABE at concentrations of 20 g/L or higher. This inhibition restricts how effectively clostridia utilize the carbon sources throughout fermentation [14]. In addition, when created at a specific quantity during fermentation, butanol is poisonous for clostridia cells. Substrate cost, butanol output and productivity, and separation and purification expenses are the three key elements that define how economically viable ABE fermentation is in producing butanol.

1.2 General Aspects of Butanol Fermentation

1.2.1 Microbes That Produce Butanol, Both in Their Wild Type and After Genetic Modification

Wild type microorganisms; The main microbe utilized to produce butanol is clostridium. Among these, *C. beijerinckii, Clostridium acetobutylicum* are often and effectively employed in laboratory applications. A gram-positive, strictly anaerobic rod-shaped bacterium called clostridia produces butanol [12, 15, 16]. *Clostridium* spp. that produce ABE are capable of fermenting a wide variety of carbon sources, including fructose, galactose, glucose, glycerol, inulin, mannose, sucrose, xylose etc. [10, 17]. Feedstocks for fermentation production employing butanol-producing clostridia frequently consist of a combination of natural carbohydrates and sugars such as sweet sorghum [18], and a variety of agricultural [19, 20] and residential wastes [21, 22]. These cheap and inedible feedstocks could make butanol fermentation more profitable.

Genetically Modified microorganisms; Improving a microbe's output is achieved mostly through adjusting metabolic activity in order to more efficiently collect the target products and generate fewer unwanted byproducts [22]. Improving butanol synthesis through genetic alteration of ABE-producing clostridia often involves the insertion of heterogeneous genes, the amplification of genetic alterations, or gene deletion and reduction. Overexpression of the expression cassette and thiolase genes led to a considerable increase in butanol synthesis (17.4 g/L) when Wang *et al.* (2017) [23] altered a critical gene in *Clostridium saccharoperbutylacetonicum* N1–4. Recently, improved butanol synthesis and protection against the "acid crash" were seen after overexpressing *adh*E2 and *ctf*AB in *C. beijerinckii*

CC101-SV6. Inhibitor product resistance and butanol concentration are both expressed in *C. beijerinckii* CC101-SV6 [10]. In addition, a number of metabolic pathways have been screwed with by the application of integrational plasmid technology [24]. Due of the intricacy of clostridia genes, researchers are now studying other species (such *Escherichia*) as suitable hosts for butanol-producing genes [25]. In order to prevent byproducts like pyruvate and butanoate while enhancing AdhE2 activity, highest butanol percentage (18.3 g/L) was achieved using metabolic engineering of *Escherichia coli* [26]. It was investigated whether metabolic engineering of *C. acetobutylicum* might transform acetone into isopropanol during ABE fermentation since isopropanol, as opposed to the acetone formed by *C. acetobutylicum* during ABE fermentation, and could be utilized as fuel [27]. That is why researchers endeavored to optimize the *C. acetobutylicum* DSM 792 strain [28]. Using genetically modified microorganisms to improve butanol fermentation is difficult, despite numerous experiments in the field [25]. These findings offer crucial guidelines for further investigation into strains that produce genetically engineered hyper butanol (Table 1.1).

1.3 *Clostridium* Species That Produce ABE and Their Respective Metabolic Characteristics

The clostridia that make ABE need both carbon and electrons to function biologically. The Emden–Meyerhof–Parnas (EMP) pathway transforms glucose into pyruvate. Two molecules of ATP are produced during glycolysis [29]. Carboxylic acid first forms, followed by the formation of pyruvate. The two distinct metabolic processes in clostridia that primarily result in the production of ABE are acidogenesis, in which a variety of organic acids are produced, and solventogenesis, in which solvents are polymerized while also recycling the derived organic acids. Both of these processes are accompanied by an increase in the pH of the surrounding environment that is fueled by acid consumption [30]. A rapid alteration in the patterns of gene expression causes the two phases to switch (Figure 1.1). Using energy produced during the acidogenesis phase by the synthesis of organic acids, primarily butyrate and acetate, which are formed from butyryl-CoA and acetyl-CoA, respectively, by CoA transferase or the inverse processes of each organic acid's biosynthesis, the cells quickly enlarge while focusing on the flow of carbon. The pH of the culture broth is lowered by these organic acids. Something occurs as the cell growth nears a standstill phase. The decrease in organic acids causes the pH of the culture broth to increase. Finally, certain dehydrogenases would change acetyl-CoA to ethanol and

Table 1.1 Genetically modified butanol-producing strains.

Strains	Substrate	Remark	The temperature and time parameters are as follows	Butanol concentration (g/l)	Total solvent concentration (g/l)	References
C. acetobutylicum ATCC 824	Glucose	Expression of adh from C. beijerinckii NRRL B593	Fed-batch fermentation mode for 48 hours at 37°C with gas stripping	25.10	35.60	[27]
C. acetobutylicum DSM 792			Batch fermentation mode for 210 hours at 37°C	10.80	18.00	[28]
C. beijerinckii CC101	Glucose	Overexpressing adhE2 and ctfAB	Batch fermentation mode for 72 hours at 37°C with CaCO3.	12.00	No Data	[10]

(Continued)

Table 1.1 Genetically modified butanol-producing strains. (Continued)

Strains	Substrate	Remark	The temperature and time parameters are as follows	Butanol concentration (g/l)	Total solvent concentration (g/l)	References
C. pasteurianum	Glucose	hydA, rex, and dhaCBE were removed out	Batch fermentation mode for 210 hours at 37°C	9.80	No Data	[90]
C. saccharoperbutylacetonicum N1-4	Glucose	Increased levels of EC expression (thl-hbd-crt-bcd)	Batch fermentation mode for 72 hours at 30°C	17.40	30.60	[23]
	Glucose	The butR smr gene was introduced to control adhE and bdhA expression.	Batch fermentation mode for 72 hours at 30°C	16.50	No Data	[91]
	Sucrose	The impact of CCR was reduced after scrR was deleted.	Batch fermentation mode for 96 hours at 35°C	17.00	26.50	[92]

(Continued)

Table 1.1 Genetically modified butanol-producing strains. (*Continued*)

Strains	Substrate	Remark	The temperature and time parameters are as follows	Butanol concentration (g/l)	Total solvent concentration (g/l)	References
C. tyrobutyricum	Glucose	The *cat1* gene was switched out for the *adhE1/adhE2* genes.	Batch fermentation mode for 72 hours at 20°C	26.20	35.10	[93]
	Sucrose, sugarcane	*ScrB, ScrA,* and *ScrK* Overexpression together along with adhesion to *adhE2*	Batch fermentation mode for 6 days at 30°C	16.00	21.80	[94]

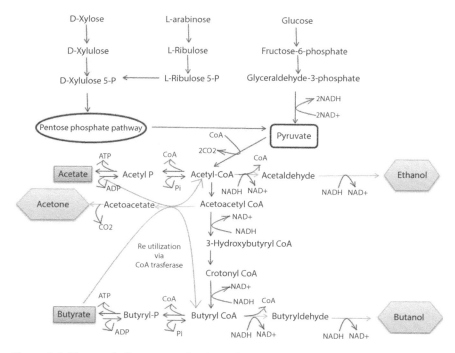

Figure 1.1 The metabolic process of ABE-producing clostridia is depicted below; the violet region explains the production of organic acids during acidogenesis, and the green region depicts the production of solvents during solventogenesis.

butyryl-CoA to butanol. Ferredoxin, a protein that functions as an electron transporter, is crucial for measuring the electron flux. During the process of acidogenesis, which results in the production of hydrogen molecules, extra electrons are transferred from reduced ferredoxin to hydrogen atoms [31]. Solventogenesis, which results in an excess of nicotinamide adenine dinucleotide being generated and employed as a reduction in absorbance in the formation of ethanol and butanol, considerably slows down this process. In conclusion, both carbon and electron transport regulate the metabolism of clostridia that produce ABE.

1.4 Traits of the Molecularly Developed Strain and the ABE-Producing Clostridia

Clostridium saccharoperbutylacetonicum N1-4 (ATCC 13564), *C. acetobutylicum* ATCC 824T, and *C. beijerinckii* NCIMB 8052 are the three main wild-type strains of the clostridia that produce the chemical and are frequently employed in studies on ABE fermentation [32–34]. These wild-type strains, among other characteristics, have a poor tolerance to butanol, a low yield, and a weak butanol titer. Mutagenesis and genetic engineering are two methods being researched to increase butanol output. The *C. acetobutylicum* strain's genome was described by Nolling *et al.* in 2001 [35], which sparked interest in using genetic engineering methods on the strain. To boost butanol synthesis by 32% over the strain, Tomas *et al.* [36] induced the expression of the groESL operon beneath the thiolase promoter of clostridium to produce *C. acetobutylicum* ATCC 824 (pGROE1). N-methyl-N-nitro-N-nitrosoguanidine and specific enrichment on the glucose analogue 2-deoxyglucose were used by Annous *et al.* [37] to transform *C. beijerinckii* NCIMB 8052 into a *C. beijerinckii* BA101 strain that produces hyper butanol. The highest concentration of butanol (20.9 g/L) has been discovered to be produced by the BA101 strain [38]. To combat issues like butanol tolerance and inadequate butanol output, genetic engineering is now used.

1.5 Substrate for ABE Fermentation in Research

Due to glucose's highly effective use by clostridia that produce ABE, it has been used frequently as a feedstock for ABE fermentation. Recent research has shown that even while amylases saccharify starch to glucose, starch can still be employed directly as a substrate for ABE fermentation [39, 40]. A favorable substrate, on the other hand, has been suggested for lignocellulosic biomass, which includes agricultural wastes like wheat and rice straw because of its abundance and inedibility [41]. Nevertheless, biomass that also contains cellulose and hemicellulose is where lignocellulose is found. Because of this, the ultimate product of saccharified lignocellulose would be a mixture of hexoses (such as glucose and fructose) and pentoses (such as xylose and arabinose), both of which are fermentable sugars. In order to establish a highly effective butanol synthesis method using heterogeneous sugars as substrates, current research into ABE fermentation is described in Sections 1.8 and 1.10.

1.6 Problem and Limitation of ABE Fermentation

The following issues arise in typical batch ABE fermentation and limit the industrial use of butanol as a fuel substitute for fossil fuels [41, 42]:

- Reduced butanol concentration levels (nearby 20 g/L) due to butanol feedback inhibition;
- Minimal butanol yields (nearby 0.35 g/g) due to hetero fermentation;
- Low volumetric butanol efficiency (nearby 0.5 g/L/h) due to cell reduced titre;
- Massive price of butanol recovery (traditional distillation is energy required);

The few occurrences of ABE fermentation using lignocellulosic biomass as the fermentative substrate come through after-enzymatic saccharification, even though the majority of clostridia that produce ABE are unable to use lignocellulose directly [33, 43, 44]. Despite being a practical method for turning lignocellulosic biomass into fermentable sugars, enzymatic saccharification still faces challenges due to the high cost of the various enzymes, such as cellulases and hemicellulases, used in the process. The efficient utilization of lignocelluloses as a feedstock for ABE fermentation requires the development of fermentative techniques and less expensive substrate conversion. To overcome these difficulties, the researchers focused on creating fermentation technology for butanol production as well as genetically engineered metabolic butanol production from planned and modified biomass.

1.7 The Development of Butanol from Designed and Modifying Biomass

Biomass with cosubstrates like acid compounds and other low-cost products generated during hydrolysis is called "designed biomass." With this modified biomass, butanol production increases while fermentation expenses decrease. Hence, during the process of being developed, utilizing acetic acid as the cosubstrate with glucose promotes metabolic efficiency and enhances solvents generation [24]. The addition of acetic acid greatly increases butanol and ABE synthesis, as discovered by Gao et al. (2016) [45]. When lactic acid was included in the specified biomass, *C. saccharoperbutylacetonicum* N1-4 was capable of producing butanol [46, 47]. Table 1.2 compiles the results of several research projects into butanol synthesis from engineered biomass.

Table 1.2 Research on butanol fermentation using designed and modified substrates.

Strains	Substrates	The temperature and time parameters are as follows	Butanol concentration (g/l)	Total solvent concentration (g/l)	Ref
Clostridium saccharoperbutyl-acetonicum N1-4	Glucose & acetic acid	pH stat batch fermentation mode for 48 hours at 30°C	15.13	24.37	[45]
		Fed-batch fermentation mode for 72 hours at 30°C	13.90	No Data	[11]
	Glucose & butyric acid	Batch fermentation mode for 120 hours at 30°C	17.76	23.51	[95, 96]
		Batch fermentation mode at 30°C	13.00	15.40	[97]
	Glucose & lactic acid	pH stat batch fermentation mode for 48 hours at 30°C	15.50	19.90	[46]
	Pentose & lactic acid	Fed-batch fermentation mode for 72 hours at 30°C	15.60	19.33	[47]

1.8 Butanol Production Enhancement Using Advanced Technology

Many studies have concentrated on increasing concentration, productivity, and yield while using various culture methods and environmental conditions, including batch fermentation, continuous fermentation, fed-batch fermentation and butanol production processes combined with butanol recovery. The production of butanol in batch, continuous culture, fed-batch modes while utilizing integrated butanol recovery processes has been the main emphasis of this section.

1.8.1 Batch Fermentation

Since batch culture is the most straightforward of the three culture types, it has undergone extensive research for a number of factors, including pH regulation, the ratio of carbon to nitrogen sources (C/N ratio), partial pressures of hydrogen or carbon monoxide in the fermenter headspace, and the addition of electron carriers (Table 1.3). At comparatively more and less pH, *C. acetobutylicum* and *C. beijerinckii* played a crucial role in acidogenesis and solventogenesis, respectively. The C. acetobutylicum ATCC 824T strain produced just 1 g/L of ABE while producing nearly 20 g/L of total organic acids at a pH of 6.0. The organism produced around 17 g/L of ABE while only 3 g/L of total organic acids were synthesized at a pH of 4.5.

On the other hand, it has been thought that the partial pressure of CO and H2 in the fermentor headspace, along with the introduction of electron carriers, might significantly alter the electron flow. In batch cultures of *C. acetobutylicum* ATCC 824T [48], *C. acetobutylicum* ATCC 4295 [49], and *C. saccharoperbutylacetonicum* ATCC 27021 [50] increasing the pressure of CO or H2 in the fermentor headspace enhanced the butanol yield, sometimes with lower yields of acetone or acids. This is because carbon monoxide (CO) and hydrogen (H2) have been implicated as hydrogenase inhibitors that produce excess electron evolution. Besides, it has been demonstrated that a number of electron carriers, including methylene blue and methyl viologen (MV), work as reducing agents for clostridia that make ABE and boost ABE production and yields. Previous research indicated that MV shifted metabolic flow away from acetone production and towards butanol production, indicating that the extra reducing power supplied may be used to produce butanol. Transcriptome explanation of batch fermentation by *C. acetobutylicum* ATCC 824T with 1 mM MV showed that MV might decrease the expression of the sol operon, which includes

Table 1.3 Examining the impact of temperature, pH, and other environmental factors on solventogenic and acidogenic acetone-butanol-ethanol (ABE) fermentation in batch cultures.

Strains	Factors	Substrate	pH	Conditions	Trait	Ref
Clostridium acetobutylicum ATCC 824T	Addition of methyl viologen (MV)	Glucose	5.0	When MV is not supplied	Minimal butanol yield (0.41mol/mol)	[98]
				When 1mM MV is added	Maximum butanol production (0.58mol/mol)	
				When MV is not supplied	The presence of a significant amount of acetic acid (73mM) Minimal butanol concentration (<110mM)	[51]
				When 1mM MV is added after 7.5h	Concentrations of acetic acid that are not too high (<16mM) Maximum butanol concentration (150mM)	
	Addition of methylene blue (MB)		-	When MB is not added	Minimal concentrations of ABE (13.4g/l)	[99]
				When 4g/l MB is added	Maximum concentrations of ABE (23.1g/l)	

(Continued)

Table 1.3 Examining the impact of temperature, pH, and other environmental factors on solventogenic and acidogenic acetone-butanol-ethanol (ABE) fermentation in batch cultures. (*Continued*)

Strains	Factors	Substrate	pH	Conditions	Trait	Ref
	Carbon : Nitrogen (C/N Ratio)		-	With a molar ratio of ammonium to glucose of 0.16, the C/N ratio is 6.25.	Maximum yield of ABE (0.545 mol/mol) Minimal acid yield (0.550 mol/mol)	[100]
				With a molar ratio of ammonium to glucose of 1.52, the C/N ratio is 0.658.	Minimal yield of ABE (0.211 mol/mol) Maximum acid yield (0.884 mol/mol)	
	Hydrogen challenge		-	With Hydrogen challenge at 274–1479 kPa	Maximum yield of acetic acid (0.168 C-mol/C-mol) Minimal yield of butanol (0.298 C-mol/C-mol)	[48]
				Without Hydrogen challenge	Maximum yield of butanol (0.345-0.361 C-mol/C-mol) Minimal yield of acetic acid (0.091-0.130 C-mol/C-mol)	

(*Continued*)

Table 1.3 Examining the impact of temperature, pH, and other environmental factors on solventogenic and acidogenic acetone-butanol-ethanol (ABE) fermentation in batch cultures. (*Continued*)

Strains	Factors	Substrate	pH	Conditions	Trait	Ref
Clostridium acetobutylicum ATCC 824T	pH	Glucose	4.5	-	High concentration of ABE (17 g/L) Low concentration of acid (<3g/l)	[101]
			6.0	-	High concentration of acid (17 g/L) Low concentration of ABE (<3g/l)	
Clostridium beijerinckii NCIMB 8052		Lactose	5.0	-	High concentration of ABE (5.1 g/L) Low concentration of acid (3.4g/l)	[102]
			7.0		High concentration of acid (6.3g/L) Low concentration of ABE (0.075g/l)	

*ctf*AB (encoding CoA transferase), *adh*E1 (encoding aldehyde/alcohol dehydrogenase), *adc* (encoding acetoacetate decarboxylase), whereas The expression of *adh*E2 (encoding aldehyde/alcohol dehydrogenase) was increased by more than a factor of 100 during MV-supplemented batch fermentation [51, 52].

Batch cultures have been demonstrated to benefit from a variety of crucial properties, such as low pH, a reasonable C/N ratio, an increase in CO and H_2 partial pressures, and the presence of electron carriers, as indicated in the opening sentence of article. Three problems continue to limit the amount of butanol that can be produced and how productively it can be produced in batch cultures: substrate inhibition [53, 54], carbon catabolite repression (CCR) of consumption of less preferred sugars caused by a more preferred sugar, and product inhibition by acquired butanol at preliminary good level [55].

1.8.2 Fed-Batch Fermentation

Benefits of fed-batch cultures over batch cultures include the removal of substrate inhibition due to factors like large initial intensity, nutrient supplementation during fermentation, the use of elevated substrates (such as starch), and the avoidance of CCR due to the use of further preferable sugars (such as glucose). To take advantage of these benefits, a number of fed-batch cultures have been developed for ABE fermentation.

Sugars like glucose and xylose usually present in higher concentrations to cause substrate inhibition, result in a prolonged lag phase and decreased butanol production [53, 54]. The ideal sugar concentration is normally between 50g/L and 60g/L, even though a butanol concentration of more than 10g/L generated may stop cell proliferation, sugar ingestion, and further butanol production. This prompted in-depth study of fed-batch cultures for the generation of butanol, which led to the invention of fresh methods for butanol recovery to prevent substrate inhibition and for sugars and other nutrients to release butanol inhibition, as mentioned in Section 1.8.4. However, a number of organic acids, including acetic acid [56], butyric acid [57], and lactic acid [58, 59], as well as cosubstrates like glucose or arabinose, are feasible substrates for butanol synthesis when taking into account the reutilization pathways used by ABE-producing clostridia. However, it has been shown that high concentrations of these organic acids at pH values of 4.76, 4.82, and 3.79, respectively, result in substrate inhibition. The pH-stat is helpful because it enables continuous management of the levels of organic acids in the broth by only measuring the pH. The *C. saccharoperbutylacetonicum* N1-4 (ATCC 13564) strain was able to produce approximately 16 g/L of butanol by converting butyric acid [57] or lactic acid [58] to butanol in a fed-batch growth method.

Batch ABE fermentation of the *C. acetobutylicum* ATCC 824T strain is reported to exhibit CCR of xylose intake (a less preferred sugar) when more than 15 g/L glucose (a higher preferred sugar) is present [55]. For their study, Fond *et al.* used a fed-batch culture in which the *C. acetobutylicum* ATCC 824T strain was provided 15 g of glucose and xylose per day at a consistent rate. While simultaneously producing 12 g/L butanol, this fed-batch culture was able to consume carbohydrates at concentrations of less than 0.1 g/L glucose and 0.7 g/L xylose [60]. Many fed-batch cultures, as detailed in the preceding and following paragraph, can increase butanol synthesis by utilizing organic acids and emitting CCR to prevent substrate inhibition. Moreover, fed-batch culture using butanol recovery techniques has been found to be more effective than fed-batch culture without them for butanol production.

1.8.3 Continuous Fermentation

The disadvantage of butanol inhibition can be mitigated by diluted butanol in the broth with fresh medium in a continuous culture system, making the process more conducive to high butanol production. Continuous culture is preferable than batch and fed-batch cultures in terms of operational stability, even if the butanol concentration in continuous culture should be lower than those in those two types of cultures. As a result, much effort is put into researching continuous cultures in order to build a highly productive butanol manufacturing process.

It is not uncommon for a continuous-chemostat culture to reach a steady state after three or more medium changes, with respect to cell, substrate, and product concentrations in the broth. Some of the variables that have been studied in relation to ABE fermentation in continuous-chemostat cultures include: CO challenge, dilution rate, electron carriers, nutrients, pH, substrate concentration (Table 1.4). High butanol productivity highly depends on pH regulation. Continuous cultures of *C. acetobutylicum* using glucose as the feedstock showed metabolic flows towards butanol formation and acid generation at pH ranging from as 4.3 to 6.0 [61, 62]. However, *C. beijerinckii* and *C. saccharoperbutylacetonicum* from glucose and xylose respectively [63, 64], generated 0.34 g/L/h and 0.529 g/L/h of butanol at moderate pH values of 5.5 and 5.6, and 0.14 g/L/h and 0.199 g/L/h of butanol at low pH values of 5.0 and 4.6. Additionally, it has been observed that employing *C. acetobutylicum* and *C. beijerinckii*, CO challenge, yeast extract, ρ-aminobenzoic acid, NH_4Cl, and greater glucose concentrations can boost butanol or ABE productivity from glucose [62, 63, 65, 66].

Table 1.4 Investigating the effects of temperature, pH, and other environmental factors on solventogenic and acidogenic continuous cultures of acetone-butanol-ethanol (ABE) fermentation.

Strains	Factors	Substrates	D/h	pH control	Other conditions	Traits	Ref.
C. acetobutylicum ATCC 824T	Dilution rate	Glucose (40g/L)	0.06	4.8	-	Productivity of ABE is high (0.75 g/L/h) Productivity of acid is low (0.15 g/L/h)	[118]
			0.1			Productivity of ABE is low (0.194 g/L/h) Productivity of acid is high (0.29 g/L/h)	
	Addition of yeast extract, CO challenge, Glucose concentration, Dilution rate, pH	Glucose (44 mM)	0.25	4.5	Without CO challenge	Productivity of Butyric acid and acetic acid are high (0.626g/L/h), (0.290g/L/h). Productivity of butanol and acetone are low (0.0853g/L/h), (0.0231g/L/h).	[119]

(*Continued*)

Table 1.4 Investigating the effects of temperature, pH, and other environmental factors on solventogenic and acidogenic continuous cultures of acetone-butanol-ethanol (ABE) fermentation. (*Continued*)

Strains	Factors	Substrates	D/h	pH control	Other conditions	Traits	Ref.
		Glucose (139 mM)	0.125		With CO challenge	Productivity of butanol is high (0.211g/L/h) Productivity of Butyric acid, acetic acid, acetone are low (0.147 g/L/h), (0.138g/L/h), (0.146g/L/h).	
					Without CO challenge	Productivity of butanol and acetic acid are high (0.341 g/L/h), (0.0529g/L/h). Productivity of Butyric acid and acetic acid are low (0.229g/L/h), (0.202g/L/h).	

(*Continued*)

Table 1.4 Investigating the effects of temperature, pH, and other environmental factors on solventogenic and acidogenic continuous cultures of acetone-butanol-ethanol (ABE) fermentation. (*Continued*)

Strains	Factors	Substrates	D/h	pH control	Other conditions	Traits	Ref.
		Glucose (139 mM)	0.25	4.5	Without CO challenge, With 5 g/L yeast extract	Productivity of butanol, butyric acid, acetic acid and acetone are high (0.514g/L/h), (0.667g/L/h), (0.537g/L/h), (0.143g/L/h).	
				6.0	Without CO challenge	Productivity of Butyric acid and acetic acid are high (1.54g/L/h), (0.738g/L/h). Productivity of butanol, acetic acid, acetone are low (0.09 g/L/h), (0g/L/h)	

(*Continued*)

Table 1.4 Investigating the effects of temperature, pH, and other environmental factors on solventogenic and acidogenic continuous cultures of acetone-butanol-ethanol (ABE) fermentation. (*Continued*)

Strains	Factors	Substrates	D/h	pH control	Other conditions	Traits	Ref.
		Glucose (44 mM)				Productivity of butanol, butyric acid, acetic acid and acetone are low (0g/L/h), (0.601g/L/h). (0.201g/L/h), (0g/L/h).	
C. acetobutylicum ATCC 824T	Addition of ρ-aminobenzoic acid in synthetic medium	Glucose (51.7 g/L)	0.28	4.4	No ρ-amino-benzoic acid	Productivity of ABE and cell concentration is low (0.09g/L/h), (0.32g/L/h)	[65]
					8.0 g/L ρ-amino-benzoic acid	Productivity of ABE and cell concentration is high (1.89 g/L/h), (2.07g/L/h)	

(*Continued*)

Table 1.4 Investigating the effects of temperature, pH, and other environmental factors on solventogenic and acidogenic continuous cultures of acetone-butanol-ethanol (ABE) fermentation. (*Continued*)

Strains	Factors	Substrates	D/h	pH control	Other conditions	Traits	Ref.
Clostridium beijerinckii NCIMB 8052	pH, Dilution rate	Glucose (50 g/L)	0.0610	5.0	-	Productivity of Butyric acid and acetic acid are high (0.091/L/h), (0.12 g/L/h). Productivity of butanol, and acetone are low (0.14 g/L/h), (0.058g/L/h)	[120]
			0.0610	5.5		Productivity of butanol, and acetone are high (0.34 g/L/h), (0.16g/L/h) Productivity of Butyric acid and acetic acid are low (0.059/L/h), (0.082 g/L/h).	

(*Continued*)

Table 1.4 Investigating the effects of temperature, pH, and other environmental factors on solventogenic and acidogenic continuous cultures of acetone-butanol-ethanol (ABE) fermentation. (*Continued*)

Strains	Factors	Substrates	D/h	pH control	Other conditions	Traits	Ref.
			0.158	5.5		Productivity of butanol, butyric acid, acetic acid and acetone are high (0.35 g/L/h), (0.10 g/L/h). (0.249g/L/h), (0.14g/L/h).	
		Xylose (50 g/L)	0.20			Productivity of butanol is low (0.272g/L/h)	

(*Continued*)

Table 1.4 Investigating the effects of temperature, pH, and other environmental factors on solventogenic and acidogenic continuous cultures of acetone-butanol-ethanol (ABE) fermentation. (*Continued*)

Strains	Factors	Substrates	D/h	pH control	Other conditions	Traits	Ref.
Clostridium beijerinckii NCIMB 8052	NH4Cl concentration	Glucose (20 g/L)	0.06	5.5	0.24 g/L NH4Cl	Productivity of Butyric acid is high (0.352g/L/h) Productivity of butanol and acetic acid are low (0.031/L/h), (0.0013 g/L/h).	[120]
					0.72 g/L NH4Cl	Productivity of butanol and acetic acid are high (0.085/L/h), (0.0097 g/L/h). Productivity of Butyric acid is low (0.158g/L/h)	

Continuous cultures with high concentrations of cells have been produced as a means of overcoming the major problem of low cell concentration in chemostat cultures. The enhanced butanol productivity, reduced reactor sizes, and accessibility of medium made for production rather than growth are some advantages of this approach. Cell immobilization on a variety of carriers (Table 1.5) and membrane-based cell recycling (Table 1.6) are two techniques claimed to increase cell density for butanol synthesis and cell recycling using membranes. Several carriers, including brick,

Table 1.5 Using immobilised cells during continuous, high-density fermentation of acetone, butanol, and ethanol (ABE).

Carrier	Strain	Dilution rate/h	ABE productivity (g/L/h)	Ref.
Brick	Clostridium beijerinckii BA101	2	15.8	[103]
	Clostridium acetobutylicum BCRC10639	0.054	0.48	[104]
Bonechar	Clostridium saccharobutylicum NCP 262T	1	4.1	[105]
Ca-alginate	Clostridium saccharobutylicum spoA2	0.196	3.02	[106]
	Clostridium acetobutylicum DSM 792	1.02	4.02	[107]
Coke	Clostridium acetobutylicum ATCC 824T	0.1	1.12	[108]
Fibrous bed	Clostridium acetobutylicum ATCC 55025	0.6	4.6	[109]
Sponge segments	Clostridium acetobutylicum ATCC 824T	0.272	4.2	[110]

Table 1.6 ABE fermentation refers to the process of producing acetone, butanol, and ethanol in a continuous, high-density culture by recycling the cells.

Membrane	Strain	Dilution rate of cell (recycling)	Dilution rate of cell (bleeding)	ABE productivity (g/L/h)	Ref.
Microfiltration membrane	Clostridium acetobutylicum ATCC 824T	0.64	-	5.4	[111]
	Clostridium saccharobutylicum NCP 262T	0.39	0.02	4.06	[112]
	Clostridium saccharoperbutylacetonicum N1–4 (ATCC 13564)	0.71	0.11, 0.14, 0.16	7.55	[67]
Ultrafiltration membrane	Clostridium acetobutylicum ATCC 824T	0.44	0.065	6.5	[113]
	Clostridium saccharobutylicum NCP 262T	0.4	-	4.1	[114]

calcium alginate, and κcarrageenan, have been explored for their potential to immobilize cells. *C. beijerinckii* BA101 has the highest ABE production at 16.2 g/L/h and can work at a high dilution rate of up to 2/h when utilizing brick as the carrier. Nevertheless, ultrafiltration and microfiltration membranes can execute cell recycling at dilution rates greater than 0.3/h (Table 1.4). Tashiro *et al.* concentrated 4 L of the broth to 0.4 L, and therefore reached a high cell concentration of 20 g/L in only 12 h, despite the fact that this is typically a time-consuming process [67]. It has also been demonstrated that cell-bleeding technique at dilution rates of 0.11 to 0.16/h maintains a consistent cell concentration in the fermentor, providing greater operating stability (207 h) than typical (48 h) without any cell bleeding. With a total dilution rate of 0.85/h, including 0.11 to 0.16/h for cell bleeding, *C. saccharoperbutylacetonicum* N1-4 (ATCC 13564) obtained the highest ABE productivity to date, 7.55 g/L/h [67]. However, in batch, fed-batch, and continuous cultures, a high butanol concentration should impede cell growth and butanol synthesis. To successfully produce higher quantities of butanol, as detailed in Section 1.8.4, it is therefore important to research both fermentation technology and butanol removal methods.

1.8.4 ABE Fermentation with Butanol Elimination

Pervaporation, gas stripping, liquid–liquid extraction, and liquid–membrane extraction are a few techniques for separating butanol (Table 1.7). Also, it has been shown that fed-batch fermentation, when compared to batch cultures, is substantially more successful at increasing butanol synthesis. Pervaporation is the act of regaining and gathering a volatile substance that has been selectively separated across a membrane under less pressure. By using fed-batch culture along with pervaporation as opposed to conventional batch culture (12.8 g/L), Qureshi *et al.* were able to produce higher butanol percentages (105 g/L) [68]. However, the cells and culture broth had to be separated because pervaporation demands a higher temperature than fermentation. Liu, however, was able to do fed-batch cultivation with *in situ* pervaporation at 37 degrees Celsius using a polydimethyl polyxan-ceramic membrane [69]. Moreover, the process of "gas stripping" uses aeration to transform components from the liquid phase into the gas phase. When fed-batch culture was employed in conjunction with gas stripping as opposed to batch culture (18.6 g/L), butanol production was increased by Ezeji *et al.* (81.3 g/L) [70]. Moreover, Grobben reported a butanol concentration of 27 g/L and a yield of 0.32 g/g in his fed-batch culture studies employing liquid–membrane extraction with methyl fatty acid [71]. Researchers also developed methods for avoiding butanol feedback

Table 1.7 ABE fermentation with butanol elimination.

Strains	Elimination method	ABE			Butanol			Ref.
		Concentration g/L	Productivity g/L/h	Yields g/g	Concentration g/L	Productivity g/L/h	Yields g/g	
Clostridium acetobutylicum ATCC 824T	Pervaporation	155	0.18	0.348	105	0.121	0.237	[68]
Clostridium acetobutylicum JB200	Gas stripping	109	0.41	0.32	76.4	0.29	0.23	[44]
Clostridium beijerinckii BA101	Pervaporation	51.5	0.69	0.42	No Data	No Data	No Data	[115]
	Gas stripping	81.3	0.593	0.360	56.2	0.410	0.249	[70]
Clostridium saccharoper-butylacetonicum N1-4 (ATCC 13564)	Liquid–liquid extraction	29.8	0.784	0.400	20.9	0.55	0.281	[116]
	Liquid membrane	No Data	No Data	No Data	20.1	0.394	0.234	[117]

inhibition through the use of different extraction techniques for high titer and pure butanol production.

1.9 Utilizing Pre-Treatment and Saccharification to Produce Butanol from Lignocellulosic Biomass

Global environmental problems and unexpected increases in fuel price are direct results of widespread usage of fossil fuels in both developing and industrialized nations [72]. The America and Brasil have increased their use of biodiesel made from food biomasses like sugarcane, corn as a response to this [73]. But this has contributed to a scarcity of agricultural supplies, driving up the cost of food. Thus, there is a lot of interest in developing methods to create biofuels from nonfood sources such lignocellulosic materials. The higher order crystal structure of the cellulose and hemicellulose that make up lignocellulosic biomass makes it challenging for the most common strains employed in biofuel production to directly consume it [74]. To get fermentable sugars out of lignocellulosic biomass, various preprocessing and saccharification processes have been devised. Physical, chemical, and biological methods are the primary categories of lignocellulosic biomass utilization processes. Steam explosion [75] is one example of a physical method, while concentrated sulphuric acid and alkaline treatment [76, 77] are the most common chemical methods. However, cellulase and hemicellulase are just two of the enzymes used in the biological perspective. Despite its imperfections (such as fuel consumption, hazardous wastes, and expensive demand), these methods are necessary for the effective exploitation of lignocellulosic biomasses and will be further developed.

1.10 Eliminating CCR to Produce Butanol

The processing and saccharification of lignocellulosic materials release various amounts of fermentable sugars, such as arabinose, glucose, and xylose. Yet, when bacteria are cultured in a solution containing glucose and other carbs, glucose inhibits the catabolism of the bacteria. Glucose causes the Carbon Catabolite Repression (CCR) phenomenon, which limits the microorganism's utilization of other sugars and results in inefficient ABE fermentation. This behavior has been linked to ABE-producing clostridia as well as other bacteria and yeast [78]. Metabolic engineering, mixed culture, using an ABE-producing strain in a mixed sugar fermentation system with the addition of exogenous trace elements, and using a semihydrolysis method for lignocellulosic biomass have all been investigated and shown to be efficient ways to get around this restriction [79]. Recently, single and dual strains were metabolically engineered to consume glucose and glycerol simultaneously, and both cultures thrived when using this sugar combination. To create up to 6 g/L of butanol, the single strain in particular used all of the glycerol and glucose available [80]. Bruder et al. (2015) [81] recommended employing both glucose and xylose at the same time to get around CCR's inhibitory effects; a modified *C. acetobutylicum* utilized 30% of the xylose. Following 48 hours of fermentation, glucose enabled the production of *C. aacetobutylicum*, which had a 7.5-fold higher yield than the strain's wild counterpart. With moderate zinc treatment, it may be possible to increase ABE fermentation from xylose/glucose sugar combinations. Wu et al.'s use of xylose and glucose resulted in 11.5 g/L of butanol at a specific xylose consumption rate of 0.3 g g-DCW/h [82]. Noguchi et al. (2013) [83] fermented mixed sugars (xylose/cellobiose) without CCR and produced significant amounts of butanol (16 g/L) and ABE (23 g/L) using *C. saccharoperbutylacetonicum* N1-4. After 72 hours of fermentation, the productivity of xylose and cellobiose increased by 1.9 and 1.8 g/L/h, respectively. As was previously mentioned, semihydrolysis of lignocellulosic biomass can stop CCR. Zhao et al. (2018) [79] created a semihydrolysate with minimum enzyme loading using pre-treatment rice straw and H_2SO_4 to minimize CCR and increase butanol fermentation efficiency. In addition, the output of butanol increased from 0.0628 to 0.265 g/L/h.

1.11 Butanol Production from Alternative Substrate to Sugar

Using ABE-producing clostridia, no one has yet impact on daily ABE fermentation from lignocellulosic biomass before even treating it to processing and saccharification methods. We are currently looking at various substrates for ABE fermentation that can be derived from lignocellulosic biomass through other fermentation methods including lactate fermentation [58, 59]. While DL-lactic acid cannot be used as feedstock for poly-lactic acid, Oshiro et al. [58] used pH stat fed-batch culture to efficiently utilize lactate with sugar as the cosubstrate, resulting in high butanol synthesis (15.5 g/L). Besides that, Yoshida et al. studied at arabinose as such cosubstrate in replacement of glucose and gained immense butanol synthesis (15.6 g/L) without using a pH controller [59]. Utilizing various substrates will become a major aspect of any approach for optimizing lignocellulosic biomass and establishing biorefinery-based civilizations.

1.12 Economics of Biobutanol

Manufacturing on a broad scale of biobutanol is hindered by factors such as the low butanol fermentation yield and the common citizenry's lack of knowledge about the benefits of biobutanol [84]. Since the productivity level of biobutanol via ABE fermentation from biomass is still questionable in comparison to the petrochemical route, its branding and contribution to the biofuel industry are constrained. This ambiguity is caused by the multiple sugar types present in the biomass, the poor performance of infomercial microbes in regards to the biobutanol, and the ineffectiveness of current separation techniques. Research on wheat straw, hydrolysate distiller's dried grains with solubles (DDGS), *C. beijerinki* BA101, and *C. acetobutylicum* P260 indicates that industrial synthesis of biobutanol from agricultural residues is conceivable [85].

On the other hand, the economy of the ABE fermentation process is closely related to the mass and energy balance [86]. The mass and energy balance of ABE fermentation using wheat straw is shown in Figure 1.2 [87]. The investigation on ABE fermentation was carried out using Life Cycle Inventory (LCI) data sets that were located in Ecoinvent. To produce biobutanol efficiently and economically on a large scale, significant advancements in microbial culture, fermentation, and waste feedstock are needed. As a result, a large number of research institutions linked to global corporations are continuously looking for solutions to these issues. Extensive research and development activities in England, France, China, Switzerland, and the United States have led to technological advances in the production pathways of biobutanol from residual crops, such as cereal straws. China (45%) and North America (23%), which are shown to have the highest production and consumption rates of butanol, respectively, in Figure 1.3. There are currently reports of some technoeconomic study on the production of butanol from biomass. According to Okoli and Adams (2014) [88], the minimum butanol selling price (MBSP) of biobutanol production from lignocellulosic bio material across the thermochemical range is between $0.55 and $1.17/L, which puts it in the same price range as ABE butanol ($0.59 to $1.05/L) and petrol ($0.82 to Lbeq). From a financial perspective, fermenting cornflour to produce biobutanol only makes sense if the resulting n-biobutanol can be sold as a premium commodity.

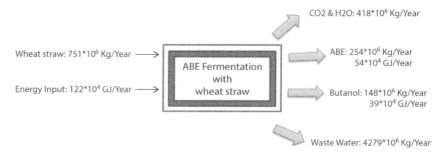

Figure 1.2 Balance of mass and energy during the fermentation of wheat straw to produce ABE.

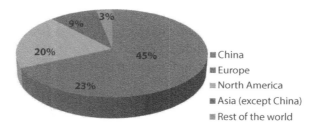

Figure 1.3 During the fermentation process of making ABE from wheat straw, the mass and energy balance are monitored.

With a final MBSP of $1.58/L, ABE fermentation using 2-ethyl-hexanol as an extractant may be the most cost-effective alternative [89].

Requirement for n-butanol as a, chemical intermediary, coating, solvent, for butyl acetate, ethers, and other compounds is increasing rapidly because of its widespread application in the construction sector. The manufacturing and construction industries are the largest consumers of n-butanol due to its use in a variety of products such paints and coatings, lubricants, varnishes. Increased demand for paints and coatings is expected to drive growth in the n-butanol market over the next few years, as building and infrastructure projects continue to mushroom around the world. By 2032, analysts predict that the worldwide n-butanol market will have expanded to 82,000 metric tons (Figure 1.4).

When broken down by geography, it's clear that Asia Pacific is where n-butanol really shines. This area accounted for over 35% of worldwide n-butanol consumption in 2021. Construction activity in fast developing nations like India, China, and Japan will increase, generating a slew of new industrial projects that in turn will drive up demand for paints and

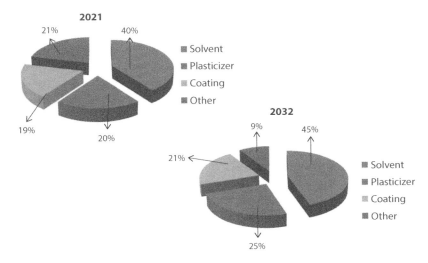

Figure 1.4 Volume-based market shares of n-various butanol's applications in 2021 and 2032.

coatings for use in the region's buildings and other infrastructure, driving up prices for n-butanol.

The global n-butanol market is divided into several submarkets based on application: solvent, plasticizer, coating, and others. In 2021, the Solvent sector alone will account for almost 40% of the global n-butanol market. N-butyl acetate, made from n-butanol, is a clear solvent that works well with a wide variety of materials, including cellulose nitrate, hydrocarbons, plastics, polymers. The chemical, ink, leather, paint industries all use it. N-butyl acrylate, another derivative of n-butanol, is utilized in a broad variety of industries, including the coatings industry, to increase low-temperature and chemical resistance. It is also found in adhesives, fiber, inks paints, paper, rubber, textiles.

OQ Chemicals, Formosa Plastics Corporation, Sasol Limited, INEOS Group Limited, The Dow Chemical Company, BASF SE, Yancon Cathay Coal Chemicals Co. Ltd., Lihuayi Weiyuan Chemical Co. Ltd., Eastman Chemical Company etc are among the world's largest manufacturers of n-butanol.

1.13 Future Prospects

Biobutanol is a second-generation alcoholic fuel used in the biofuel industry that has reduced volatility and a higher energy density than ethanol.

The commercial development of biobutanol is the primary objective of numerous businesses. The food supply is unaffected by biobutanol, a new form of biofuel that can compete with oil. Advanced technologies are being made to improve the glucose extraction, energy outputs, and sugar mixture fermentation of lignocellulosic biomass on a widespread scale. Although there has been a lot of research on fermenting lignocellulosic biomass to produce butanol, this process is still hindered by the necessity of a pre-treatment step. Biorefineries, unlike oil refineries, will need to process a wider variety of raw materials, necessitating the use of numerous pre-treatment processes. However, lignocellulosic biomass hinders biodegradation due to its stiff and complicated structure and variable chemical composition, while agricultural biomass as substrate (e.g., cornflour) has proven technology to manufacture these large value byproducts. Nonlignocellulosic biomass, such food waste, should be the subject of research since it contains a lot of starch, which clostridia can readily devour to make butanol. Furthermore, because starchy food waste was already gelatinized during cooking, it does not require pre-treatment with gelatinization to dissolve the intermolecular bonds of starch molecules. Another strategy is to use "designed biomass" as the butanol synthesis substrate. This sort of modified biomass increases butanol production while decreasing the cost of the fermentation process since it uses cosubstituents that are less expensive, including acid compounds.

The need for fuel is growing, which has stoked researchers' interest in discovering alternative, sustainable, and renewable energy sources. By utilizing renewable feedstock, petrol that is both commercially feasible and environmentally responsible can be distributed throughout the world. The promising renewable biofuel biobutanol could replace nonrenewable fossil fuels for both present and future generations.

ABE fermentation, which uses agricultural waste to produce biobutanol, is a green technology that is less harmful to the environment than traditional technology. Despite the limitations of the technology, such as low productivity and yield brought on by heterofermentation and product inhibition, these problems can be genetically fixed by using methods like genetic engineering and mutagenesis. Continuous culture rather than batch culture can be used to regulate product inhibition, which eventually lowers the cost of butanol recovery. The method can be made less expensive by manufacturing crude enzymes from various microbial sources and employing a biorefinery approach. The technique is typically costly since it requires pre-treatment of the biomass or substrate and additional saccharification using commercial enzymes. The cost of producing butanol would decrease if detoxification was enhanced, microbial strains were more efficient, processes were optimized, and more renewable sources were utilized. The focus of future research might be on the processes of bacterial genetic modification and adaptability that raise butanol titers.

The problem brought on by the high cost of raw materials might be resolved by the cheap and abundant lignocellulosic biomass. Corn stalk is one of the least expensive lignocellulosic substrates for solvent production in most nations with an agricultural economy. High inhibitor yield and insufficient butanol synthesis may be addressed by a biphasic pre-treatment approach that uses hydrothermal followed by diluted acid and/or alkaline, depending on the process parameters. In this circumstance, using large particle size may be one of the better options for use in industrial scales. Therefore, future study should concentrate on creating powerful biological pre-treatment techniques.

Detoxification of the hydrolysate may be necessary, depending on how the hydrolysate is utilized and how the pre-treatment method is applied. It is advisable to lessen inhibitor synthesis during the pre-treatment stage. Detoxification techniques that are physical, chemical, or biological have all been researched with the aim of lowering inhibitory effects on butanol fermentation.

The simplest physical detoxification technique for removing weak acids, furfural, and 5-hydroxymethylfurfural is evaporation (HMF). Biological detoxification involves the removal of phenolic compounds via peroxidase and laccase. To lessen the concentration of acetic acid, furfural, and phenolic compounds, processing with activated carbon as well as neutralization/alkalinization with calcium hydroxide, calcium oxide, calcium carbonate, ethyl acetate, potassium hydroxide, sodium hydroxide, and calcium hydroxide may be helpful. Scientists must take into account the salts produced during detoxification that also act as *Clostridium* species inhibitors.

Due to its lower rate of sugar utilization, decreased ability to tolerate high butanol concentrations, and inadequate productivity with inhibitors, a pure culture of the clostridium strain does not appear to be able to tackle the issue of low solvent output from maize stover. The issues related to the pre-treatment procedure may be resolved by using combination of different microbial strains culture and their respective growth media. A clostridium strain-based mixed culture is used to accelerate the fermentation process and maximize the butanol percentage of solvents. There is currently no commercially viable microbial strain that can efficiently ferment a mixture of different lignocellulosic hydrolysate into butanol. Future research must focus on developing highly butanol-producing, inhibitor-tolerant clostridium strains via gene and metabolic engineering techniques.

It is possible to steer fermentation pathways towards the production of butanol rather than other acids and alcohols by manipulating crucial enzymes, which is a crucial step towards commercially viable butanol fermentation from maize stover. By changing their structural makeup and/or adding a cosubstrate to their active side, enzymes involved in the breakdown of ABE fermentation byproducts can have their activity modulated. High hydrogen partial pressure is one explanation, but there are still additional ones that should be investigated.

Without producing a lot of acetone or ethanol, butanol can be produced by two-stage fermentation techniques that first convert fermentable carbohydrates into butyric acid and then further into butanol. In addition to these, the butanol fermentation is significantly impacted by buffer solution concentrations, mineral, redox balance, vitamin, and, as well as ecological factors including agitation with hydrogen, carbon dioxide, carbon monoxide, pH, and temperature. There is still a need to standardize these variable process parameters. All agriculturally based nations may successfully implement this standard technique for the commercial synthesis of biobutanol, which will ultimately decrease wealth export and aid in reducing the fossil fuel problem.

In the end, it is anticipated that a profitable and sustainable method of producing butanol will be developed soon.

1.14 Conclusion

Researchers are interested in investigating renewable, sustainable, and ecologically acceptable energy sources because of the rising demand for fuel. Using renewable feedstock yields fuel that is both environmentally responsible and commercially viable for all nations. Future generations may benefit from using biobutanol as an alternative to finite fossil fuels. Butanol has lately gained widespread interest as a promising liquid biofuel due to the advantages it offers over conventional alcohol-based biofuels at a time when both oil supplies and environmental concerns are dwindling. Nonetheless, butanol can be utilized as a liquid transportation biofuel in most cars with no adaptations needed. Hence, this chapter has covered the batch fermentation, continuous fermentation, and fed-batch fermentation procedures that are presently being used by various researchers for biobutanol production. There was also an increase in biobutanol output and production rates due to the incorporation of microbes into the fermentation process. Via either the overexpression or interruption of specific genes, the genetic alteration of the butanol-producing strain described in this article was able to increase the amount of butanol produced. Continuous fermentation that maintains a high cell concentration, on the other hand, has been a widely utilized for some time, as opposed to engineering strains. Furthermore, pre-treatment and saccharification, in conjunction with inhibitor removal methods, are utilized in order to break down lignocelluloses into cellulose, hemicelluloses, and lignin since these methods result in a reduction in the level of toxicity produced. Thus, future developments in fermentative biobutanol synthesis will be illuminated by the ideal mix of the prospective substrate, bacteria, and technology.

A lot of work remains to be done before the deficit in understanding regarding biobutanol production's efficiency is overcome. In spite of this, the manufacturing of biobutanol continues to be a source of concern from the point of view of profitable industry. The development of a potential microorganism to enhance or boost biobutanol production at the commercialization level is still under investigation. Wild-type strains with desired properties are being redesigned using a wide variety of molecular techniques. In conclusion, the manufacturing of biobutanol at an industrial scale will offer employee privileges as well as significantly expand the possibilities for growth in the economy.

References

1. Baral, N.R. and Shah, A., Techno-economic analysis of cellulosic butanol production from corn stover through acetone-butanol-ethanol fermentation. *Energy Fuel*, 30, 7, 5779–5790, 2016.
2. Lee, S.Y., Park, J.H., Jang, S.H., Nielsen, L.K., Kim, J., Jung, K.S., Fermentative butanol production by Clostridia. *Biotechnol. Bioeng.*, 101, 2, 209–228, 2008.
3. Ellis, J.T., Hengge, N.N., Sims, R.C., Miller, C.D., Acetone, Butanol, and Ethanol Production from Wastewater Algae. *Bioresour. Technol.*, 111, 491–495, 2012.
4. Li, H.G., Luo, W., Gu, Q.Y., Wang, Q., Hu, W.J., Yu, X.B., Acetone, butanol, and ethanol production from cane molasses using *Clostridium beijerinckii* mutant obtained by combined low-energy ion beam implantation and N-methyl-N-nitro-N-nitrosoguanidine induction. *Bioresour. Technol.*, 137, 254–260, 2013.
5. Li, S., Guo, Y., Lu, F., Huang, J., Pang, Z., High-level butanol production from Cassava Starch by a newly isolated *Clostridium acetobutylicum*. *Appl. Biochem. Biotechnol.*, 177, 4, 831–841, 2015.
6. Becerra, M., Cerdán, M.E., González-Siso, M., II, Biobutanol from cheese whey. *Microb. Cell Fact.*, 14, 27, 2015.

7. Gao, K., Li, Y., Tian, S., Yang, X., Screening and characteristics of a butanol-tolerant strain and butanol production from enzymatic hydrolysate of NaOH-pretreated corn stover. *World J. Microbiol. Biotechnol.*, 28, 10, 2963–2971, 2012.
8. Sarchami, T. and Rehmann, L., Optimizing enzymatic hydrolysis of inulin from Jerusalem artichoke tubers for fermentative butanol production. *Biomass Bioenergy*, 69, 175–182, 2014.
9. Al-Shorgani, N.K.N., Kalil, M.S., Ali, E., Hamid, A.A., Yusoff, W.M.W., The use of pretreated palm oil mill effluent for acetone-butanolethanol fermentation by *Clostridium saccharoperbutylacetonicum* N1-4. *Clean Technol. Environ. Policy*, 14, 5, 879–887, 2012.
10. Lu, C., Yu, L., Varghese, S., Yu, M., Yang, S.T., Enhanced robustness in acetone-butanol-ethanol fermentation with engineered *Clostridium beijerinckii* overexpressing adhE2 and ctfAB. *Bioresour. Technol.*, 243, 1000–1008, 2017.
11. Gao, M., Tashiro, Y., Yoshida, T., Zheng, J., Wang, Q., Sakai, K., Sonomoto, K., Metabolic analysis of butanol production from acetate in *Clostridium saccharoperbutylacetonicum* N1-4 using13C tracer experiments. *RSC Adv.*, 5, 11, 8486–8495, 2015.
12. Tashiro, Y., Yoshida, T., Noguchi, T., Sonomoto, K., Recent advancesand future prospects for increased butanol production by acetonebutanol-ethanol fermentation. *Eng. Life Sci.*, 13, 5, 432–445, 2013.
13. Visioli, L.J., Enzweiler, H., Kuhn, R.C., Recent advances on biobutanol production. *Sustain. Chem. Processes*, 2, 15, 2014.
14. Qureshi, N. and Eller, F., Recovery of butanol from *Clostridium beijerinckii* P260 fermentation broth by supercritical CO2 extraction. *J. Chem. Technol. Biotechnol.*, 93, 4, 1206–1212, 2018.
15. Karimi, K., Tabatabaei, M., Horváth, I.S., Kumar, R., Recent trends in acetone, butanol, and ethanol (ABE) production. *Biofuel Res.*, 2, 4, 301–308, 2015.
16. Lütke-Eversloh, T., Application of new metabolic engineering tools for *Clostridium acetobutylicum*. *Appl. Microbiol. Biotechnol.*, 98, 5823–5837, 2014.
17. Mao, S., Luo, Y., Zhang, T., Li, J., Bao, G., Zhu, Y., Chen, Z., Zhang, Y., Li, Y., Ma, Y., Proteome reference map and comparative proteomic analysis between a wild type *Clostridium acetobutylicum* DSM 1731 and its mutant with enhanced butanol tolerance and butanol yield. *J. Proteome Res.*, 9, 6, 3046–3061, 2010.
18. Rochon, E., Ferrari, M.D., Lareo, C., Integrated ABE fermentation-gas stripping process for enhanced butanol production from sugarcanesweet sorghum juices. *Biomass Bioenergy*, 98, 153–160, 2017.
19. Khedkar, M.A., Nimbalkar, P.R., Gaikwad, S.G., Chavan, P.V., Bankar, S.B., Sustainable biobutanol production from pineapple waste by using *Clostridium acetobutylicum* B 527: Drying kinetics study. *Bioresour. Technol.*, 225, 359–366, 2017.

20. Nimbalkar, P.R., Khedkar, M.A., Chavan, P.V., Bankar, S.B., Biobutanol production using pea pod waste as substrate: Impact of drying on saccharification and fermentation. *Renewable Energy*, 117, 520–529, 2018.
21. Claassen, P.A., Budde, M.A., López-Contreras, A.M., Acetone, butanol and ethanol production from domestic organic waste by solventogenic clostridia. *J. Mol. Microbiol. Biotechnol.*, 2, 1, 39–44, 2000.
22. Lütke-Eversloh, T. and Bahl, H., Metabolic engineering of *Clostridium acetobutylicum*: Recent advances to improve butanol production. *Curr. Opin. Biotechnol.*, 22, 5, 634–647, 2011.
23. Wang, S., Dong, S., Wang, Y., Enhancement of solvent production by overexpressing key genes of the acetone-butanol-ethanol fermentation pathway in *Clostridium saccharoperbutylacetonicum* N1-4. *Bioresour. Technol.*, 245, Pt A, 426–433, 2017.
24. Zheng, J., Tashiro, Y., Wang, Q., Sonomoto, K., Recent advances to improve fermentative butanol production: Genetic engineering and fermentation technology. *J. Biosci. Bioeng.*, 119, 1, 1–9, 2015.
25. Kumar, M. and Gayen, K., Developments in biobutanol production: New insights. *Appl. Energy*, 88, 6, 1999–2012, 2011.
26. Ohtake, T., Pontrelli, S., Laviña, W.A., Liao, J.C., Putri, S.P., Fukusaki, E., Metabolomics-driven approach to solving a CoA imbalance for improved 1-butanol production in *Escherichia coli*. *Metab. Eng.*, 41, 135–143, 2017.
27. Lee, J., Jang, Y.S., Choi, S.J., Im, J.A., Song, H., Cho, J.H., Seung, do Y., Papoutsakis, E.T., Bennett, G.N., Lee, S.Y., Metabolic engineering of *Clostridium acetobutylicum* ATCC 824 for isopropanol-butanol-ethanol fermentation. *Appl. Environ. Microbiol.*, 78, 5, 1416–1423, 2012.
28. Bankar, S.B., Jurgens, G., Survase, S.A., Ojamo, H., Granström, T., Genetic engineering of *Clostridium acetobutylicum* to enhance isopropanol-butanol-ethanol production with an integrated DNA technology approach. *Renewable Energy*, 83, 1076–1083, 2015.
29. Ezeji, T., Milne, C., Price, N.D., Blaschek, H.P., Achievements and perspectives to overcome the poor solvent resistance in acetone and butanol-producing microorganisms. *Appl. Microbiol. Biotechnol.*, 85, 6, 1697–1712, 2010.
30. Jones, D.T. and Woods, D.R., Acetone-butanol fermentation revisited. *Microbiol. Rev.*, 50, 4, 484–524, 1986.
31. Girbal, L., Croux, C., Vasconcelos, I., Soucaille, P., Regulation of metabolic shifts in *Clostridium acetobutylicum* ATCC 824. *FEMS Microbiol. Rev.*, 17, 287–297, 1995.
32. Thang, V.H., Kanda, K., Kobayashi, G., Production of acetone– butanol–ethanol (ABE) in direct fermentation of cassava by *Clostridium saccharoperbutylacetonicum* N1–4. *Appl. Biochem. Biotechnol.*, 161, 1-8, 157–170, 2010.
33. Shah, M.M. and Lee, Y.Y., Process improvement in acetonebutanol production from hardwood by simultaneous saccharification and extractive fermentation. *Appl. Biochem. Biotechnol.*, 45, 585–597, 1994.

34. Qureshi, N., Saha, B.C., Hector, R.E., Dien, B., Hughes, S., Liu, S., Iten, L., Bowman, M.J., Sarath, G., Cotta, M.A., Production of butanol (a biofuel) from agricultural residues. II. Use of corn stover and switchgrass hydrolysates. *Biomass Bioenergy*, 34, 566–571, 2010.
35. Nölling, J., Breton, G., Omelchenko, M.V., Makarova, K.S., Zeng, Q., Gibson, R., Lee, H.M., Dubois, J., Qiu, D., Hitti, J., Wolf, Y., II, Tatusov, R.L., Sabathe, F., Doucette-Stamm, L., Soucaille, P., Daly, M.J., Bennett, G.N., Koonin, E.V., Smith, D.R., Genome sequence and comparative analysis of the solvent-producing bacterium *Clostridium acetobutylicum*. *J. Bacteriol.*, 183, 16, 4823–4838, 2001.
36. Tomas, C.A., Welker, N.E., Papoutsakis, E.T., Overexpression of groESL in *Clostridium acetobutylicum* results in increased solvent production and tolerance, prolonged metabolism, and changes in the cell's transcriptional program. *Appl. Environ. Microbiol.*, 69, 8, 4951–4965, 2003.
37. Annous, B.A. and Blaschek, H.P., Isolation and characterization of *Clostridium acetobutylicum* mutants with enhanced amylolytic activity. *Appl. Environ. Microbiol.*, 57, 9, 2544–2548, 1991.
38. Chen, C.K. and Blaschek, H.P., Acetate enhances solvent production and prevents degeneration in *Clostridium beijerinckii* BA101. *Appl. Microbiol. Biotechnol.*, 52, 2, 170–173, 1999.
39. Ezeji, T.C., Qureshi, N., Blaschek, H.P., Continuous butanol fermentation and feed starch retrogradation: Butanol fermentation sustainability using *Clostridium beijerinckii* BA101. *J. Biotechnol.*, 115, 2, 179–187, 2005.
40. Madihah, M.S., Ariff, A.B., Sahaid, K.M., Suraini, A.A., Direct fermentation of gelatinized sago starch to acetone–Butanol–Ethanol by *Clostridium acetobutylicum*. . *World J. Microbiol. Biotechnol.*, 17, 567–576, 2001.
41. Qureshi, N. and Blaschek, H.P., Recent advances in ABE fermentation: Hyper-butanol producing *Clostridium beijerinckii* BA101. *J. Ind. Microbiol. Biotechnol.*, 27, 5, 287–291, 2001.
42. Tashiro, Y. and Sonomoto, K., Advances in butanol production by clostridia, in: *Current Research, Technology and Education Topics in Applied Microbiology and Microbial Biotechnology (Microbiology Book #2)*, A.M. Vilas (Ed.), pp. 1383–1394, Formatex Research Center, Badajoz, 2010.
43. Liu, Z., Ying, Y., Li, F., Ma, C., Xu, P., Butanol production by *Clostridium beijerinckii* ATCC 55025 from wheat bran. *J. Ind. Microbiol. Biotechnol.*, 37, 5, 495–501, 2010.
44. Lu, C., Zhao, J., Yang, S.T., Wei, D., Fed-batch fermentation for n-butanol production from cassava bagasse hydrolysate in a fibrous bed bioreactor with continuous gas stripping. *Bioresour. Technol.*, 104, 380–387, 2012.
45. Gao, M., Tashiro, Y., Wang, Q., Sakai, K., Sonomoto, K., High acetone-butanol-ethanol production in pH-stat co-feeding of acetate and glucose. *J. Biosci. Bioeng.*, 122, 2, 176–182, 2016.
46. Oshiro, M., Hanada, K., Tashiro, Y., Sonomoto, K., Efficient conversion of lactic acid to butanol with pH-stat continuous lactic acid and glucose

feeding method by *Clostridium saccharoperbutylacetonicum*. *Appl. Microbiol. Biotechnol.*, 87, 3, 1177–1185, 2010.
47. Yoshida, T., Tashiro, Y., Sonomoto, K., Novel high butanol production from lactic acid and pentose by *Clostridium saccharoperbutylacetonicum*. *J. Biosci. Bioeng.*, 114, 5, 526–530, 2012.
48. Yerushalmi, L., Volesky, B., Szczesny, T., Effect of increased hydrogen partial pressure on the acetone-butanol fermentation by *Clostridium acetobutylicum*. *Appl. Microbiol. Biotechnol.*, 22, 103–107, 1985.
49. Kim, B.H., Bellows, P., Datta, R., Zeikus, J.G., Control of Carbon and Electron Flow in *Clostridium acetobutylicum* Fermentations: Utilization of Carbon Monoxide to Inhibit Hydrogen Production and to Enhance Butanol Yields. *Appl. Environ. Microbiol.*, 48, 4, 764–770, 1984.
50. Brosseau, J.D., Yan, J.Y., Lo, K.V., The relationship between hydrogen gas and butanol production by *Clostridium saccharoperbutylacetonicum*. *Biotechnol. Bioeng.*, 28, 3, 305–310, 1986.
51. Hönicke, D., Janssen, H., Grimmler, C., Ehrenreich, A., Lütke-Eversloh, T., Global transcriptional changes of *Clostridium acetobutylicum* cultures with increased butanol: Acetone ratios. *New Biotechnol.*, 29, 4, 485–493, 2012.
52. Chauvatcharin, S., Siripatana, C., Seki, T., Takagi, M., Yoshida, T., Metabolism analysis and on-line physiological state diagnosis of acetone-butanol fermentation. *Biotechnol. Bioeng.*, 58, 6, 561–571, 1998.
53. Shinto, H., Tashiro, Y., Yamashita, M., Kobayashi, G., Sekiguchi, T., Hanai, T., Kuriya, Y., Okamoto, M., Sonomoto, K., Kinetic modeling and sensitivity analysis of acetone-butanol-ethanol production. *J. Biotechnol.*, 131, 1, 45–56, 2007.
54. Shinto, H., Tashiro, Y., Kobayashi, G., Sekiguchi, T., Hanai, T., Kuriya, Y., Okamoto, M., Sonomoto, K., Kinetic study of substrate dependency for higher butanol production in acetone–butanol–ethanol fermentation. *Process Biochem.*, 43, 12, 1452–1461, 2008.
55. Fond, O., Engasser, J.M., Matta-El-Amouri, G., Petitdemange, H., The acetone butanol fermentation on glucose and xylose. I. Regulation and kinetics in batch cultures. *Biotechnol. Bioeng.*, 28, 2, 160–166, 1986.
56. Cho, D.H., Shin, S.J., Kim, Y.H., Effects of acetic and formic acid on ABE production by *Clostridium acetobutylicum* and *Clostridium beijerinckii*. *Biotechnol. Bioprocess Eng.*, 17, 270–275, 2012.
57. Tashiro, Y., Takeda, K., Kobayashi, G., Sonomoto, K., Ishizaki, A., Yoshino, S., High butanol production by *Clostridium saccharoperbutylacetonicum* N1-4 in fed-batch culture with pH-Stat continuous butyric acid and glucose feeding method. *J. Biosci. Bioeng.*, 98, 4, 263–268, 2004.
58. Oshiro, M., Hanada, K., Tashiro, Y., Sonomoto, K., Efficient conversion of lactic acid to butanol with pH-stat continuous lactic acid and glucose feeding method by *Clostridium saccharoperbutylacetonicum*. *Appl. Microbiol. Biotechnol.*, 87, 3, 1177–1185, 2010.

59. Yoshida, T., Tashiro, Y., Sonomoto, K., Novel high butanol production from lactic acid and pentose by *Clostridium saccharoperbutylacetonicum*. *J. Biosci. Bioeng.*, 114, 5, 526–530, 2012.
60. Fond, O., Engasser, J.M., Matta-El-Amouri, G., Petitdemange, H., The acetone butanol fermentation on glucose and xylose. II. Regulation and kinetics in fed-batch cultures. *Biotechnol. Bioeng.*, 28, 2, 167–175, 1986.
61. Bahl, H., Andersch, W., Gottschalk, G., Continuous production of acetone and butanol by *Clostridium acetobutylicum* in a two-stage phosphate limited chemostat. *Eur. J. Appl. Microbiol. Biotechnol.*, 15, 201–205, 1982.
62. Mermelstein, L.D., Papoutsakis, E.T., Petersen, D.J., Bennett, G.N., Metabolic engineering of *Clostridium acetobutylicum* ATCC 824 for increased solvent production by enhancement of acetone formation enzyme activities using a synthetic acetone operon. *Biotechnol. Bioeng.*, 42, 9, 1053–1060, 1993.
63. Hartmanis, M.G.N., Åhlman, H., Gatenbeck, S., Stability of solvent formation in *Clostridium acetobutylicum* during repeated subculturing. *Appl. Microbiol. Biotechnol.*, 23, 369–371, 1986.
64. Zheng, J., Tashiro, Y., Yoshida, T., Gao, M., Wang, Q., Sonomoto, K., Continuous butanol fermentation from xylose with high cell density by cell recycling system. *Bioresour. Technol.*, 129, 360–365, 2013.
65. Soni, B.K., Soucaille, P., Goma, G., Continuous acetone-butanol fermentation: Influence of vitamins on the metabolic activity of *Clostridium acetobutylicum*. *Appl. Microbiol. Biotechnol.*, 27, 1–5, 1987.
66. Meyer, C.L., Roos, J.W., Papoutsakis, E.T., Carbon monoxide gasing leads to alcohol production and butyrate uptake without acetone formation in continuous cultures of *Clostridium acetobutylicum*. *Microbiol. Biotechnol.*, 24, 159–167, 1986.
67. Tashiro, Y., Takeda, K., Kobayashi, G., Sonomoto, K., High production of acetone-butanol-ethanol with high cell density culture by cell-recycling and bleeding. *J. Biotechnol.*, 120, 2, 197–206, 2005.
68. Qureshi, N., Meagher, M.M., Huang, J., Hutkins, R.W., Acetone butanol ethanol (ABE) recovery by pervaporation using silicalite-silicone composite membrane from fed-batch reactor of *Clostridium acetobutylicum*. *J. Membr. Sci.*, 187, 93–102, 2001.
69. Liu, G., Wei, W., Wu, H., Dong, X. et al., Pervaporation performance of PDMS/ceramic composite membrane in acetone butanol ethanol (ABE) fermentation–PV coupled process. *J. Membr. Sci.*, 373, 121–129, 2011.
70. Ezeji, T.C., Qureshi, N., Blaschek, H.P., Production of acetone butanol (AB) from liquefied corn starch, a commercial substrate, using *Clostridium beijerinckii* coupled with product recovery by gas stripping. *J. Ind. Microbial. Biotechnol.*, 34, 12, 771–777, 2007.
71. Grobben, N.G., Eggink, G., Cuperus, F.P., Huizing, H.J., Production of acetone, butanol and ethanol (ABE) from potato wastes: Fermentation with integrated membrane extraction. *Appl. Microbiol. Biotechnol.*, 39, 494–498, 1993.

72. Nigam, P.S. and Singh, A., Production of liquid biofuels from renewable resources. *Prog. Energy Combust. Sci.*, 37, 52–68, 2011.
73. Kumar, M. and Gayen, K., Developments in biobutanol production: New insights. *Appl. Energy*, 88, 1999–2012, 2011.
74. Brodeur, G., Yau, E., Badal, K., Collier, J., Ramachandran, K.B., Ramakrishnan, S., Chemical and physicochemical pre-treatment of lignocellulosic biomass: A review. *Enzyme Res.*, 2011, 1–17, 2011.
75. Qureshi, N., Saha, B.C., Hector, R.E., Cotta, M.A., Removal of fermentation inhibitors from alkaline peroxide pretreated and enzymatically hydrolyzed wheat straw: Production of butanol from hydrolysate using *Clostridium beijerinckii* in batch reactors. *Biomass Bioenergy*, 32, 1353–1358, 2008.
76. Ezeji, T., Qureshi, N., Blaschek, H.P., Butanol production from agricultural residues: Impact of degradation products on *Clostridium beijerinckii* growth and butanol fermentation. *Biotechnol. Bioeng.*, 97, 6, 1460–1469, 2007.
77. Martinez, A., Rodriguez, M.E., Wells, M.L., York, S.W., Preston, J.F., Ingram, L.O., Detoxification of dilute acid hydrolysates of lignocellulose with lime. *Biotechnol. Progr.*, 17, 2, 287–293, 2001.
78. Grimmler, C., Held, C., Liebl, W., Ehrenreich, A., Transcriptional analysis of catabolite repression in *Clostridium acetobutylicum* growing on mixtures of D-glucose and D-xylose. *J. Biotechnol.*, 150, 3, 315–323, 2010.
79. Zhao, T., Tashiro, Y., Zheng, J., Sakai, K., Sonomoto, K., Semi-hydrolysis with low enzyme loading leads to highly effective butanol fermentation. *Bioresour. Technol.*, 264, 335–342, 2018.
80. Saini, M., Lin, L.J., Chiang, C.J., Chao, Y.P., Effective production of n-butanol in *Escherichia coli* utilizing the glucose–glycerol mixture. *J. Taiwan Inst. Chem. Eng.*, 81, 134–139, 2017.
81. Bruder, M.R., Moo-young, M., Chung, D.A., Chou, C.P., Elimination of carbon catabolite repression in *Clostridium acetobutylicum*—A journey toward simultaneous use of xylose and glucose. *Appl. Microbiol. Biotechnol.*, 99, 7579–7588, 2015.
82. Wang, Y., Guo, W., Cheng, C.L., Ho, S.H., Chang, J.S., Ren, N., Enhancing bio-butanol production from biomass of *Chlorella vulgaris* JSC-6 with sequential alkali pre-treatment and acid hydrolysis. *Bioresour. Technol.*, 200, 557–564, 2016.
83. Noguchi, T., Tashiro, Y., Yoshida, T., Zheng, J., Sakai, K., Sonomoto, K., Efficient butanol production without carbon catabolite repression from mixed sugars with *Clostridium saccharoperbutylacetonicum* N1-4. *J. Biosci. Bioeng.*, 116, 6, 716–721, 2013.
84. Pfromm, P.H., Amanor-Boadu, V., Nelson., R., Vadlani., P., Madl, R., Biobutanol vs. bio-ethanol: A technical and economic assessment for corn and switchgrass fermented by yeast or *Clostridium acetobutylicum*. *Biomass Bioenergy*, 34, 4, 515–524, 2010.
85. Bharathiraja., B., Jayamuthunagai, J., Sudharsanaa, T., Bharghavi, A., Praveenkumar, R., Chakravarthy, M., Yuvaraj, D., Biobutanol—an impending

biofuel for future: A review on upstream and downstream processing techniques. *Renewable Sustain. Energy Rev.*, 68, 788–807, 2017.
86. Ranjan, A. and Moholkar, V.S., Biobutanol: Science, engineering andeconomics. *Int. J. Energy Res.*, 2012, 36, 277–323, 2012.
87. Brito, M. and Martins, F., Life cycle assessment of butanol production. *Fuel*, 208, 476–482, 2017.
88. Okoli., C.O. and Adams, T.A., Techno-economic analysis of a thermochemical lignocellulosic biomass-to-butanol process. *Comput. Aided Chem. Eng.*, 33, 1681–1686, 2014.
89. Dalle, Ave, G. and Adams, T.A., Techno-economic comparison of acetone-butanol-ethanol fermentation using various extractants. *Energy Convers. Manag.*, 156, 288–300, 2018.
90. Schwarz, K.M., Grosse-Honebrink, A., Derecka, K., Rotta, C., Zhang, Y., Minton, N.P., Towards improved butanol production through targeted genetic modification of *Clostridium pasteurianum*. *Metab. Eng.*, 40, 124–137, 2017.
91. Tanaka, Y., Kasahara, K., Hirose, Y., Morimoto, Y., Izawa, M., Ochi, K., Enhancement of butanol production by sequential introduction of mutations conferring butanol tolerance and streptomycin resistance. *J. Biosci. Bioeng.*, 124, 4, 400–407, 2017.
92. Zhang, J., Wang, P., Wang, X., Feng, J., Sandhu, H.S., Wang, Y., Enhancement of sucrose metabolism in *Clostridium saccharoperbutylacetonicum* N1-4 through metabolic engineering for improved acetone-butanol-ethanol (ABE) fermentation. *Bioresource Technol.*, 270, 430–438, 2018.
93. Zhang, J., Zong, W., Hong, W., Zhang, Z.T., Wang, Y., Exploiting endogenous CRISPR-Cas system for multiplex genome editing in *Clostridium tyrobutyricum* and engineer the strain for high-level butanol production. *Metab. Eng.*, 47, 49–59, 2018.
94. Zhang, J., Yu, L., Lin, M., Yan, Q., Yang, S.T., n-Butanol production from sucrose and sugarcane juice by engineered *Clostridium tyrobutyricum* overexpressing sucrose catabolism genes and adhE2. *Bioresour. Technol.*, 233, 51–57, 2017.
95. Al-Shorgani, N.K.N., Ali, E., Kalil, M.S., Yusoff, W.M.W., Bioconversion of butyric acid to butanol by *Clostridium saccharoperbutylacetonicum* N1-4 (ATCC 13564) in a limited nutrient medium. *BioEnergy Research*, 5, 2, 287–293, 2012.
96. Al-Shorgani, N.K., Kalil, M.S., Yusoff, W.M., Biobutanol production from rice bran and de-oiled rice bran by *Clostridium saccharoperbutylacetonicum* N1-4. *Bioprocess. Biosyst. Eng.*, 35, 5, 817–826, 2012.
97. Al-Shorgani, N.K.N., Ali, E., Kalil, M.S., Yusoff, W.M.W., Bioconversion of butyric acid to butanol by *Clostridium saccharoperbutylacetonicum* N1-4 (ATCC 13564) in a limited nutrient medium. *Bioenergy Res.*, 5, 2, 287–293, 2012.

98. Peguin, S., Delorme, P., Goma, G., Soucaille, P., Enhanced alcohol yields in batch cultures of *Clostridium acetobutylicum* using a three-electrode potentiometric system with methyl viologen as electron carrier. *Biotechnol. Lett.*, 16, 269–274, 1994.
99. Ballongue, J., Amine, J., Petitdemange, H., Gay, R., Enhancement of solvents production *by Clostridium acetobutylicum* cultivated on a reducing compounds depletive medium. *Biomass*, 10, 121–129, 1986.
100. Roos, J.W., McLaughlin, J.K., Papoutsakis, E.T., The effect of pH on nitrogen supply, cell lysis, and solvent production in fermentations of *Clostridium acetobutylicum*. *Biotechnol. Bioeng.*, 27, 5, 681–694, 1985.
101. Monot, F., Engasser, J.M., Petitdemange, H., Influence of pH and undissociated butyric acid on the production of acetone and butanol in batch cultures of *Clostridium acetobutylicum*. *Appl. Microbiol. Biotechnol.*, 19, 422–426, 1984.
102. Holt, R.A., Stephens, G.M., Morris, J.G., Production of Solvents by *Clostridium acetobutylicum* Cultures Maintained at Neutral pH. *Appl. Environ. Microbiol.*, 48, 6, 1166–1170, 1984.
103. Lienhardt, J., Schripsema, J., Qureshi, N., Blaschek, H.P., Butanol production by *Clostridium beijerinckii* BA101 in an immobilized cell biofilm reactor. *Biotechnol. Appl. Biochem.*, 98, 591–598, 2002.
104. Yen, H.-W., Li, R.-J., Ma, T.-W., The development process for a continuous acetone–butanol–ethanol (ABE) fermentation by immobilized *Clostridium acetobutylicum*. *J. Taiwan Inst. Chem. Eng.*, 42, 902–907, 2011.
105. Qureshi, N., Paterson, A.H.J., Maddox, I.S., Model for continuous production of solvents from whey permeate in a packed bed reactor using cells of Clostridium acetobutylicum immobilized by adsorption onto bonechar. *Appl. Microbiol. Biotechnol.*, 29, 323–328, 1988.
106. Largier, S.T., Long, S., Santangelo, J.D., Jones, D.T., Woods, D.R., Immobilized Clostridium acetobutylicum P262 Mutants for Solvent Production. *Appl. Environ. Microbiol.*, 50, 2, 477–481, 1985.
107. Frick, C. and Schügerl, K., Continuous acetone-butanol production with free and immobilized *Clostridium acetobutylicum*. *Appl. Microbiol. Biotechnol.*, 25, 186–193, 1986.
108. Welsh, F.W., Williams, R.E., Veliky, I.A., Solid carriers for a Clostridium acetobutylicum that produces acetone and butanol. *Enzyme Microb. Technol.*, 9, 500–502, 1987.
109. Huang, W.-C., Ramey, D.E., Yang, S.-T., Continuous production of butanol by Clostridium acetobutylicum immobilized in a fibrous bed bioreactor. *Biotechnol. Appl. Biochem.*, 115, 887–898, 2004.
110. Park, C.-H., Okos, M.R., Wankat, P.C., Acetone-butanolethanol (ABE) fermentation in an immobilized cell trickle bed reactor. *Biotechnol. Bioeng.*, 34, 18–29, 1989.
111. Afschar, A.S., Biebl, H., Schaller, K., Schugerl, K., Production of acetone and butanol by *Clostridium acetobutylicum* in continuous culture with cell recycle. *Appl. Microbiol. Biotechnol.*, 22, 394–398, 1985.

112. Ennis, B.M. and Maddox, I.S., Production of solvents (ABE fermentation) from whey permeate by continuous fermentation in a membrane bioreactor. *Bioprocess. Eng.*, 4, 27–34, 1989.
113. Pierrot, P., Fick, M., Engasser, J.M., Continuous acetone-butanol fermentation with high productivity by cell ultrafiltration and recycling. *Biotechnol. Lett.*, 8, 253–256, 1986.
114. Schlote, D. and Gottschalk, G., Effect of cell recycle on continuous butanol-acetone fermentation with *Clostridium acetobutylicum* under phosphate limitation. *Appl. Microbiol. Biotechnol.*, 24, 1–5, 1986.
115. Qureshi, N. and Blaschek, H.P., Production of acetone butanol ethanol (ABE) by a hyper-producing mutant strain of *Clostridium beijerinckii* BA101 and recovery by pervaporation. *Biotechnol. Progr.*, 15, 4, 594–602, 1999.
116. Ishizaki, A., Michiwaki, S., Crabbe, E., Kobayashi, G., Sonomoto, K., Yoshino, S., Extractive acetone-butanol-ethanol fermentation using methylated crude palm oil as extractant in batch culture of *Clostridium saccharoperbutylacetonicum* N1-4 (ATCC 13564). *J. Biosci. Bioeng.*, 87, 3, 352–356, 1999.
117. Tanaka, S., Tashiro, Y., Kobayashi, G., Ikegami, T., Negishi, H., Sakaki, K., Membrane-assisted extractive butanol fermentation by *Clostridium saccharoperbutylacetonicum* N1-4 with 1-dodecanol as the extractant. *Bioresour. Technol.*, 116, 448–452, 2012.
118. Fick, M., Pierrot, P., Engasser, J.M., Optimal conditions for long-term stability of acetone-butanol production by continuous cultures of *Clostridium acetobutylicum*. *Biotechnol. Lett.*, 7, 503–508, 1985.
119. Mermelstein, L.D., Papoutsakis, E.T., Petersen, D.J., Bennett, G.N., Metabolic engineering of *Clostridium acetobutylicum* ATCC 824 for increased solvent production by enhancement of acetone formation enzyme activities using a synthetic acetone operon. *Biotechnol. Bioeng.*, 42, 9, 1053–1060, 1993.
120. Stephens, G.M., Holt, R.A., Gottschal, J.C., Morris, J.G., Studies on the stability of solvent production by *Clostridium acetobutylicum* in continuous culture. *J. Appl. Bacteriol.*, 59, 597–605, 1985.

2

Recent Trends in the Pre-Treatment Process of Lignocellulosic Biomass for Enhanced Biofuel Production

Nikita Bhati, Shreya and Arun Kumar Sharma*

Department of Bioscience and Biotechnology, Banasthali Vidyapith, Rajasthan, India

Abstract

Rise in demand for fossil resources and upsurge in energy requirements have raised the necessity to switch to more cleaner and environment friendly sources. Lignocellulosic biomass (LCB), used as organic matter for the advanced biofuel production, serves in the regulation of global carbon cycle and in sustainable development processes by limiting the usage of fossil fuels, further delaying the drastic changes in climate by reduction in emission of greenhouse gases. However, the presence of lignin limits the solubilization of LCB, posing as a major obstacle in the biofuel production. Thus, pre-treatment of LCB is vital for its disintegration, increasing solubility and surface area, as well as to lessen the cellulose crystallinity. Usual pre-treatment methods pose several setbacks like high operation cost, maintenance of equipment and production of unwanted products. To overcome these drawbacks, various ecofriendly processes are being developed for the effective conversion of biomass to certify high yields of biofuels as a profitable solution. This chapter presents recent advances in the pre-treatment of lignocellulosic biomass along with hydrolysis of lignin, cellulose and hemicellulose for higher yield of biofuel. Certain principal processes of pre-treatment are summarized for guiding the further research in the cost-effective biofuel production.

Keywords: Lignocellulosic biomass, pre-treatment, biofuel production, cellulose, lignin

---

Corresponding author: arun.k.sharma84@gmail.com

2.1 Introduction

Consumption of fossil fuel at an alarming rate along with rise in energy requirement is detrimental to the environment as it causes increase in emission of greenhouse gases giving rise to global warming hence, causing concern globally. Spike in temperature is reported due to global warming affecting health of population overall. It is reported that heatwave like conditions intensified since the beginning of this century and is likely to worsen in upcoming years causing serious health issues among population [1]. The solution to these problems is switching to cleaner and ecofriendly fossil fuels. The advent of lignocellulosic biomass as renewable source of energy has gained indispensable reception due to its widespread occurrence and potential to produce value added products like biofuel in an economically beneficial way [2–4]. LCB consists of three integral parts—cellulose, hemicellulose and lignin along with small quantity of protein, extractives, pectin and inorganic compounds minerals [5].

Constitution of cellulose and hemicellulose is in majority about 70% to 80% (dry weight), whereas lignin is comprises about 10% to 25% [6]. Lignin acting as recalcitrant in the LCB complex needs certain pre-treatment to breakdown the cellulose and hemicellulose part into simple sugars [7]. Pre-treatment of LCB can be classified as physical, biological, dilute acid hydrolysis, alkaline pre-treatment, hydrothermal treatment, ammonia fiber explosion pre-treatment and novel ecofriendly approaches, such as supercritical fluids and ionic liquids [8–10].

Pre-treatment processes enhance the surface area and offers easy accessibility for enzymes to binding sites [11]. The conventional pre-treatment methods faced substantial shortcomings in the aim to attain economic, large scale, environment friendly production for sustainable development. Whereas, new methods of pre-treatment includes low temperature steep delignification (LTSD), supercritical fluid based, cosolvent enhanced lignocellulosic fractionation (CELF) and ionic liquids causes improved sugar yields with minimal generation of by-products, hence are deemed to be the most advanced approaches [12, 13].

The sugars (pentose and hexose) released are utilized to produce alcohols, polyols, bioplastics, fatty acids and organic acids by several microorganisms [14]. Several side products like weak acids (formic acid and acetic acid), furan aldehydes (furfural, 5-hydroxy methyl furfural (HMF)), phenolics (vanillin), and simple sugars are released during the pre-treatment process resulting in inhibited microbial growth due to accumulation of reactive oxygen species, intracellular acidification as well as energy

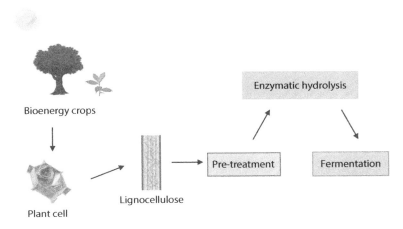

Figure 2.1 A schematic diagram of steps in biofuel production.

drainage [15]. Thus, detoxification of biomass hydrolysate is necessary prior to exposure to microbial fermentation to evade the harmful effects during the process of fermentation.

Several approaches of detoxification such as biological (enzymatic and microbial), physical (membrane filtration, evaporation), and chemical (active carbon adsorption, resin exchange, alkali detoxification) are used to get rid of the inhibitors [16]. Studies are being done in the direction of improving plants for less lignin content and use of recombinant strategies on microbes to tolerate several inhibitors [17]. The transformation of LCB into biofuel and many more value based products appears to be more economical and viable through simultaneous detoxification and fermentation processing (Figure 2.1) [18].

2.2 Composition of Lignocellulosic Biomass

The understanding of the structure-based intricacy and the recalcitrant nature of a typical LCB comes from the core knowledge of the individual constituents existing in any biomass type. Generally, LCB is composed of three major constituents cellulose (35–50%), hemicellulose (20–35%), and lignin (5–30%) [19]. Different biomass feedstock may have varying percent arrangement of the three components (Table 2.1) among which cellulose constituent is found to be higher in percent in nearly all the LCBs as mentioned in Table 2.1.

Table 2.1 LCB composition from varying sources.

Source of LCB	Cellulose (%)	Hemicellulose (%)	Lignin (%)	Reference
Rice straw	35.8	21.5	24.4	[20]
Rice husk	37.1	29.4	24.1	[21]
Corn stalks	50	20	30	[22]
Bamboo	41.8	18	29.3	[23]
Corn stover	38.4	22.9	20.1	[24]
Sugarcane tops	43	27	17	[25]
Chili postharvest residue	39.9	17.8	25.3	[26]
Water hyacinth	24.5	34.1	8.6	[27]

Cellulose is a polysaccharide consisting various monomer units of glucose adjoined by β-1,4 glycosidic linkages [28]. LCB has an intricated structure and is recalcitrant in nature due to its high degree of polymerization (approx. 10,000 units), exclusive crystalline structure as well as existence of an composite linkages of intermolecular and intramolecular hydrogen bond attached to hydroxyl groups in cellulose [19]. When compared with composite fractions of cellulose, hydrophilic and amorphous parts of hemicellulose in LCB have low degree of polymerization which results in insignificant contribution of hemicellulose moieties towards recalcitrant property of LCB.

Hemicellulose (being a branched heteropolymer) consists of five or six-carbon sugars connected by β-1,4 glycosidic bonds and its each segment functions in holding together the cellulose and lignin parts which overall increases the rigidity of biomass matrix [29]. Cellulose and hemicellulose both have reducing sugars thus, gaining importance as a basis of various commercially profitable chemicals [30]. In contrast, the lignin part is composed of extremely complex methoxylated phenylpropanoid units which make the LCB recalcitrant in nature.

Lignin is a 3-D macromolecule biosynthesized from three commonly occurring aromatic monolignol precursors includes coniferyl alcohol, p-coumaryl alcohol and sinaphyl alcohol which are generally referred as G, H and S units respectively and are associated via C–C as well as C–O–C linkages. Several high-end aromatics are obtained by the catalysis of the lignin fraction [31].

LCB recalcitrance serves as the natural defence mechanism in plant cells against animal attacks, microbial deterioration along with other physical factors, however other conditions such as the existence of proteins and acetyl groups as well as the porous nature of biomass also influences the recalcitrance. Deacetylation of LCB might raise lignocellulosic breakdown up to five to seven times as the acetyl groups use covalent ester linkages to bind hemicellulose. This recalcitrant nature serves hindrance in the industrial application of LCB and several pre-treatment methods are used to control this obstacle (Table 2.2). Recalcitrance is also influenced by proteins both positively and negatively, a few proteins improves degradation by disrupting hydrogen bonds between polysaccharides, whereas several proteins supress the function of many hydrolases [32].

Use of dried LCB in the biorefinery is done to dry and store the biomass denature proteins to regulate the inhibitory effect of various proteins. The physical assembly of the material like particle size, pore volume along with accessible surface area (ASA) performs a vital part in making biomass recalcitrant. High ASA offers additional surface area for enzyme during the process of hydrolysis. Refined grinding forms particle with small sizes causing alterations in polymerization and porosity, thus improves enzyme hydrolysis. Pore size of biomass plays a crucial part as the enzymes enter the pores of certain size only [33].

2.3 Insight on the Pre-Treatment of LCB

Breakdown of the three constituents of LCB after pre-treatment is very crucial for their further conversion to bioenergy with high efficiency. Pre-treatment process is influenced by the parameters such as energy uptake and its economic value which serves as the defining step for industrial feasibility of this process. In the pre-treatment method, it is important that the inherent assembly of the constituents of LCB remain intact. In case of lignin its native structure is altered in the process of delignification, hence, making lignin prone to valorization [40]. During the pre-treatment process the decrease in crystallinity affects the enzyme accessibility to cellulosic part [41]. For the development of optimal approach of pre-treatment it is necessary to consider the intricacy and variability of the LCB so that it does not cause hindrance to the structural and constitutional features

Table 2.2 Impact of LCB constitution and structure on recalcitrance.

Method of pre-treatment	Constituent	Part in recalcitrance	Reference
Physical structure	Particle size	Reduced size of particle increases reaction surface area as well as minimizes degree of polymerization and crystallinity.	[34]
	Surface area	Higher ASA offers additional contact with enzyme and upsurges rate of hydrolysis.	[35]
	Crystallinity	Amorphous form of cellulose has rate of hydrolysis approx. 30 times more than crystalline form of cellulose.	[36]
	Pore size	It constitutes essential aspect in the process of hydrolysis as cellulase can pass through pores larger than 5.1 nm only.	[37]
	Degree of polymerization	Depolymerization process causes hydrolysis of biomass. Reduced degree of polymerization offers additional availability to enzymes thus, improving hydrolysis.	[36]

(*Continued*)

Table 2.2 Impact of LCB constitution and structure on recalcitrance. (*Continued*)

Method of pre-treatment	Constituent	Part in recalcitrance	Reference
Chemical constitution	Hemicellulose	Hemicellulose functions as a physical hurdle and inhibits the approachability of cellulase enzyme to cellulose.	[38]
	Acetyl group	It prevents hydrogen bonding among catalytic domain of cellulase and cellulose hence, leading to interference in enzyme recognition	[39]
	Lignin	It also functions as a physical hurdle making cellulose inaccessible. Derivatives of lignin have repressive effect on hydrolases.	[39]

of the individual components. The efficiency of a pre-treatment approach relies on its potential to:

- Keeping the native lignin structure intact while delignifying the LCB
- Less energy uptake
- Decrease the crystallinity index of cellulose
- Cost-efficient operation
- Improvement in the surface area for enhanced enzymatic hydrolysis due to reduced particle size of LCB
- Evade the production of enzyme inhibitors
- usage of biodegradable chemicals
- Pretreat various types of LCB feedstocks

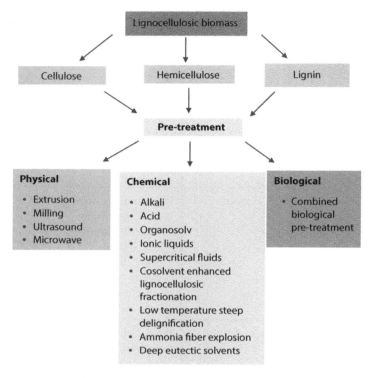

Figure 2.2 Overview of different LCB pre-treatment processes.

Overall, the pre-treatment methods are classified as physical, chemical, and physicochemical approaches (Figure 2.2). In the following segments, we have discussed current developments in techniques for the LCB pre-treatment with their pros and cons.

2.4 Physical Pre-Treatment Method

This method of pre-treatment is utilized to minimize the particle size of the biomass. Physical pre-treatment involves usage of grinders, millers, UV or microwave radiations. Furthermore, to aid the process utilization of secondary or tertiary origin like solvents, enzymes, or chemicals is done.

2.4.1 Extrusion Method

In this method, usage of extruder screws contoured in a fitted barrel is required. When the biomass is packed in the extruder, the extruder screws spin to produce higher shear force among the barrel, biomass, as well as the screw. These shear forces created along the extruder is a thermomechanical procedure that causes rise in temperature and pressure. The higher temperature and pressure through the process alters the physical such as shortening of fibre and defibrillation along with chemical structure of the biomass which raises the surface area and hence, improves the enzyme availability for enabling hydrolysis [42, 43]. Maximum output during the process is effected by many significant parameters including rotation speed of the screw, type of screw, moisture, temperature, etc.

2.4.2 Milling Method

This method of pre-treatment helps in reduction of particle size of biomass. It is carried out in many forms such as chipping, milling, grinding, shredding, etc. Though, the type of method used influences the final particle size for instance size of 0.2 to 2 mm is obtained while milling or grinding of biomass whereas chipping results in particle size of 10 to 30 mm [29]. Benefits associated with this method are:

- Decrease in crystallinity index of cellulose
- Enhancement in the existing surface for enzyme hydrolysis
- Minimal degree of cellulose polymerization
- Reduction in the particle size results in improved mass transfer

2.4.3 Ultrasound Method

Ultrasound is a form of acoustic energy that travels as waves having frequencies over the hearing range (>20 kHz). These waves can propagate in any medium which generates the acoustic cavitation causing initiation of spontaneous development of microbubbles and the collapse of the latter results in creation of shockwaves/local hotspots that releases high pressure of ~50 MPa and temperature of ~5000°C [44]. During the encounter with the shockwaves at high temperature and pressure, disintegration of crystalline structure of the solid particles occurs. According to one hypothesis it is believed that the acoustic oscillations generates high frequency which results in the breakdown of the water molecules producing oxidizing radicals; H- and OH- that are highly reactive in nature and break the glycosidic linkages existing in the complex alignment of LCB [45, 46]. Recently this approach has appeared as the pacesetter in the LCB pre-treatment. Usage of this technique helps control the LCB recalcitrance; thus catalyzing broad array of reactions for the commercialization of biomass [47]. Advantages of the ultrasound waves are high activation energy, lower residence time and adequate mass transfer for efficient LCB breakdown.

2.4.4 Microwave Method

This method uses microwaves that are nonionizing electromagnetic radiations that enable the disturbance of the recalcitrant structures induced by blasts within the particles in a material. On the basis of green chemistry principles, the emergence of microwave as a thermal procedure to pretreat LCB has gained wide reception in biorefinery industries. LCB serves as biological conductor due to the occurrence of crystalline ionic along with nonionic amorphous sections [48]. Heat produced by microwaves cause development of hot spots in the biomass particles because of existence of polar groups in the cellulosic constituents of a biomass [49]. Relocation of the crystalline structure is accelerated due to constant blast within the particles [50].

In another theory [51], pressure is created in the hydrated and nonhydrated components of biomass due to microwave heating. Benefits offered by the usage of microwave heating are rapid and uniform heating, economically feasible, low reaction activation energy, environment friendly, small reaction time, less side products, higher product yield and energy saving [52]. Hence, the usage of microwave heating has led to the advancement of progressive pre-treatment technologies for LCB since the prior 30 years. Microwave method of pre-treatment is significant as developing system for biomass processing having many benefits as discussed above. These emerging techniques do undergo from some drawbacks like high capital investment, operational costs at high loads and pressure, energy consumption, moisture content of biomass sample and many more are needed to be considered for their industrial setup.

2.5 Chemical Pre-Treatment Methods

2.5.1 Alkali Method

In this method pre-treatment of biomass is done using the hydroxides of sodium, ammonium, potassium, calcium [29]. Out of these hydroxides, sodium hydroxide is the most prevalent [53]. Alkali substance disrupts the side chain of esters and glycosides leading to structural alteration of lignin, swelling of cellulosic fibre, heating of swollen cellulose resulting in decrystallization and hemicellulose solvation [54–56]. This method of pre-treatment takes place in mild environment, few require ambient temperature, as proven by soaking of biomass in sodium hydroxide (NaOH), or in ammonium hydroxide (NH_4OH). It can minimize the requirement for costly materials and special strategies in handling of corrosion and drastic reaction conditions. Furthermore, when acetyl groups and uronic acid substitutions in hemicelluloses are removed in the midst of alkali pre-treatment process it raises the availability of carbohydrates to enzyme hydrolysis [43, 57].

When the parameters of alkali pre-treatment of sugarcane bagasse and filter mud were optimized to improve biomethanation, it resulted in the elimination of lignin by 86.27% via 1% of NaOH at 100°C for about 3 hours thus, increasing methane yield by 82.20% [58]. However, this method is more in favour of biomass with lower content of lignin for instance agricultural remains and herbaceous crops and less beneficial for hardwoods.

2.5.2 Acid Method

This approach is generally used for removal of hemicellulose that leads to better availability of enzymes for cellulose [29]. Acetic acid, phosphoric acid, sulphuric acid are mostly required for pre-treatment. Acid pre-treatment goes two ways—pre-treatment by dilute acid (0.1%) at elevated temperature (less than 200°C), pre-treatment by concentrated acid (about 30–70%) at comparatively low temperature; both approaches having their own set of pros and cons [59, 60]. The acid consumption is less in the dilute acid method but the energy required for complete procedure is high because of elevated temperatures. In contrast, energy uptake in concentrated acid approach is less due to the low temperature requirement however, fermentation inhibitors (furfural, 5-hydroxymethylfurfural, phenolic acids, and aldehydes) are produced due to high acid uptake.

Microbes used for the fermentation procedure are severely impacted by the presence of inhibitors which includes DNA breakdown and decreased synthesis of RNA leading to hurdles in enzyme activity [61]. Besides, high concentration of acid is more vulnerable to corrosion of the reaction vessel. In few instances, enzyme hydrolysis step can be evaded as acid itself converts biomass to fermentable sugars but, thorough cleaning is required to get rid of acid prior to fermentation of sugars [62]. Many types of reactors have been developed such as plug flow, batch, percolation, counter current reactors, shrinking-bed, and flow-through. To make this approach economically feasible, retrieval of concentrated acid after hydrolysis is necessary.

2.5.3 Organosolv Method

In this approach organic solvents are used such as ethanol, ethylene glycol, tetrahydrofurfuryl alcohol, acetone, methanol is involved for biomass pre-treatment and occasionally organic acids (oxalic, salicylic, acetylsalicylic) as well as bases (lime, sodium hydroxide) are utilized as catalysts [63]. In this process, due to solubilization linkages between lignin and hemicellulose are cleaved resulting in enhanced overall surface area of cellulose for enzymatic hydrolysis [43]. This process requires occurrence of acid, base or salt catalyst for the reaction to proceed [64]. In organosolv pre-treatment, dependence on biomass type and catalyst influences the temperature range which may reach 200°C. Lignin being a value based product is mainly extracted from this method, along with cellulose portion and hemicellulose mixture of C5 and C6 sugars. Lignin removal from biomass centres the cellulosic fibres for enzyme hydrolysis causing maximum conversion of biomass [65].

In pretreated biomass, physical parameters such as degree of cellulose polymerization, crystallinity, fiber length etc. relies on variable aspects like solvent concentration, reaction time, catalyst used and temperature. Fermentation inhibitors are formed due to high concentration of acid, lengthy reaction time and elevated temperatures. Impact of various catalysts (NaOH, $MgSO_4$ and H_2SO_4) on pine was studied, concluding H_2SO_4 as the highly efficient catalyst for ethanol yield while, in respect of digestibility, NaOH was reported to be efficient especially at increased concentration of 2% [66]. H_2SO_4 though being a strong catalyst due to its high reactivity is also inhibitory in nature, toxic and corrosive and has major drawback in the form of high solvent cost that can reduced by the recovery and recycle of solvents through evaporation and condensation. Solvent removal is necessary due to its negative effect microbial growth, enzyme hydrolysis and fermentation [65]. This is less preferred approach because handling of harsh organic solvents involves high risk as they are highly flammable and lack of appropriate measures for safety may lead to grave damage causing huge fire explosions. Treatment of biomass with combination of phenol, HCl, and water at pressure 1 atm and temperature of about 100 °C is called battelle (a type of organosolv approach) [67].

2.5.4 Ionic Liquids

Ionic liquids (ILs) bearing unique solvation properties are classified as green solvents (Table 2.3). It needs low vapor pressure and exhibits huge thermal stabilities and lower toxicity. ILs particularly remove biomass parts

Table 2.3 List of ILs used for LCB delignification.

ILs	Biomass type	Conditions	Removal of lignin (%)	Reference
1-Butylimidazolium hydrogen sulphate [C4Him][HSO$_4$]	Miscanthus	[C4Him][HSO4] 80%, 120°C, 4 hours	81.14	[72]
	Miscanthus	[C4Him][HSO4] 95%, 120°C, 20 h	92.83	[72]
1,3-Dimethylimidazolium methyl sulphate [C4Clim][MeSO$_4$]	Miscanthus	[C4Clim][MeSO4] 80%, 120°C, 2 hours	27.2	[72]
1-Butyl-3-methylimidazolium acetate [C4Clim]MeCO$_2$]	Miscanthus	[C4Clim]MeCO$_2$] 80%, 120°C, 22 hours	26.23	[72]
	Switchgrass	[C4Clim]MeCO2] 100%, 160°C, 3 hours	65	[73]
	Oak	[C4Clim]MeCO2 100%, 110°C, 16 hours	34.9	[73]
1-Butyl-3-methylimidazolium hydrogen sulfate [C4Clim][HSO$_4$]	Willow	[C4Clim][HSO4] 80%, 120°C, 22 hours	85.07	[72]
	Miscanthus	[C4Clim][HSO$_4$] 80%, 120°C, 22 hours	92.84	[72]

lignin and hemicellulose to offer pure cellulose for effective hydrolysis and this method can be performed more orderly in continuous method with high input of biomass [68, 69]. Setbacks associated with this method of pre-treatment are pH compatibility, process intricacy, ILs toxicity and high cost [10]. Hemicellulose and lignin derivatives are used for the synthesis of cost-effective and ecofriendly ILs [70].

ILs such as [VanEt2NH][H2PO4], [p-AnisEt2NH][H2PO4], and [FurEt2NH][H_2PO_4] are generated by reduced amination of lignin monomers vanillin, p-anisaldehyde, and furfural, respectively, with subsequent treatment of phosphoric acid. Enzyme hydrolysis of pretreated biomass via freshly synthesized three ionic liquid and [C2mim][OAc] resulted in enhanced sugar yields. ILs causes breakup of cellulose because of good solubilities of about 5% to 20% and this cellulosic breakdown is associated with strong hydrogen bonds among anions of ILs and equatorial hydroxyl groups of cellulose. Two processes are extensively utilized for complete biomass solubilization; in first process use of acidified or acidic LCB dissolving ILs is done whereas, in second process use of ILs that target dissolution of lignin and part hemicellulose while keeping cellulose fraction intact. IL pre-treatment has attained 15% to 92% of lignin removal [69]. Industrial application of ILs depends upon its cost-effectiveness, recovery and reuse of ILs. Addition of glycerol to carboxylate ILs led to about full recovery of ILs [71]. Use of ILs at industrial level for biomass pre-treatment is limited due to huge expenses of imidazolium cations. Hence, renewable sources are used to develop novel cations based ILs.

2.5.5 Supercritical Fluids

Supercritical fluids (SCFs) technique are generally attributed as sustainable, alternative, and favorable ecofriendly approach for LCB pre-treatment. They possess exclusive features of liquid and gas, involving viscosities and diffusivity like gases and density akin to liquids [74]. It consists of exceptional physicochemical parameters like less viscosity and high diffusivity that permits easy movement of these nontoxic and relatively cheap solvents through solid materials. Industrial applications of SCFs are widely acknowledged due to less maintenance costs and higher-energy productivity thus, they are also endorsed as future green solvents [75].

Frequently used two SCFs are supercritical water (SCW) and supercritical CO_2 (ScCO_2) both of them are nonflammable, thermodynamically stable and noncarcinogenic [76]. This technique has appeared as an advanced and potent approach for valorization of biomass. They can further be engaged in pre-treatment techniques for biomass and for effectual extraction of value-based products from LCB. Normally, treatment with solvent at high temperature and pressure alters the intricated physical appearance of a regular LCB which results in widening of the fibrous micropores and raises fibre porosity hence, making solvent diffusion in the biomass more effective [77]. The bonds existing in the matrix of lignocellulose are broken due to increase in exchanges among solid biomass particles and SCF at high pressure [78].

Pre-treatment of corn stover using $ScCO_2$ resulted in the higher glucose yield of 30% at 150°C and 3500 psi for 1 h [79]. Usage of $ScCO_2$ improves penetration and lowers the viscosity thus increasing mass transfer. The reaction amongst water and CO_2 existing in the biomass forms carbonic acid thus, creating weak acidic environment which increases the mass transfer and the hydrolysis of hemicellulose during the pre-treatment. The potential of $ScCO_2$ is also increased due to presence of cosolvents in the reaction medium as they perform a crucial part in improving the polarity and solvation abilities [80]. Combined pre-treatment due to the collaborative action on biomass treatment has attained improved glucan retrieval and enzyme based conversion of sugars [81].

2.5.6 Cosolvent Enhanced Lignocellulosic Fractionation

This process practices a mixture of the organic compound tetrahydrofuran (THF), water, and dilute sulfuric acid for partition of lignin from LCB [82]. The higher sugar yields at lower dose of enzyme is obtained using THF; a multipurpose renewable solvent which solubilize acetylated lignin from biomass as well as encourages hydrolysis of cellulose in water [83]. Delignification of biomass is done by using polar cosolvents like dimethyl sulfoxide, γ-valerolactone, acetone, and 1,4-dioxane [84]. This approach is used to meet the necessity to break hydrogen bonding and obstruct hydrophobic stacking exchanges in cellulose. Cosolvent enhanced lignocellulosic fractionation (CELF) pre-treatment functions at elevated temperatures and releases other side products which includes 5-hydroxymethylfurfural, levulinic acid and furfural that can further be utilized for value-based chemical production.

In comparison to conventional dilute acid pre-treatment, CELF pre-treatment in mild conditions attains over 95% recovery of fermentable sugars postenzymatic hydrolysis [85]. Poplar biomass when treated at different conditions of catalysts (0.1 M H_2SO_4), time period (15–60 min), THF:water ratios (1:1 or 7:1 (v:v)) and temperature (160–180°C) resulted in recovery of new type of lignin, CELF lignin. CELF pretreated samples have higher sugar yields because biomass structure consists of broken cell walls which causes enzymes seeking cellulose surface more easy. Changes in physiochemical nature of the lignin structure is a setback in the CELF process. The dissolution of native β-O-4 linkages causes significant reduction of molecular weight of the CELF lignin that shows clear structural alterations takes place through the CELF process [12].

2.5.7 Low Temperature Steep Delignification

Low-temperature steep delignification (LTSD) is an effective approach for LCB conversion that uses lower concentrations of nontoxic chemicals, bases and oxygen. Separation of cellulose, lignin and hemicellulose is done under mild process conditions. Almost 90% of the lignin is removed and recovered from biomass in this method. Advantages associated with LTSD are low cost of operation, recycle of chemicals, production of toxic chemicals or inhibitors is none, environmental sustainability. Bioprocess Innovation Company handling softwood biomass (miscanthus and switchgrass), woody biomass (forestry residues) and agricultural residues (wheat straw) developed LTSD [86]. It permits incorporation of pre-treatment approach with enzyme hydrolysis, fermentation as well as distillation process at a preliminary level. This approach was utilized to produce sugars from mixed hardwood chips [87].

Hydrogen peroxide, sodium chlorite and oxygen treatments were examined to study the effect of delignification on biomass. Lignin removal in mixed hardwood samples was found to be most efficient in sodium chlorite delignification (SCD) method in comparison to hydrogen peroxide delignification (HPD) methods and oxygen delignification (OD). Complete xylan conversion was obtained with 78% yield of glucan after enzyme hydrolysis when mixed hardwood samples were exposed to repetitive OD treatments [87]. Conversion of carbohydrates was only 2.5% from untreated mixed hardwood samples. Enzymatic degradation of hemicellulose was more effective by HPD method. Rise in conversion for glucan and xylan when SCD and OD process were used. SCD was found to be most efficient out of the three delignification processes, for lignin removal and to obtain high conversion yield.

2.5.8 Ammonia Fiber Explosion

Ammonia fiber explosion (AFEX) method of pre-treatment consists of LCB treatment with anhydrous liquid ammonia (1:1 w/w) at higher pressure (15–30 bar) under average reaction temperatures (60–170°C) and for a small time period (5–60 min) [88]. Many benefits are associated with this method, such as moderate temperature, high retention of cellulose/hemicellulose fraction, enhanced surface area of LCB along with availability of enzymes for hydrolysis, deacetylation of hemicellulose and convenient breakdown of LCB, insignificant inhibitor production (sugar hydrolysis products), less residence time etc. Optimization of the AFEX process is done to acquire high yields of glucan and xylan via variation in five main features—residence time, ammonia content, temperature, water content, and pressure [89].

AFEX and dilute acid process used for corn stover pre-treatment at 140°C to 160°C for 15 to 20 minutes caused reduction in acid insoluble lignin from 17.2% to 12.2% for untreated corn stover and AFEX corn stover respectively because of partial solubilization of lignin. Pretreated corn stover in case of AFEX when subjected to hydrolysis and fermentation generate ethanol yield of 20.5 kg of ethanol per kg of biomass whereas in dilute acid treated sample lower yield of 14 kg was generated [88]. AFEX-pretreated switchgrass reported ethanol yield about 2.5 times high in comparison to the untreated sample [90]. In this process, among cellulose fibers rearrangement of the hydrogen bonding takes place that causes an raise in interhydrogen linkages and a decline in intrahydrogen linkages hence, converting crystalline form of cellulose to its amorphous form [89]. Improved version of AFEX process was developed known as liquid ammonia pre-treatment (LAT) to pretreat giant reed [91]. LAT process releases ammonia slowly rather than fast release (in case of AFEX process) hence, evading explosion during process which makes this method more controlled and cost-effective. Several drawbacks are associated with this method among which the most important is the huge capital cost of equipment to resist the involvement of high pressure during the process.

The industrialization of the process are facing issues such as the huge charges of ammonia and the rise in energy necessities for recycle of ammonia [60]. Moreover, pre-treatment of biomass consisting high content of lignin has lower efficiency for the AFEX process. Higher hydrolysis yield of 90% for bagasse and bermudagrass having 15% and 5% lignin content respectively was obtained as compared to lesser hydrolysis yield of 40% to 50% was observed for AFEX-pretreated aspen chips and newspapers having 25% lignin content [43].

2.5.9 Deep Eutectic Solvents

Rendering to the basics of green chemistry, usage of environmental sustainable solvents is a crucial part in making the overall process of pre-treatment renewable. Industrial use of traditional solvents is lethal and presents grave danger to the ecosystem thus, increasing the need to develop alternate green solvents. Combination of hydrogen bond donor (HBD) and hydrogen bond acceptor (HBA) in suitable proportions was done to synthesize deep eutectic solvent (DESs) [92]. In comparison to ionic liquids (ILs) and conventional solvents, DESs are inexpensive, less lethal, and green solvents as well as its synthesis is simple and doesn't require complex steps of purification. The eutectic solvents have extensive tuneability of its physicochemical parameters comparable to ILs which makes them beneficial [93].

DESs is used for effectual delignification of diverse biomass feedstocks because of its exceptional solvent properties involving high polarity, ease in miscibility with water, and other cosolvents, etc. [94]. DESs have potential for selective lignin removal from the LCB complex hence, generates an environment sustainable and novel ways for the LCB pre-treatment in mild condition. DESs showed effectual lignin solubilization, whereas solubility of cellulose was insignificant (Table 2.4) [95]. Cellulose showed insignificant solubility (<2 wt%) in choline chloride (ChCl) based DESs at 110°C for 12 hours [96]. Cellulose possess intense cohesive energy which averts the breakup and restructuring of the hydrogen bonds that is needed for dissolution of cellulose in DESs. Strong hydrogen bonding system with cellulose is formed due to the existence of hydroxyl groups in ChCl that stabilizes the DES-cellulose system [97]. Delignification of the biomass feedstocks is severely affected by the high viscosity of the DESs because of the limitations in the associated mass transfer [98]. Reduction in DES viscosity occurs due to elevation in temperature hence increasing mass transfer in the system [99].

Presence of several HBAs and HBDs aids in the synthesis of broad range of task oriented DESs with exceptional physicochemical parameters hence, creating endless prospects for use of DESs in diverse research areas. Usage of DESs in broad scale applications is limited due to certain setbacks associated with it such as recyclability, thermal stability, high viscosity, etc. Complete perception of the DESs connections with the LCB constituents is still unknown [98].

2.6 Biological Pre-Treatment Methods

Biological method in contrast to chemical and physical methods, is regarded as an effective, ecofriendly and energy saving approach. Microbes (cellulolytic and hemicellulolytic) that can be precisely aimed for efficient pre-treatment of biomass are found in abundance in nature [102]. In this

Table 2.4 Lignin and cellulose solubility in various DESs.

DES (HBD: HBA)	Molar ratio (HBD: HBA)	Temp. (°C)	Lignin (weight %)	Cellulose (weight %)	Reference
LA[a]: Betaine	2:1	60	9	<1	[100]
LA: Histidine	9:1	60	11.8	0	[95]
LA: ChCl	2:1	60	5.3	0	[95]
Oxalic acid dihydrate: ChCl	1:1	60	3.6	0	[95]
Malic acid: ChCl	1:1	100	3.4	0	[95]
Acetic acid: ChCl	2:1	60	12	<1	[100]
LA: Glycine	9:1	60	8.7	0	[95]
PEG[b] : DBU[c]	3:1	90	41.1	Not reported	[101]

*Lactic acid (LA); polyethylene glycol 200 (PEG); 1,5-diazabicyclo[5.4.0]-5-undecene (DBU)

method microbes that breakdown hemicellulose, lignin as well as little amount of cellulose are white, brown and soft-rot fungi [103]. Occurrence of laccases and peroxidases (lignin degrading enzymes) in white-rot fungi causes lignin degradation [104]. Common white-rot fungal species are *Cyathus stercolerus, Pycnoporus cinnarbarinus, Phanerochaete chrysosporium, Pleurotus ostreaus, Ceriporia lacerata,* and *Ceriporiopsis subvermispora* used for pre-treatment. Several species of basidiomycetes such as *Fomes fomentarius, Trametes versicolor, Ganoderma resinaceum, Lepista nuda, Bjerkandera adusta, Phanerochaete chrysosporium,* and *Irpex lacteus* are reported to have high efficiency for delignification [104, 105].

2.6.1 Combined Biological Pre-Treatment

Researchers have found that biological process when combined with other pre-treatment process results in more efficient process. Enzymatic saccharification of *Populus tormentosa* was enhanced by combining biological process with liquid hot water method as it reported largest hemicellulose degradation (approx. 92.33%) thus, achieving raise in glucose yield by 2.66-fold [106]. The unique mixture of physical (ultrasound) and chemical (H_2O_2) pre-treatment method along with biological (*P. ostreatus*) method on rice husk resulted in significant lignin removal within 48 hours than using sing step pre-treatment by *P. ostreatus* for 60 days [107].

2.7 Future Prospects

Above mentioned methods of pre-treatment have progressively increased in their potential of biomass degradation. A serious assessment of pre-treatment methods points the need to develop a commercial and ecofriendly approach that can enable LCB delignification completely and valorization of by-products. The upcoming research attempts must be focused more on complete knowledge of reaction mechanism of pre-treatment to create suitable methods of pre-treatment with consideration to diverse LCB composition as well as the desired product. A considerable evaluation of pre-treatment of biomass by combining two or more approaches is also recommended for complete disintegration of biomass. Selective efforts are required to confront the necessity of biomass pre-treatment approach that is cheap and economically fruitful. Suggestion for pre-treatment method that is regulated in terms of energy expenditure is also given.

2.8 Conclusion

Development of effective and environment friendly pre-treatment techniques helps in perceiving the concept of sustainable biorefineries. Different pre-treatment methods when evaluated suggested the need for ideal approach that should improve cellulose crystallinity, control recalcitrant property of LCB and promise full recovery of sugars along with other value based products. The current advancements in pre-treatment methods on the basis of physical and chemical approaches as well as their pros and cons are discussed above. The deeper understanding of different methods of pre-treatment will speed up the progress of new profitable, less energy intensive, and environment sustainable methods also confronting the problems attached helps in enhancing the present pre-treatment methods.

References

1. Watts, N., Amann, M., Arnell, N., Ayeb-Karlsson, S., Belesova, K., Berry, H. *et al.*, The 2018 report of the Lancet Countdown on health and climate change: Shaping the health of nations for centuries to come. *Lancet*, 392, 2479–2514, 2018.
2. Mazarji, M., Kuthiala, S., Tsapekos, P., Alvarado-Morales, M., Angelidaki, I., Carbon dioxide anion radical as a tool to enhance lignin valorization. *Sci. Total Environ.*, 682, 47–58, 2019.
3. Luo, H., Zheng, P., Bilal, M., Xie, F., Zeng, Q., Zhu, C., Yang, R., Wang, Z., Efficient bio-butanol production from lignocellulosic waste by elucidating the mechanisms of *Clostridium acetobutylicum* response to phenolic inhibitors. *Sci. Total Environ.*, 710, 136399, 2020.
4. Toor, M., Kumar, S.S., Malyan, S.K., Bishnoi, N.R., Mathimani, T., Rajendran, K., Pugazhendhi, A., An overview on bioethanol production from lignocellulosic feedstocks. *Chemosphere*, 242, 125080, 2020.
5. Kavitha, S., Kannah, R.Y., Kasthuri, S., Gunasekaran, M., Pugazhendi, A., Rene, E.R., Pant, D., Kumar, G., Banu, J.R., Profitable biomethane production from delignified rice straw biomass: The effect of lignin, energy and economic analysis. *Green Chem.*, 22, 8024–8035, 2020.
6. Liguori, R. and Faraco, V., Biological processes for advancing lignocellulosic waste biorefnery by advocating circular economy. *Bioresour. Technol.*, 215, 13–20, 2016.
7. Nanda, S., Mohanty, P., Pant, K.K., Naik, S., Kozinski, J.A., Dalai, A.K., Characterization of North American lignocellulosic biomass and biochars in terms of their candidacy for alternate renewable fuels. *Bioenergy Res.*, 6, 663–677, 2013.

8. Kumar, A.K. and Sharma, S., Recent updates on different methods of pre-treatment of lignocellulosic feedstocks: A review. *Bioresour. Bioprocess.*, 4, 1–19, 2017.
9. Sankaran, R., Parra Cruz, R.A., Pakalapati, H., Show, P.L., Ling, T.C., Chen, W.-H., Tao, Y., Recent advances in the pre-treatment of microalgal and lignocellulosic biomass: A comprehensive review. *Bioresour. Technol.*, 298, 122476, 2019.
10. Singh, S., Designing tailored microbial and enzymatic response in ionic liquids for lignocellulosic biorefineries. *Biophys. Rev.*, 10, 911–913, 2018.
11. Karthikeyan, O.P., Trably, E., Mehariya, S., Bernet, N., Wong, J.W., Carrere, H., Pre-treatment of food waste for methane and hydrogen recovery: A review. *Bioresour. Technol.*, 249, 1025–1039, 2018.
12. Meng, X., Parikh, A., Seemala, B., Kumar, R., Pu, Y., Christopher, P., Wyman, C.E., Cai, C.M., Ragauskas, A.J., Chemical Transformations of Poplar Lignin during Cosolvent Enhanced Lignocellulosic Fractionation Process. *ACS Sust. Chem. Eng.*, 6, 8711–8718, 2018.
13. Sorn, V., Chang, K.-L., Phitsuwan, P., Ratanakhanokchai, K., Dong, C.-D., Effect of microwave-assisted ionic liquid/acidic ionic liquid pre-treatment on the morphology, structure, and enhanced delignification of rice straw. *Bioresour. Technol.*, 293, 121929, 2019.
14. Jagtap, S.S., Bedekar, A.A., Liu, J.-J., Jin, Y.-S., Rao, C.V., Production of galactitol from galactose by the oleaginous yeast *Rhodosporidium toruloides* IFO0880. *Biotechnol. Biofuels.*, 12, 250, 2019.
15. Moreno, A.D., Carbone, A., Pavone, R., Olsson, L., Geijer, C., Evolutionary engineered *Candida intermedia* exhibits improved xylose utilization and robustness to lignocellulose-derived inhibitors and ethanol. *Appl. Microbiol. Biotechnol.*, 103, 1405–1416, 2019.
16. Farmanbordar, S., Amiri, H., Karimi, K., Simultaneous organosolv pre-treatment and detoxification of municipal solid waste for efficient biobutanol production. *Bioresour. Technol.*, 270, 236–244, 2018.
17. Shafrin, F., Ferdous, A.S., Sarkar, S.K., Ahmed, R., Hossain, K., Sarker, M., Rencoret, J., Gutiérrez, A., Jose, C., Sanan-Mishra, N., Modification of monolignol biosynthetic pathway in jute: Different gene, different consequence. *Sci. Rep.*, 7, 39984, 2017.
18. Hazeena, S.H., Nair Salini, C., Sindhu, R., Pandey, A., Binod, P., Simultaneous saccharification and fermentation of oil palm front for the production of 2,3-butanediol. *Bioresour. Technol.*, 278, 145–149, 2019.
19. Kassaye, S., Pant, K.K., Jain, S., Synergistic effect of ionic liquid and dilute sulphuric acid in the hydrolysis of microcrystalline cellulose. *Fuel Process. Technol.*, 148, 289–294, 2016.
20. Imman, S., Arnthong, J., Burapatana, V., Champreda, V., Laosiripojana, N., Fractionation of rice straw by a single-step solvothermal process: Effects of solvents, acid promoters, and microwave treatment. *Renew. Energy*, 83, 663–673, 2015.

21. Kalita, E., Nath, B.K., Deb, P., Agan, F., Islam, M.R., Saikia, K., High quality fluorescent cellulose nanofibers from endemic rice husk: Isolation and characterization. *Carbohydr. Polym.*, 122, 308–313, 2015.
22. Christopher, M., Mathew, A.K., Kiran Kumar, M., Pandey, A., Sukumaran, R.K., A biorefinery-based approach for the production of ethanol from enzymatically hydrolysed cotton stalks. *Bioresour. Technol.*, 242, 178–183, 2017.
23. Ma, Y., Tan, W., Wang, J., Xu, J., Wang, K., Jiang, J., Liquefaction of bamboo biomass and production of three fractions containing aromatic compounds. *J. Bioresour. Bioprod.*, 5, 114–123, 2020.
24. Wan, C. and Li, Y., Microbial pre-treatment of corn stover with *Ceriporiopsis subvermispora* for enzymatic hydrolysis and ethanol production. *Bioresour. Technol.*, 101, 6398–6403, 2010.
25. Sindhu, R., Kuttiraja, M., Binod, P., Sukumaran, R.K., Pandey, A., Physicochemical characterization of alkali pretreated sugarcane tops and optimization of enzymatic saccharification using response surface methodology. *Renew. Energy*, 62, 362–368, 2014.
26. Sindhu, R., Binod, P., Mathew, A.K., Abraham, A., Gnansounou, E., Ummalyma, S.B., Thomas, L., Pandey, A., Development of a novel ultrasound-assisted alkali pre-treatment strategy for the production of bioethanol and xylanases from chili post harvest residue. *Bioresour. Technol.*, 242, 146–151, 2017.
27. Ruan, T., Zeng, R., Yin, X.-Y., Zhang, S.-X., Yang, Z.-H., Water hyacinth (*Eichhornia crassipes*) biomass as a biofuel feedstock by enzymatic hydrolysis. *BioResources*, 11, 2372–2380, 2016.
28. Dora, S., Bhaskar, T., Singh, R., Naik, D.V., Adhikari, D.K., Effective catalytic conversion of cellulose into high yields of methyl glucosides over sulfonated carbon based catalyst. *Bioresour. Technol.*, 120, 318–321, 2012.
29. Veluchamy, C., Kalamdhad, A.S., Gilroyed, B.H., Advanced pre-treatment strategies for bioenergy production from biomass and biowaste. *Handb. Environ. Mater. Manag.*, 1–19, 2018.
30. Quereshi, S., Ahmad, E., Pant, K.K., Dutta, S., Insights into microwave-assisted synthesis of 5-ethoxymethylfurfural and ethyl levulinate using tungsten disulfide as a catalyst. *ACS Sustain. Chem. Eng.*, 8, 1721–1729, 2020.
31. Zhang, C. and Wang, F., Catalytic lignin depolymerization to aromatic chemicals. *Acc. Chem. Res.*, 53, 470–484, 2020.
32. Zhao, X., Zhang, L., Liu, D., Biomass recalcitrance. Part I: The chemical compositions and physical structures affecting the enzymatic hydrolysis of lignocellulose. *Biofuels. Bioprod. Biorefin.*, 6, 465–482, 2012.
33. Zhang, H., Chen, L., Lu, M., Li, J., Han, L., A novel film–pore–surface diffusion model to explain the enhanced enzyme adsorption of corn stover pre-treated by ultrafine grinding. *Biotechnol. Biofuels.*, 9, 181, 2016.
34. Zhai, Q., Li, F., Wang, F., Feng, J., Jiang, J., Xu, J., Ultrafine grinding of poplar biomass: Effect of particle morphology on the liquefaction of biomass for methyl glycosides and phenolics. *Cellulose*, 26, 3685–3701, 2019.

35. Cho, E.J., Trinh, L.T.P., Song, Y., Lee, Y.G., Bae, H.-J., Bioconversion of biomass waste into high value chemicals. *Bioresour. Technol.*, 298, 122386, 2019.
36. Liu, Y., Nie, Y., Lu, X., Zhang, X., He, H., Pan, F., Zhou, L., Liu, X., Ji, X., Zhang, S., Cascade utilization of lignocellulosic biomass to high-value products. *Green. Chem.*, 21, 3499–3535, 2019.
37. Ponnusamy, V.K., Nguyen, D.D., Dharmaraja, J., Shobana, S., Banu, J.R., Saratale, R.G., Chang, S.W., Kumar, G., A review on lignin structure, pre-treatments, fermentation reactions and biorefinery potential. *Bioresour. Technol.*, 271, 462–472, 2019.
38. Kumar, R., Bhagia, S., Smith, M.D., Petridis, L., Ong, R.G., Cai, C.M., Mittal, A., Himmel, M.H., Balan, V., Dale, B.E., Ragauskas, A.J., Cellulose–hemicellulose interactions at elevated temperatures increase cellulose recalcitrance to biological conversion. *Green. Chem.*, 20, 921–934, 2018.
39. Wang, H.L., Pu, Y.Q., Ragauskas, A., Yang, B., From lignin to valuable products-strategies, challenges, and prospects. *Bioresour. Technol.*, 271, 449–461, 2019.
40. Zakaria, M.R., Fujimoto, S., Hirata, S., Hassan, M.A., Ball milling pre-treatment of oil palm biomass for enhancing enzymatic hydrolysis. *Appl. Biochem. Biotechnol.*, 173, 1778–1789, 2014.
41. Zhao, C., Ding, W., Chen, F., Cheng, C., Shao, Q., Effects of compositional changes of AFEX-treated and H-AFEX-treated corn stover on enzymatic digestibility. *Bioresour. Technol.*, 155, 34–40, 2014.
42. Duque, A., Manzanares, P., Ballesteros, M., Extrusion as a pre-treatment for lignocellulosic biomass: Fundamentals and applications. *Renew. Energy*, 114, 1427–1441, 2017.
43. Kumar, B., Bhardwaj, N., Agrawal, K., Chaturvedi, V., Verma, P., Current perspective on pre-treatment technologies using lignocellulosic biomass: An emerging biorefinery concept. *Fuel Process. Technol.*, 199, 106244, 2020.
44. Ashokkumar, M., The characterization of acoustic cavitation bubbles-an overview. *Ultrason. Sonochem.*, 18, 864–872, 2011.
45. Luo, J., Fang, Z., Smith, R.L., Ultrasound-enhanced conversion of biomass to biofuels. *Prog. Energy Combust. Sci.*, 41, 56–93, 2014.
46. Velmurugan, R. and Muthukumar, K., Ultrasound-assisted alkaline pre-treatment of sugarcane bagasse for fermentable sugar production: Optimization through response surface methodology. *Bioresour. Technol.*, 112, 293–299, 2012.
47. Cho, H.M., Gross, A.S., Chu, J.W., Dissecting force interactions in cellulose deconstruction reveals the required solvent versatility for overcoming biomass recalcitrance. *J. Am. Chem. Soc*, 133, 14033–14041, 2011.
48. Budarin, V.L., Clark, J.H., Lanigan, B.A., Shuttleworth, P., Macquarrie, D.J., Microwave assisted decomposition of cellulose: A new thermochemical route for biomass exploitation. *Bioresour. Technol.*, 101, 3776–3779, 2010.
49. Shi, J., Pu, Y., Yang, B., Ragauskas, A., Wyman, C.E., Comparison of microwaves to fluidized sand baths for heating tubular reactors for hydrothermal

and dilute acid batch pre-treatment of corn stover. *Bioresour. Technol.*, 102, 5952–5961, 2011.
50. Hu, Z. and Wen, Z., Enhancing enzymatic digestibility of switchgrass by microwave-assisted alkali pre-treatment. *Biochem. Eng. J.*, 38, 369–378, 2008.
51. Palav, T. and Seetharaman, K., Impact of microwave heating on the physico-chemical properties of a starch-water model system. *Carbohydr. Polym.*, 67, 596–604, 2007.
52. Li, H., Qu, Y., Yang, Y., Chang, S., Xu, J., Microwave irradiation-a green and efficient way to pretreat biomass. *Bioresour. Technol.*, 199, 34–41, 2016.
53. Kumar, R. and Wyman, C.E., Effects of cellulase and xylanase enzymes on the deconstruction of solids from pre-treatment of poplar by leading technologies. *Biotechnol. Prog.*, 25, 302–314, 2009.
54. Olejnik, K., Skalski, B., Stanislawska, A., Wysocka-Robak, A., Swelling properties and generation of cellulose fines originating from bleached kraft pulp refined under different operating conditions. *Cellulose*, 24, 3955–3967, 2017.
55. Zhang, X., Qu, T., Mosier, N.S., Han, L., Xiao, W., Cellulose modification by recyclable swelling solvents. *Biotechnol. Biofuels*, 11, 1–12, 2018.
56. Sills, D.L. and Gossett, J.M., Assessment of commercial hemicellulases for saccharification of alkaline pretreated perennial biomass. *Biores. Technol.*, 102, 1389–1398, 2011.
57. Maurya, D.P., Singla, A., Negi, S., An overview of key pre-treatment processes for biological conversion of lignocellulosic biomass to bioethanol. *3 Biotech.*, 5, 597–609, 2015.
58. Talha, Z., Ding, W., Mehryar, E., Hassan, M., Bi, J., Alkaline pre-treatment of sugarcane bagasse and filter mud codigested to improve biomethane production. *BioMed. Res. Int.*, 2016, 1–10, 2016.
59. Den, W., Sharma, V.K., Lee, M., Nadadur, G., Varma, R.S., Lignocellulosic biomass transformations via greener oxidative pre-treatment processes: Access to energy and value added chemicals. *Front. Chem.*, 6, 1–23, 2018.
60. Rezania, S., Oryani, B., Cho, J., Talaiekhozani, A., Sabbagh, F., Hashemi, B., Rupani, P.F., Mohammadi, A.A., Different pre-treatment technologies of lignocellulosic biomass for bioethanol production: An overview. *Energy*, 199, 117457, 2020.
61. Lorenci Woiciechowski, A., Dalmas Neto, C.J., de Souza, P., Vandenberghe, L., de Carvalho Neto, D.P., Novak Sydney, A.C., Letti, L.A.J., Karp, S.G., Zevallos Torres, L.A., Soccol, C.R., Lignocellulosic biomass: Acid and alkaline pre-treatments and their effects on biomass recalcitrance–Conventional processing and recent advances. *Bioresour. Technol.*, 304, 122848, 2020.
62. Sassner, P., Martensson, C.G., Galbe, M., Zacchi, G., Steam pre-treatment of H_2SO_4 impregnated salix for the production of bioethanol. *Biores. Technol.*, 99, 137–145, 2008.
63. Cheah, W.Y., Sankaran, R., Show, P.L., Ibrahim, T.N.B.T., Chew, K.W., Culaba, A., Chang, J.S., Pre-treatment methods for lignocellulosic biofuels

production: Current advances, challenges and future prospects. *Biofuel Res. J.*, 7, 1115–1127, 2020.
64. Bajpai, P., *Pre-treatment of lignocellulosic biomass for biofuel production*, pp. 17–70, Springer, Singapore, 2016.
65. Agbor, V.B., Cicek, N., Sparling, R., Berlin, A., Levin, D.B., Biomass pre-treatment: Fundamentals toward application. *Biotechnol. Adv.*, 29, 675–685, 2011.
66. Park, N., Kim, H.Y., Koo, B.W., Yeo, H., Choi, I.G., Organosolv pre-treatment with various catalysts for enhancing enzymatic hydrolysis of pitch pine (*Pinus rigida*). *Bioresour. Technol.*, 101, 7046–7053, 2010.
67. Villaverde, J.J., Ligero, P., De Vega, A., Miscanthus x giganteus as a source of biobased products through organosolv fractionation: A mini review. *Open Agric. J.*, 4, 102–110, 2010.
68. Morais, A.R.C., da Costa Lopes, A.M., Bogel-Łukasik, R., Carbon dioxide in biomass processing: Contributions to the green biorefinery concept. *Chem. Rev.*, 115, 3–27, 2015.
69. Brandt, A., Ray, M.J., To, T.Q., Leak, D.J., Murphy, R.J., Welton, T., Ionic liquid pre-treatment of lignocellulosic biomass with ionic liquid–water mixtures. *Green. Chem.*, 13, 2489–2499, 2011.
70. Socha, A.M., Parthasarathi, R., Shi, J., Pattathil, S., Whyte, D., Bergeron, M., George, A., Tran, K., Stavila, V., Venkatachalam, S., Hahn, M.G., Simmons, B.A., Singh, S., Efficient biomass pre-treatment using ionic liquids derived from lignin and hemicellulose. *Proc. Nat. Acad. Sci.*, 111, E3587–E3595, 2014.
71. Clough, M.T., Griffith, J.A., Kuzmina, O., Welton, T., Enhancing the stability of ionic liquid media for cellulose processing: Acetal protection or carbene suppression. *Green. Chem.*, 18, 3758–3766, 2016.
72. Gschwend, F.J.V., Malaret, F., Shinde, S., Brandt-Talbot, A., Hallett, J.P., Rapid pre-treatment of Miscanthus using the low-cost ionic liquid triethylammonium hydrogen sulfate at elevated temperatures. *Green. Chem.*, 20, 3486–3498, 2018.
73. Williams, C.L., Li, C., Hu, H., Allen, J.C., Thomas, B.J., Three way comparison of hydrophilic ionic liquid, hydrophobic ionic liquid, and dilute acid for the pre-treatment of herbaceous and woody biomass. *Front. Energy. Res.*, 6, 1–12, 2018.
74. Mani Rathnam, V. and Madras, G., Conversion of *Shizochitrium limacinum* microalgae to biodiesel by non-catalytic transesterification using various supercritical fluids. *Bioresour. Technol.*, 288, 121538, 2019.
75. Brunner, G., Applications of supercritical fluids. *Annu. Rev. Chem. Biomol. Eng.*, 1, 321–342, 2010.
76. Parhi, R. and Suresh, P., Supercritical fluid technology: A review. *J. Adv. Pharm. Sci. Technol.*, 1, 13–36, 2013.
77. Raud, M., Olt, J., Kikas, T., N_2 explosive decompression pre-treatment of biomass for lignocellulosic ethanol production. *Biomass Bioenergy*, 90, 1–6, 2016.

78. Pasquini, D., Pimenta, M.T.B., Ferreira, L.H., Curvelo, A.A.S., Sugar cane bagasse pulping using supercritical CO_2 associated with co-solvent 1-butanol/water. *J. Supercrit. Fluids*, 34, 125–131, 2005.
79. Narayanaswamy, N., Faik, A., Goetz, D.J., Gu, T., Supercritical carbon dioxide pre-treatment of corn stover and switchgrass for lignocellulosic ethanol production. *Bioresour. Technol.*, 102, 6995–7000, 2011.
80. De Melo, M.M.R., Silvestre, A.J.D., Silva, C.M., Supercritical fluid extraction of vegetable matrices: Applications, trends and future perspectives of a convincing green technology. *J. Supercrit. Fluids*, 92, 115–176, 2014.
81. Zhang, H. and Wu, S., Pre-treatment of eucalyptus using subcritical CO_2 for sugar production. *J. Chem. Technol. Biotechnol.*, 90, 1640–1645, 2015.
82. Keating, D.H., Zhang, Y., Ong, I.M., McIlwain, S., Morales, E.H., Grass, J.A., Tremaine, M., Bothfeld, W., Higbee, A., Ulbrich, A., Aromatic inhibitors derived from ammonia-pretreated lignocellulose hinder bacterial ethanologenesis by activating regulatory circuits controlling inhibitor efflux and detoxification. *Front. Microbiol.*, 5, 402, 2014.
83. Nguyen, T.Y., Cai, C.M., Kumar, R., Wyman, C.E., Overcoming factors limiting high-solids fermentation of lignocellulosic biomass to ethanol. *Proc. Natl. Acad. Sci.*, 114, 11673–11678, 2017.
84. Petridis, L. and Smith, J.C., Molecular-level driving forces in lignocellulosic biomass deconstruction for bioenergy. *Nat. Rev. Chem.*, 2, 382–389, 2018.
85. Nguyen, T.Y., Cai, C.M., Kumar, R., Wyman, C.E., Co-solvent pre-treatment reduces costly enzyme requirements for high sugar and ethanol yields from lignocellulosic biomass. *ChemSusChem*, 8, 1716–1725, 2015.
86. Inc, B.-P.I., BPI announces startup of pilot plant featuring LTSD technology, 2015. Ethanol producer magazine.
87. Park, J., Shin, H., Yoo, S., Zoppe, J.O., Park, S., Delignification of Lignocellulosic Biomass and Its Effect on Subsequent Enzymatic Hydrolysis. *BioResources*, 10, 12, 2015.
88. Uppugundla, N., Da Costa Sousa, L., Chundawat, S.P.S., Yu, X., Simmons, B., Singh, S., Gao, X., Kumar, R., Wyman, C.E., Dale, B.E., Balan, V., A comparative study of ethanol production using dilute acid, ionic liquid and AFEXTM pretreated corn stover. *Biotechnol. Biofuels*, 7, 1–14, 2014.
89. Da Costa Sousa, L., Jin, M., Chundawat, S.P.S., Bokade, V., Tang, X., Azarpira, A., Lu, F., Avci, U., Humpula, J., Uppugundla, N., Gunawan, C., Pattathil, S., Cheh, A.M., Kothari, N., Kumar, R., Ralph, J., Hahn, M.G., Wyman, C.E., Singh, S., Simmons, B.A., Dale, B.E., Balan, V., Next-generation ammonia pre-treatment enhances cellulosic biofuel production. *Energy Environ. Sci.*, 9, 1215–1223, 2016.
90. Alizadeh, H., Teymouri, F., Gilbert, T.I., Dale, B.E., Pre-treatment of switchgrass by Ammonia fiber explosion (AFEX). *Appl. Biochem. Biotechnol.*, 121, 1133–1141, 2005.
91. Zhao, C., Cao, Y., Ma, Z., Shao, Q., Optimization of liquid ammonia pre-treatment conditions for maximizing sugar release from giant reed (*Arundo donax* L.). *Biomass Bioenergy*, 98, 61–69, 2017.

92. Abbott, A.P., Capper, G., Davies, D.L., Rasheed, R.K., Tambyrajah, V., Novel solvent properties of choline chloride/urea mixtures. *Chem. Commun.*, 70–71, 2003.
93. Chen, Y. and Mu, T., Application of deep eutectic solvents in biomass pre-treatment and conversion. *Green Energy Environ.*, 4, 95–115, 2019.
94. Pandey, A., Dhingra, D., Pandey, S., Hydrogen bond donor/acceptor cosolvent-modified choline chloride-based deep eutectic solvents. *J. Phys. Chem. B*, 121, 4202–4212, 2017.
95. Francisco, M., Van Den Bruinhorst, A., Kroon, M.C., New natural and renewable low transition temperature mixtures (LTTMs): Screening as solvents for lignocellulosic biomass processing. *Green. Chem.*, 14, 2153–2157, 2012.
96. Zhang, Q., Benoit, M., Dea Oliveiraa Vigier, K., Barrault, J., Jerŏme, F., Green and inexpensive choline-derived solvents for cellulose decrystallization. *Chem. A Eur. J.*, 18, 1043–1046, 2012.
97. Kumar, A.K. and Parikh, B.S., Natural deep eutectic solvent mediated pre-treatment of rice straw: Bioanalytical characterization of lignin extract and enzymatic hydrolysis of pretreated biomass residue. *Environ. Sci. Pollut. Res.*, 23, 9265–9275, 2016.
98. Pandey, A., Mankar, A.R., Ahmad, E., Pant, K.K., Deep eutectic solvents: A greener approach towards biorefineries, in: *Biomass, Biofuels, Biochemicals*, T. Bhaskar and A. Pandey (Eds.), pp. 193–219, Elsevier, 2021.
99. Li, Y., Liu, H., Song, C., Gu, X., Li, H., Zhu, W., Yin, S., Han, C., The dehydration of fructose to 5-hydroxymethylfurfural efficiently catalyzed by acidic ion-exchange resin in ionic liquid. *Bioresour. Technol.*, 133, 347–353, 2013.
100. Lynam, J.G., Kumar, N., Wong, M.J., Deep eutectic solvents 'ability to solubilize lignin, cellulose, and hemicellulose; thermal stability; and density. *Bioresour. Technol.*, 238, 684–689, 2017.
101. Liu, Q., Mou, H., Chen, W., Zhao, X., Yu, H., Xue, Z., Mu, T., Highly efficient dissolution of lignin by eutectic molecular liquids. *Ind. Eng. Chem. Res.*, 58, 23438–23444, 2019.
102. Vats, S., Maurya, D.P., Shaimoon, M., Negi, S., Development of a microbial consortium for the production of blend enzymes for the hydrolysis of agricultural waste into sugars. *J. Sci. Ind. Res.*, 72, 585–590, 2013.
103. Sánchez, C., Lignocellulosic residues: Biodegradation and bioconversion by fungi. *Biotechnol. Adv.*, 27, 185–194, 2009.
104. Kumar, P., Barrett, D.M., Delwiche, M.J., Stroeve, P., Methods for pre-treatment of lignocellulosic biomass for efficient hydrolysis and biofuel production. *Ind. Eng. Chem. Res.*, 48, 3713–3729, 2009.
105. Shi, J., Chinn, M.S., Sharma-Shivappa, R.R., Microbial pre-treatment of cotton stalks by solid state cultivation of Phanerochaete chrysosporium. *Bioresour. Technol.*, 99, 6556–6564, 2008.

106. Wang, W., Yuan, T., Wang, K., Cui, B., Dai, Y., Combination of biological pre-treatment with liquid hot water pre-treatment to enhance enzymatic hydrolysis of Populus tomentosa. *Bioresour. Technol.*, 107, 282–286, 2012.
107. Yu, J., Zhang, J., He, J., Liu, Z., Yu, Z., Combinations of mild physical or chemical pre-treatment with biological pre-treatment for enzymatic hydrolysis of rice hull. *Bioresour. Technol.*, 100, 903–908, 2009.

3
Current Status of Enzymatic Hydrolysis of Cellulosic Biomass

Ram Bhajan Sahu, Janki Pahlwani and Priyanka Singh[*]

Institute of Allied Medical Science and Technology, NIMS University, Rajasthan, India

Abstract

In current scenario, biofuel has strong marketing values due to continuous depletion of fossil fuels. Lignocellulosic biomass as renewable resource is commercially used as alternative energy resources. The production of ethanol from these biomasses has major limitation of release of recalcitrant by products under mild conditions. The pre-treatment methodologies have been effectively employed to delignify a diverse portfolio of lignocellulosic biomass feedstocks. The efficient production of bioethanol at low cost will depend on the involvement of suitable pre-treatment system. Specific pre-treatment technique could not be employed effectively for different sources of lignocellulosic biomass due to exhibiting different structure and chemical composition of cell walls in biomass feedstocks. In this book chapter, the different sources for lignocellulosic feedstocks, their structure, and chemical composition have been described in detail. The pre-treatment methodologies in terms of efficient downstream biocatalytic hydrolysis of various lignocellulosic biomass materials have been simultaneously analyzed.

Keywords: Bioethanol, biodiesel, lignocellulosic biomass, pre-treatment methodology

3.1 Introduction

Scientists are continuously exploring the low-cost biological methodology for achieving production of biofuel. In this concern, renewable sources have been preferred for improving production of different generation of biofuels. These resources are being used as alternative sources for production

---

Corresponding author: priyay20@gmail.com

Arindam Kuila and Mainak Mukhopadhyay (eds.) Production of Biobutanol from Biomass, (77–104) © 2024 Scrivener Publishing LLC

of biofuels by reducing cost of fuel and inhibiting the emission of greenhouse gases in environment [1]. The targeted crop for production of first-generation liquid biofuels are generally food crops like sugarcane, oil seeds, and cereals. Their competence for cultivated agriculture land and high production cost has limited their demand for production of biofuel [2]. Cellulosic polymers from corn starch have been hydrolyzed into monomeric glucose units by specific pre-treatment methodologies for production of first-generation bioethanol. The great demand of cellulose for production of ethanol has been fulfilled by using cheap feedstock of biological waste products from Agriculture sectors and forest. But their recalcitrant structure limits their application for bioconversion into bioethanol by physiochemical approaches [1]. Cellulose polymer consist of linear chain of D-anhydroglucopyranose monomers interconnected by β-1, 4-glycosidic bonds [3, 4]. Linear chains of D-cellobiose repeating units are interconnected by intermolecular and intermolecular hydrogen bonding. The recalcitrant nature of cellulose may be due to exhibiting fibrous structure of cellulose polymer [5]. The use of Lignocellulosic biomass has improved the production of ethanol up to 50-80 % with low degree of emission of green-house gases (GHG). The extraction process for Lignocellulosic derivatives for bioconversion into biofuel has major limitation like inherent inefficiency of extracting lignin and release of highly recalcitrant polymer by-product [6]. It is necessary to remove the lignin for successful enzymatic scarification of cellulose and hemicellulose products. Pre-treatment methodologies have been employed for delignification and then delignified products have been used for enzyme saccharification process. The treatment with specific enzyme mixture results the hydrolysis of hemicellulose and improved the efficiency of cellulose hydrolysis. The overall crystallinity of partially hydrolyzed products have been improved by specific physical treatment methodology and major fractions of undesirable components has been removed by chemical treatment process. These pre-treatment methodologies are subsequently applied for removal of lignin and hemicelluloses which may decrease crystallinity of cellulose [7, 8]. Acid hydrolysis as simple pre-treatment methodology have been mainly applied for conversion of polymeric chain of cellulose into glucose monomer using dilute acid or concentrated acid at high temperature (120-180°C) and pressure (8-10atm) [9]. The acidic treatment of cellulosic biomass with 0.7 % sulphuric acid (w/v) at 190°C for 3 min has result their bioconversion up to 50% glucose yield. The bioconversion efficiency with 90% glucose has been improved after acid treatment with concentrated acid at moderate temperature and pressure [10].

The pre-treatment of oilseed crops has been carried out at high temperature for hydrolysis of cellulosic components in their cell wall structure.

During heat treatment, cell wall of seed has been degraded with coagulation of protein and denaturation of enzyme inside the cell membrane [11, 12]. Physiochemical methodology like soaking, solvent extraction, irradiation and enzymatic treatment has been also employed for cellulosic hydrolysis in cell wall [13]. The pre-treatment methodology of soaking and fermentation has significantly reduced the level of phytate, glycosides and tannin in cell wall. The oil of safflower seeds has been used as alternative feedstock for the production of biodiesel after Tran's esterification process. The viscosity of extracted biodiesel has been reported as closer to the petroleum diesel and its calorific value has been found 5.5% lesser than diesel [14].

The biological pre-treatment methodologies will subsequently result the effective delignification and polymerization of cellulose [15]. Microalgae has been reported as significant biological agent for production of biohydrogen, diesel fuel. In current scenario, genetically improved strains for microalgae have been developed by engineering metabolic pathways for lipids, polysaccharides and other hydrocarbons. These metabolic approaches improved the yield of desired biofuels at a greater extent.

This book chapter provides information for technology related implantations for production of significant biofuel. It has highlighted the process for production of all generation biofuels with their negative consequences for sustainability. The different pre-treatment methodologies for hydrolysis of Lignocellulosic feedstocks have been compared for their bioconversion process. The fermentative production of bioethanol using different microbial strains and their physiochemical parameters has been subsequently discussed in concern of current status for cellulosic hydrolysis approaches.

3.2 Overview on Biofuels and Its Classification

3.2.1 First-Generation Biofuels

Biofuels are center of attraction in worldwide as potential alternative for fossil derived petroleum transportation fuel. These fossil-based transport fuels result energy cost of oil crisis and global warming environmental issues. Ethanol as first-generation biofuel has been mainly produced by the fermentation of sugarcane, sugar beet potato, maize, wheat. These feedstocks mainly processed for production of bioethanol by Trans esterification, fermentation and distillation process. Biodiesel based on vegetable oils, alcohol based on food crop, and biogas are considered as most common first-generation biofuel [16–21].

The use of vegetable oil for production of biodiesel has economic, environmental and energy related benefits. The consumption of vegetable oils has generated approximately 90% heat equivalent to heat released due to consumption of diesel fuel. Vegetable oils extracted from oilseed crop can be easily transformed into biodiesel after Tran's esterification process. In current scenario, waste cooking oils and vegetable oils are considered as alternative fuels for diesel engines [22]. The continuous utilization of vegetable oil at large scale for production of biofuel can result the crisis of valuable oilseed crop in global sector. The substitution of vegetable oil with unrefined oil can damage the efficiency of engine through deposition of carbon after their combustion process. The commercial use of palm oil for production of biofuel has caused serious concern for global warming by increasing rate of emission of greenhouse gases. The fermentation process of cane sugars and starches result the production of butanol, ethanol. Ethanol has been considered as significant first-generation biofuel [23]. The corn has been used as main feedstock for the production of bioethanol in United States of America [16]. The emission of greenhouse gases has been reduced by blending ethanol with small percentage of gasoline. These blended biofuels have been categorized as E85 (85 % ethanol & 15% gasoline), E10 (10 % ethanol & 90% gasoline) [24]. The ethanol was recovered from fermented product using low-cost dry milling process. This process results the removal of 90 % phosphorus from fermentable starch with undesirable residues containing protein and oil [25]. Wet milling process was used for the production of highly refined pure glucose. This process has high cost due to requirement of substantial capital and operating investment [26].

Dry-mill corn fractionation process has separated low-cost food and animal feed-quality fiber from fractionated corn mash [27]. Ethanol has been considered as highly inflammable with reduce high octane additives to diminish ozone layer of the environment. Fuels with >10% of ethanol are not compatible with non-ready fuel scheme and may possibly reduce the content of ferrous components [28]. These fuels are not considered compatible with capacitance fuel plane gauging indicators. Soya bean is highly used for the production of Biodiesel rather than ethanol.

Biodiesel (serene of mono alkyl esters containing long chain fatty acids) has been extracted from vegetable oil or animal fats. The content of biodiesel has been classified into four categories such as edible vegetable oil, nonedible vegetable oil, waste or recycled oil and animal fats [29]. These oil derivatives could be effectively used along with the normal diesel as mixture. They are biodegradable, less toxic due to zero percentage of sulfur. The consumption of agricultural feedstocks for their production has increased its capital cost than normal diesel [30]. The use of biodiesel as transport fuel can pollute environment by releasing nitrogen oxide during combustion process. The waste products released due to human activities by agriculture, food, cosmetic, textile, pharmaceutics sectors may contain polymers of harmful toxicants. These polymers can be both natural (such as carbohydrates, proteins, and lipids) and Synthetic (such as plastic, polyester). Lignocelluloses have been considered as the main constituent of biological waste products. These can be hydrolyzed for the production of methane biogas and ethanol after fermentation process [30]. Methano genesis is the anaerobic process in which organic harmful toxic substrates are converted into methane in presence of methanogens. These biological waste residues will release more than 25% of the energy from organic food and solid waste residues from agricultural sectors which can be utilized for generation of electricity. The biogas mainly comprised of 40% to 65% methane, 30-40% carbon dioxide and small percentage of hydrogen sulphide, ammonia [31, 32]. Biogas produced from waste water have higher percentage of methane than biogas produced in landfill.

3.2.1.1 Advantage of First-Generation Biofuel

Biofuel production is becoming the key to open up to new markets in energy production. Agricultural sector will play the most important role in both food production as well as energy production. Biofuel is a lipid derived fuel which benefits the small and medium scale producer companies. Ethanol can be used in bagasse (the fibrous residues after pressing of sugarcane). In Brazil, this bagasse is burned and id used for the distillation process and electricity production. Glycerin is also an important by-product from industrial point of view.

3.2.1.2 Limitation of First-Generation Biofuel

The main substrate of the first-generation fuels is feedstock of cultivated food crops. The continuous utilization of theses feedstocks for production of biofuel has limited their food availability for common people in global sectors. The cost of these food crops have been raised due to the following factors:

- More use of biofuels is increasing the demand of water supply
- The cultivation of feedstock for biofuel production is increasing the land demand

- The first-generation fuels are more expensive than the gasoline, petrol.
- The use of vegetable oils for production of Biodiesel has limited their availability in whole world

3.2.2 Second-Generation Lignocellulosic Biofuel

Second-generation fuels are more industrially developed than first-generation fuels and are known as "Advanced biofuels" because unlike first-generation biofuels, these are produced from feedstock that is not utilized by human. The extraction of these biofuels is more difficult. Non-food feedstock of second-generation biofuels include wood, food wastes, crop enriched with cellulose, hemicelluloses and lignin [33]. The main derived second-generation biofuel include BTL-diesel (biomass to liquids) and lignocelluloses ethanol. The second-generation biofuels are mainly produced in North America, Europe, Brazil, China, India and Thailand [34]. Fischer-Tropsch fuels as second-generation fuels are obtained by biological reactions for converting mixture of carbon dioxide and hydrogen gas into liquid hydrocarbons [35, 36]. Lignocellulosic biomass has been consisted of cellulose, hemicelluloses, and lignin. Lignin has an aromatic functionality with three different phenyl propane monomers. Thermochemical and biochemical processing has been employed for hydrolysis of these Lignocellulosic biomass to produce ethanol biofuel. Synthetic gas of bio-SNG as another second-generation biofuel has been obtained from straw and other plant residues containing gaseous mixture of carbon monoxide, hydrogen and other hydrocarbons [37]. Hydrogen is used as a fuel whereas the other hydrocarbons are used for the production of gas oil.

3.2.2.1 Different Types of Feedstocks for Second-Generation Biofuels

In some oilseed crop, vegetable oil has been secreted as waste product (no nutritive component), which are being preferred for production of biofuel. Some engines are designed in such a manner that the biofuel can be obtained directly without blending or refining. There are many advantages of using waste vegetable oil such as it does not demand land for cultivation and also does not release sulfur that causes acid rain. Some of the limitations are that the engines can get damaged due to avoid of the pretreatment of this waste vegetable oil.

Miscanthus grasses and switch grasses are being used in South-East Asia and United States respectively for production of second-generation biofuels. It is very advantageous as it grows quickly, can be yielded multiple times a year, require a smaller number of fertilizers for cultivation. The major disadvantages for growing grasses for biofuels are as follows:

- Sometimes grass biomass can be used directly but biofuel may not be favorable.
- Bioconversion of solid residues of grasses into alcohol is more complicated than other processes.
- Although they are easy to cultivate, they are required to be protected from growth of unwanted weeds.
- They can grow only in areas of substantial humidity levels not in arid soils.
- No supplements can yield the dense biomass within the first few years of cultivation.

The biomass obtained from seed crops can possess comparatively less biofuel energy than soybean biomass. The popularity of this method has been decreased due to many difficulties in cultivation of soybean. The biomass utilized for production of biofuels include plant solid waste, human waste, grass, landfill components. The following fuels are listed as second-generation biofuels:

- Cellulosic ethanol – It is derived from the fermentation of sugars from cellulose and hemicelluloses fractions of Lignocellulosic biomass.
- Biobutanol – It is also produced by fermentation of agricultural solid residues in presence of specific microorganisms. The yield of biobutanol has been s reported as lower than the production of ethanol, therefore, biobutanol can be considered as a replacement of gasoline without blending. BTL (biomass to liquids) technology has been started with the production of synthetic gas (syngas) followed by Fischer-Tropsch process to process for production of gasoline, diesel and jet fuel.

Methanol, dimethyl ether (DME) and other mixed alcohols can be produced from syngas through catalytic approach. Although, alcohols are also obtained by the fermentation of syngas after inoculation with specific microorganisms. Biosynthetic gas (BIO-SNG) can be obtained by the gasification of natural gas followed by the catalytic methylation and anaerobic digestion by specific microorganisms. This gas mainly comprised of methane and carbon dioxide, which could be efficiently used as compressed natural gas (CNG) and liquefied petroleum gas (LPG) in vehicles or natural gas cylinders [38]. Hydro treated vegetable oils are considered as a substitute of diesel and possesses physical properties like high amount of ketone, non-aromatic components and does not contain sulfur. Pyrolyzed oil (biocrude) are obtained by rapid heating to about 1000° F followed by rapid cooling of ash. Refinement of these crude oils has generated liquid fuel for transportation [39].

3.2.2.2 Advantages

The second-generation biofuels are currently being preferred in compare to first-generation biofuels due to following reasons: -

- The utilization of non-food feedstock for production of second-generation biofuels
- These biofuels are effective alternatives for conventional petroleum-based fuels and can be used directly without blending and refining.
- They do not produce harmful by-products and not cause emission of greenhouse gases
- The cultivated lands are not going to be disturbed and it require less land for cultivation of these feedstock.

3.2.2.3 Disadvantages

- High cost of machinery and engines based on second-generation biofuels.
- Current harvesting, storage and transport systems are required at large scales.
- Requirement of long-term policy framework
- Agricultural lands are used for cultivation of grasses to achieve large production of second-generation fuels.
- These are more expensive than first-generation fuels.

The second-generation fuels do not race in the markets with first-generation biofuels as they are produced from different biomasses. They use limited land for cultivation of feedstock for biomass production. Cellulosic sources grow side by side along with food crops that do not disturb the policy of farmers. There is no or very less requirement of fertilizers and pesticides for the cultivation of biomass. It opens the doors of economic sector for the business of energy production. Second-generation biofuels are considered as more economical and eco-friendly sustainable approach for environmental and societal benefits [40].

3.2.3 Third-Generation Biofuels

Microalgae has efficiency to produce biofuels during their lipid metabolism process. They have tendency to produce liquid biofuels after consumption of biological waste product like animal fats, garbage, and spent cooking oil. Algal biomasses are being reported to produce third-generation biofuels, whereas Lignocellulosic biomass are being consumed for production of first- and second-generation biofuel [33, 41]. Algae has efficiency to produce biofuels with 30 times more energy per acre than the land crops [42]. Biorefinery technologies based on physiochemical and biological treatment process has been applied for conversion of algal biomass to energy sources [43]. The environmental factors like carbon dioxide, salts, and light energy source have been supplied for cultivation of autotrophic algae [41]. Microalgae with small size (few micrometer) has grown faster in the water bodies and contain more lipids than macro algae. The main advantage of cultivation of algae is the short harvesting cycle once or twice a year. These algae produced biofuels like gasoline, methane, ethanol, butanol, vegetable oil, biodiesel. The following strategies are being opted for cultivation of culture of different types of algae:

- Open ponds – Algae can easily be cultivated in open ponds supplemented with nutritive components at low cost.
- Closed-loop systems – Algae has been cultivated in water bodies maintained in closed system by inhibiting its exposure from normal atmosphere and continuous supply of carbon dioxide in bioreactor
- Photo bioreactors – These are advance level of bioreactor fitted with different photocell units for supply of artificial light. They are inoculated with different types of algal species and cultivation process has been carried out under optimum condition. These photo bioreactors are being used for large scale production of biofuel at high capital costs.

3.2.3.1 Advantages

- Algae has greater capability for growth and cultivation than other food crops

- They have high efficiency to consume high level of carbon dioxide content
- Algal cultivation process requires less consumption of water
- They have high rate of expansion capacity than other sources
- Algae are easily grown in waste water by consuming raw materials from water source
- They have capability to grow in saline, brackish water, coastal seawater
- These are efficiently secreting third-generation biofuels
- They are easily grown 10 times more per acre than other land crop
- They do not demand extra land for production of algal biomass

3.2.3.2 Disadvantages

- They require higher degree of management skill for maintenance of their specific cultivated land in compare to other traditional crops
- They require comparatively more energy input for harvesting of algae
- It is considered to be costly techniques due to involvement of Flocculation, centrifugation, sedimentation, and filtration for concentrating algal biomass [44].
- Although algae are grown in wastewater, but they require large quantity of water, light sources, nitrogen and phosphorus for their cultivation process.

The third-generation biofuels are much useful than other generation biofuels as it is more eco-friendly, high energetic, and fully renewable sources of energy. The sources of first-, second-, and third-generation biofuels are food feedstock, non-food feedstock and algae respectively. Algae has been proved to be very useful than food and non-food feedstock as it has required any land for its cultivation. Algae produces overall higher biomass than other feedstock sources and can be cultivated with the help of sewage or waste water. The algal harvesting used biorefinery but now by applying advance downstream technologies, bioreactors and other machineries are being developed. As the demands and investments are high in this sector, a sustainable approach is considered for the construction of biofuel industries.

3.2.4 Fourth-Generation Biofuels

The ecological point of view for production of the various generation biofuels are still considered insufficient on their impact of Greenhouse gases in the atmosphere. After the production of third-generation biofuel through algal biomass, it is now suitable to fulfill the needs of environmentally suitable biofuel [45]. The sources of first- and second-generation biofuels are food feedstock (sugarcane and corn) and non-food feedstock (cellulosic, lignocellulose, and hemicellulose biomass) respectively. Whereas the third- and fourth-generation biofuels are considered as "algae-to-biofuels". The production of fourth-generation biofuel through algal biomass involves oxygenic photosynthetic microorganisms as they are inexhaustible, cheap and extensively available. They efficiently convert vegetable oil and biodiesel into gasoline [33]. These fourth-generation biofuels are considered as more superior than third-generation biofuels due to exhibiting high lipid content along with their algal biomass. They have capability for capturing large amount of carbon dioxide and high efficiency for production of biofuel. The production of fourth-generation biofuel from algal biomass has initial requirement of high capital investments for cultivation of microalgae.

3.3 Pre-Treatment Methodologies for Hydrolysis of Lignocellulosic Biomass

3.3.1 Overview

Pre-treatment methodologies efficiently break cellulosic structure of solid plant waste residue for the bioconversion into biofuel. Physical methodologies have been considered as most expensive processing technique for hydrolysis of cellulosic biomass. These techniques will disrupt the plant cell wall network for partial separation of lignin, cellulose, and hemicellulose as major polymer components. Fermentation process has great potential for improving efficiency for production of biofuel at a low cost. Pre-treatment has been employed for hydrolysis of lignocellulose into biofuel and further used as raw materials for enzyme hydrolysis. These methodologies may release recalcitrant byproducts containing inhibitors and deactivators during the enzymatic hydrolysis of lignocellulose. Therefore, it should take precaution for designing specific pre-treatment technology for reducing the secretion of these recalcitrant by-products during hydrolysis of Lignocellulosic biomass.

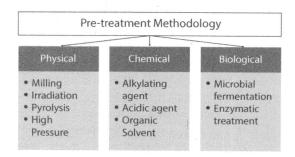

Figure 3.1 Different techniques for physical, chemical and biological pre-treatment methodologies for hydrolysis of Lignocellulosic biomaterials.

These pre-treatment methodologies have been classified as physical, chemical and microbial fermentation process as shown in Figure 3.1.

Biorefinery process for this Lignocellulosic hydrolysis are very costly and therefore, it is a major challenge for their economic management [46]. This issue has attracted the scientist to explore novel pre-treatment methodologies with low-cost value based on physical, chemical and biological methodology. The effective bioconversion of cellulose into fermentable sugars has been achieved using methodologies based on physical treatment (milling and grinding), thermochemical treatment (thermolysis, steam explosion, and wet oxidation), chemical treatment (organic solvents, alkali, dilute acid, oxidizing agents), and biological fermentation processes. The different methodologies for pre-treatment process with their specific characteristics has been discussed in Table 3.1.

Among these pre-treatment methodologies, chemical process results rapid conversion of cellulose into fermentable sugars after enzymatic hydrolytic process. The biological pre-treatment process results the bioconversion process with a very slow rate of fermentation of sugars into biofuel component. Physical pre-treatment processes have generally required high level of energy input throughout the bioconversion process.

These Lignocellulosic biomasses have been firstly pretreated by physical methodology based on pyrolysis, combustion, liquefaction, or gasification. Fractionation process has been further carried out by separating byproducts from major component of lignocellulosic biomass. Biorefinery techniques are further carried out for efficient utilization of raw materials for production of biofuel by separating waste biological products as alternative fossil fuel source such as coal, petroleum.

Table 3.1 Effect of different types of pre-treatment methodologies on hydrolysis of solid residues.

Types of chemicals used for pre-treatment process	Efficiency for extraction process	References
Acidic agents	• Requires less operation time • higher yield of sugar • Release of furfural inhibitory by-product	[6, 47, 48, 49]
Ammonia Fibre Explosion (AFEX)	• High yield of xylose sugars • Produce lignin content as inhibitory compound	[6, 15]
Alkylating agent (Lime)	• Highly effective without release of any inhibitory component • It requires long duration of operation	[6, 51, 52, 53, 54]
Organic solvent	• High yield of pentose sugar • Expensive solvent recovery system	[6, 55, 56]
Physical method for pre-treatment		
Milling -process	• Required lesser operation time • Poor yield of biofuel • Required high energy	[6]
Irradiation and ultrasonic wave treatment	• Higher yield of sugar • Need specific equipment for process • Not produce inhibitory by-product	[6, 57, 58, 59, 60]
High-pressure	• Efficient process for higher yield of biofuel • Not produce any inhibitory component • Challenge for maintain high pressure during process	[6, 61]

(*Continued*)

Table 3.1 Effect of different types of pre-treatment methodologies on hydrolysis of solid residues. (*Continued*)

Types of chemicals used for pre-treatment process	Efficiency for extraction process	References
Biological pre-treatment methodology		
Microbial fermentation process	• Required less energy • Not produce inhibitory compound • Carried out under mild operation conditions • Required long period of fermentation process	[6, 62, 63]

3.3.2 Structural Analysis for Cellulosic Hydrolysis

The different plant sources for Lignocellulosic biomass for production of biofuels has been shown in Figure 3.2.

Lignocellulosic cell biomasses are being consisted of cellulosic polymer of β-1, 4-polyacetal of cellobiose (4-O-β-D-glucopyranose-D-glucose). These cellobiose chains have been further cross linked with β, 1-4 D-glycosidic bond of glucose unit (Solomon 1988). The linkage with hydrogen bond

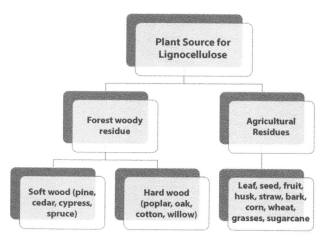

Figure 3.2 Biomass based on Lignocellulosic materials are being classified as follows: (1) agricultural wastes generated due to plant cultivations and farming, and (2) forest plant residues.

has provided crystalline fibrous structure to cellulosic polymeric matrix. Cellulosic polymers are arranged in long straight parallel chains together with hydroxyl groups across both sides of the glucose monomer. This arrangement results the formation of hydrogen bond between two hydroxyl groups of different polymer chains of cellobiose unit [64, 65].

The structure of hemicellulose differs from cellulose with lack of hydrogen bond and presence of significant number of ester functional group and salt molecule. Xylan is considered as most common hemicellulose components with linkage of xylopyranosyl units associated with α-(4-O)-methyl-D-glucurono-pyranosyl group attached to anhydrous xylose molecule. This linkage results the formation of branched polymer chain between xylose and glucose molecule.

Lignin has secondary cell wall with cross linkage of phenyl-propanoic acid units with amorphous phenolic macromolecule. These polymeric structures have provided mechanical strength to the plant tissue and individual fibers in structure of lignin. The oxidative coupling of 4-hydrophenylpropanoids during secondary cell wall deposition result high level of heterogeneity in lignin structure. The Lignocellulosic matrix has specific bonds of ether, ester, C-C bond and hydrogen bond, which provides linkage within the individual components of Lignocellulosic polymer. The interactive bond of ester group are mainly observed in lignin, polysaccharides, and hemicellulose polymer. Acetyl group has formed the ester bonding with hydroxyl group of the main chain of the polysaccharides. Ethers and C-C bonding are mainly responsible for association of building molecules within lignin polymers. Lignin polymeric matrix has 70% bonding of ether type and 30% of carbon-carbon units. These ether bonds have been observed between two aryl carbon atoms, two allyl carbon atoms, or between one allylic and one aryl carbon atom [6]. Hydrogen bond has been identified for association of cellulose with hemicellulose unit. Ester bond has connected hemicellulose with lignin and ether bond for linkage of lignin with polysaccharides [64]. The arrangement of molecules in Lignocellulosic biomass is the key parameter for selection of specific physiochemical pre-treatment methodology for bioconversion process to produce biofuels as shown in Figure 3.3.

3.3.3 Chemical Process for Pre-Treatment of Lignocellulose

3.3.3.1 *Dilute Acid Pre-Treatment Process*

The rigid cell wall structure of lignocellulose has been partially digested after pre-treatment with dilute or concentrated acids like HCl, sulfuric acid

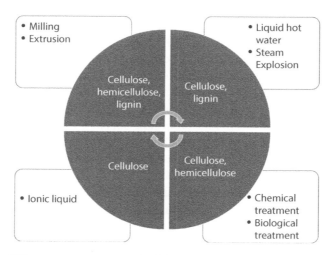

Figure 3.3 Different pre-treatment methodologies opted for hydrolysis of cellulose, lignin and hemicellulose in plant residues.

(H_2SO_4). Acid treatment subsequently with the exposure of high temperature (120-180°C) will extensively disrupt the cellular composition of outer cell wall of lignocellulose biomass [52, 66–68].

Different types of agricultural cultivated plants like grass, corn stover and poplar has been treated with dilute sulfuric acid for hydrolysis of their cellular matrix [6, 69–71]. The treatment of cellulosic materials from olive tree with sulfuric acid (1.4%) at high temperature 210°C will result 75% hydrolysis of their cell wall. The pre-treatment of cashew apple bagasse at 121°C with dilute H_2SO_4 will effectively hydrolyze them to secrete bioethanol [72, 73]. The hydrolysis with dilute acid has specific advantage than concentrated acid for pre-treatment process with higher reaction rate and shorter reaction times [74]. In case of lignin, the crystalline cellulosic polymeric structure will limit its potential for production of biofuel [75, 76]. The biomass of rapeseed has been pretreated with 1% sulfuric acid for 10 min at 180 °C for production of bioethanol. Hemicellulose could be removed from other biomass of rapeseed straw by treatment with dilute sulfuric acid (H_2SO_4) [6, 75]. Noureddini and Byun [48] has investigated the secretion of furfural byproduct after hydrolysis of corn fiber and distiller's grains to monomeric sugar molecules with dilute sulfuric acid. The higher yield of monomeric sugars has been obtained for treating substrate molecules with high concentration of dilute sulfuric acid (1.0-1.5 vol %) at 140°C temperature (Noureddini and Byun [48] for optimum time period. The formation of furfural byproduct after acid treatment at 120 °C has been obtained higher for distiller's grains in compare to corn fiber samples.

Shi et al. [77] have hydrolyzed corn straw and rice straw by treating with dilute sulfuric acid subsequently with enzymatic hydrolysis of cellulose. These straw feedstock biomasses have been pretreated at 121 °C with different concentration of sulfuric acid (1- 5%, v/v) with residence times (30-90 min). The residence time during pre-treatment process has increased the concentration of glucose monomers after hydrolysis of cellulose. Cellulase from *Trichoderma viride* has digested the cellulosic materials of acidified feedstock of corn straw and rice straw with saccharification yield as 72.38 and 82.84% respectively. The pentose sugar of xylan has been efficiently treated with dilute acid for the production of ethanol from feedstocks of corn stover. Dilute acid pre-treatment has greatly hydrolyzed xylan residues with the significant production of toxic byproducts, acetic acid for inhibiting the enzymatic process of saccharification and fermentation. These inhibitory components will decrease the bioconversion of xylose into biofuel components. The treatment with dilute acid will result the successful removal of acetyl groups from native corn stover for increasing the yield of monomeric xylose from digested Lignocellulosic biomass [6, 47, 78]. These dilute acid pre-treatment methodologies have improved saccharification process of xylan and glucan unit.

The cellulosic polymer consists of crystalline bundles surrounded by charged water molecules with dipole-dipole interactions [6, 79]. The hydrolysis reaction has been carried out with specific water molecules and H^+ ions released from acid media, which could penetrate the cellulose fiber. Acid has catalyzed the breakdown of long cellulosic polymer chains into short chain of oligomer unit. These oligomeric units have been further degraded into glucose monomers in sugar moiety. The hydrolysis of cellulose has been started due to conjugation reaction between acidic proton ions and oxygen atom linking to two glucose molecule.

3.3.4 Ionic Liquid as Pre-Treatment Agent

Ionic liquids (ILs) consisting of inorganic anion and organic molecule has been used as effective solvents for removal of lignin and cellulose content. It is effectively used as pre-treatment methodology for agricultural feedstocks for production of second-generation biofuels like bioethanol and biobutanol. These ionic liquids include quaternary ammonium ILs, N-alkylisoquinolinium ILs, N-alkyl-pyridinium ILs, and imidazolium-based ILs [6, 80]. These ionic liquids may hydrolyze lignocellulosic materials and polysaccharides (cellulose and hemicelluloses) into monomeric fermentable sugar units. This methodology has separated cellulose by incorporating hydroxyl groups in cellulose with both cationic and anionic ionic liquid. The combination of oxygen atoms of hydroxyl group and ionic liquid has been used as strong electron donors during the dissolution process of lignin content from biomass. In this process, H-atom of hydroxyl groups and ations of ionic liquid act as electron acceptors [6, 81]. The addition of organic solvents such as alcohols, ethers, or ketones has caused the regeneration of dissolved cellulosic unit. In presence of solvents, cations in ionic liquid treated cellulose matrix have formed hydrogen bonds with water molecules. These cationic molecules in IL complex have been further displaced into the aqueous phase. Cellulose associated with ionic liquids has been expelled and reunite with intra and inter molecular hydrogen bonds for precipitating out from partially digested biomass.

3.3.5 Pre-Treatment Process with Alkali Agents

Specific alkali like caustic soda, ammonium hydroxide, calcium hydroxide has been used for pre-treatment process for hydrolysis of cell wall of lignocellulose matrix. Alkali pre-treatment process has been conducted under mild conditions and effectively hydrolyzed lignin molecules for production of ethanol. Saponification of intermolecular ester bonds are the significant step for the alkaline pre-treatment process for dissociation of lignin from sugar molecules (cellulose and hemicellulose). This alkaline pre-treatment process will release major percentage of five carbon sugar in compare to hexose carbon sources after hydrolysis of Lignocellulosic biomaterials. Lime pre-treatment methodology has effectively used for partial digestion of cellulosic matrix due to utilization of less amount of energy for bioconversion of fermentable sugar into bioethanol after enzymatic hydrolysis [6, 82, 83]. During this treatment, lignin has been removed efficiently from biomass without any significant damage of structure of glucose, xylose, and arabinose carbohydrate molecule. Alkaline pre-treatments have removed acetyl groups from hemicellulose for improving saccharification by lowering content of steric hindrance of enzyme [84, 85].

The alkaline pre-treatment process has successfully digested lignocelluloses matrix and has not released furfural and methyl furfural by-products during processing [86]. Lignin has three-dimensional structure with high molecular weight (approx. 100 KD) and originated from phenyl-propanoid derivatives of vascular plants [87]. Alkaline treatment has depolymerized lignin molecules by breaking aryl ether bonds. In alkaline environment, phenolic hydroxyl groups have been reduced throughout the reaction into superoxide radical as primarily free electron radical. Delignification reactions cause the secretion of different acidic components and added hydrophilic groups onto the entire surface of lignin molecules [6, 88]. These proton ions of phenolic compound in lignin have been converted to phenolate ions under this basic environment. These phenolate ions further eliminate ether group from benzylic position to produce derivatives for highly conjugated aryl vinyl-phenol. These phenolic moieties have been oxidized to 1,4-dicarboxylic acid which are found to be soluble in basic solution. Lime pre-treatment methodology has been employed for the hydrolysis of sugarcane bagasse into bioethanol [54]. This lime pre-treatment methodology has also employed for improving enzymatic hydrolysis of bagasse and wheat straw [83, 89]. Saccharification of high dry-matter content and their subsequent fermentation process by *Saccharomyces cerevisiae* result the production of high yield of bioethanol under optimum environmental condition. Microbial fermentation process results the production of ethanol as main product about 20g/l after 50h fermentation period and also released xylan component as soluble oligomeric byproduct [89]. Feedstock of rice straw has been treated with sodium hydroxide (NaOH) and hydrated lime $Ca(OH)_2$ for production of bioethanol [90]. Lime-pre-treatment methodology has result 48% conversion of total fermentable glucose monomers with 175 mg/g dried biomass, whereas, alkaline pre-treatment results 38% bioconversion with 142 mg/g dried biomass [90]. During alkaline treatment process, phenolic hydroxyl groups has been produced after the cleavage of aryl-alkyl-ether bonds [91].

Ammonia derivatives are commonly used for pre-treatment process because of their non-toxic and inexpensive nature [92]. The pre-treatment process with aqueous ammonia cause delignification without damaging the carbohydrate contents. It is used as effective pre-treatment method for feedstocks with low content of lignin for the production of bioethanol [93]. This ammonia-based pre-treatment methodology cause the delignification activity with increased surface area of their particle size and modification of structure of cellulose and hemicellulose. Aqueous ammonia will also affect the degree of crystallinity of the Lignocellulosic biomass for enzymatic hydrolysis. The dissociation of these ammonia molecules results the release of H^+ and NH_3 species, which further induced the breakage of chemical bonds cellulosic polymeric chain [6, 94]. The dissociated ammonia ions will promote the selective cleavage of these linkages during delignification process. Ammonia fiber explosion-expansion (AFEX) has been used as advance physiochemical treatment methodology for hydrolyzing Lignocellulosic biomass into glucose monomeric units. These feedstocks are treated with liquefied anhydrous ammonia at high temperature

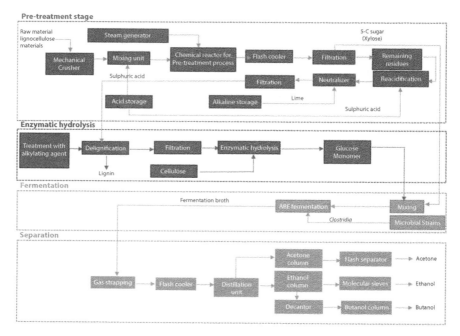

Figure 3.4 Downstream processing techniques for biosepration of ethanol and butanol from lignocellulosic biomaterials after pre-treatment, fermentation and separation methodologies.

(60-120 °C). This results the entrapment of liquid ammonia within cellular biomass to disrupt the structure of cellulose. After the operation, ammonia has been recovered at this stage as low-pressure gas and reused for treatment process. AFEX pre-treatment has increased bioconversion of glucan and xylan units into bioethanol in feedstocks like corn stover, switch grass, and bagasse. Physiochemical and biological pre-treatment process for hydrolysis of Lignocellulosic biomass into bioethanol has been shown in Figure 3.4.

3.3.6 Pre-Treatment with Ultrasonic Wave

Ultrasonic waves have different frequencies are being categorized as high frequency range (2-10 MHz), low frequency (20-100 kHz), and medium frequency (300-1000 kHz) [95]. When liquid has been exposed to high frequency (50-60 kHz), micro bubbles will be developed at the site of nucleation site. This result the increase of temperature up to 5000K due to developing of high pressure (500 atm) across the nucleation site [6, 96]. These huge pressure and temperature will result the disruption of cellulosic polymeric biomass into monomers. Other theory postulates the development of electrical double layer across the nucleation site [97]. This process results the depolymerization, emulsification, and tanning of cellulosic structure in varieties of feedstocks [87]. Ultrasonic pre-treatment process has caused 20-fold reduction in size of corn particles for facilitating the production of glucose yield by 30% after hydrolysis with specific enzymes [58]. Cassava chips has been pretreated with ultrasonic waves of maximum power output 20 KW at the frequency of 20 kHz [60]. The pre-treatment methodology of ultrasonic waves has assisted hydrogen peroxide (250 W, 30 min) and biological treatment has been employed for digestion of rice hull [6].

3.4 Conclusion

Fossil-based fuels are continuously creating new challenges for the world due to cost for oil and global warming issues. For the last few decades, scientists have been exploring alternative energy resources with eco-friendly, environmentally sustainable and renewable approaches. The first-generation biofuels have been suggested as environmentally sustainable approach to overcome the dependency of oil crisis in the world. These biofuels have major limitation for the extensive use of food crops, which has caused starving condition for millions of people across the whole world. Second-generation biofuel has some benefits for consuming agricultural solid plant residues instead of food crop. But these lignocellulosic crops have occupied major portion of lands instead of cultivation of food crop. Third-generation biofuel has suggested best possibility to get an opportunity for substitution of these fuels without competing food crop and land use. The fourth-generation biofuels using algae metabolic activities have shown more advancement in comparison to other generation biofuels group biofuels. Pre-treatment methodologies have been successfully employed for the hydrolysis and transesterification of lignocellulosic biomass materials into energy derived biofuels like bioethanol, biogas, or biodiesel. The specific pre-treatment technique should be low cost and less energy consuming process. The pre-treatment methodologies depend on the type of lignocellulosic biomaterials and reported as different for production of biogas, biodiesel and bioethanol. Saccharification and enzymatic treatment process have been subsequently employed for the hydrolysis of polymeric chain of cellulose into glucose monomers. Microbial fermentation process has been currently used as a cost-effective technique for the production of biofuel components from glucose monomers. This book chapter has described various pre-treatment methodologies for the production of different categories of biofuels.

References

1. Lee, R. and Lavoie, J.M., From first-to third-generation biofuels: Challenges of producing a commodity from a biomass of increasing complexity. *Anim. Front.*, 3, 2, 6–11, 2013.
2. Sims, R.E., Mabee, W., Saddler, J.N., Taylor, M., An overview of second-generation biofuel technologies. *Bioresour. Technol.*, 101, 6, 1570–1580, 2010.
3. Awudu, I. and Zhang, J., Uncertainties and sustainability concepts in biofuel supply chain management. *Renewable Sustain. Energy Rev.*, 16, 2, 1359–1368, 2012.
4. Taherzadeh, M.J. and Karimi, K., Acid-based hydrolysis processes for ethanol from lignocellulosic materials: Bioethanol review. *Bio-Resources*, 2, 707–738, 2007.
5. Pulidindi, N. and Gedanken, A., Biofuels and biochemicals from biomass. *Open Chem. J.*, 7, 1, 022–024, 2021. 10.17352/ojc.000024.
6. Sasmal, S. and Mohanty, K., Pre-treatment of lignocellulosic biomass toward biofuel production. *Biorefin. Biomass Biofuels, 2017*, 1107, 2017. https://doi.org/10.1007/978-3-319-67678-4_9.
7. Sun, Y. and Cheng, J., Hydrolysis of lignocellulosic materials for ethanol production: A review. *Bioresour. Technol.*, 83, 1–11, 2002.
8. El Zawawy, N.A., El-Shenody, R.A., Ali, S.S., El-Shetehy, M., A novel study on the inhibitory effect of marine macroalgal extracts on hyphal growth and biofilm formation of candidemia isolates. *Sci. Rep.*, 10, 9339, 2020.
9. Iranmahboob, J., Farhad, N., Sharareh, M., Optimizing acid-hydrolysis: A critical step for production of ethanol from mixed wood chips. *Biomass Bioenergy*, 22, 401–404, 2002.
10. Hamelinck, C., Geertje, H., Faaij, A.P.C., Ethanol from lignocellulosic biomass: Techno-economic performance in short-, middle- and long-term. *Biomass Bioenergy.*, 28, 384–410, 2005.
11. Piorreck, M., Baasch, K.H., Pohl, P., Biomass production, total protein, chlorophylls, lipids and fatty acids of freshwater and blue-green algae under different nitrogen regime. *Phytochemistry*, 23, 207–216, 1984.
12. Wakelyn, P. and Wan, P.J., Solvent extraction to obtain edible oil products, in: *Chemical Rubber Company (CRC) Handbook of Nutrition and Food*, C.C. Akoh (Ed.), pp. 89–131, CRC Press, Boca Raton, FL, 2005.
13. Bora, P., Konwar, L.J., Boro, J., Phukan, M.M., Deka, D., Konwar, B.K., Hybrid biofuels from non-edible oils: A comparative standpoint with corresponding biodiesel. *Appl. Energy*, 135, 450–460, 2014.

14. Hamamci, C., Abdurrahman, S., Tonbul, Y., Kaya, C., Kafadar, A., Biodiesel production via transesterification from safflower (Carthamus tinctorius L.) seed oil. *Energy Sources Part A: Recovery Utilization and Environmental Effects*, 512-520. doi:10.1080/15567030903096964
15. Den, W., Sharma, V.K., Lee, M., Nadadur, G., Varma, R.S., Lignocellulosic biomass transformations via greener oxidative pre-treatment processes: Access to energy and value-added chemicals. *Front. Chem.*, 27, 6, 141, 2018.
16. Roy, S., A piece of writing: Future prospect of biofuels. *J. Biofuels*, 8, 49, 2017.
17. Karaosmanoglu, F., Vegetable oil fuels: A review. *Energy Sources*, 21, 221, 1999.
18. Shah, Y.R. and Sen, D.J., Bioalcohol as green energy - A review. *Int. J. Curr. Sci. Res.*, 1, 57, 2011.
19. Hoekman, S., Broch, A., Robbins, C., Ceniceros, E., Natarajan, M., Review of biodiesel composition, properties and specifications. *Renewable Sustain. Energy Rev.*, 16, 143, 2012.
20. Balat, M. and Balat, H., Biogas as a renewable energy source-a review. *Energy Sources*, 31, 1280, 2009.
21. Dafnomilis, I., Hoefnagels, R., Pratama, W., Schott, D.L., Lodewijks, G., Junginger, M., Review of solid and liquid biofuel demand and supply in Northwest Europe towards 2030-A comparison of national and regional projections. *Renew. Sustain. Energy Rev.*, 78, 31, 2017.
22. Corsini, A., Marchegiani, A., Rispoli, F., Sciulli, F., Venturini, P., Vegetable oils as fuels in diesel engine, engine performance and emissions. *Energy Proc.*, 81, 942, 2015.
23. Hassan, M.H. and Kalam, M.A., An overview of biofuel as a renewable energy source: Development and challenges. *Proc. Eng.*, 56, 39, 2013.
24. Singh, D. and Trivedi, R.K., Biofuel from wastes an economic and environmentally feasible resource. *Energy Proc.*, 54, 634, 2014.
25. Rajagopalan, S., Ponnampalam, E., McCalla, D., Stowers, M., Enhancing profitability of dry mill ethanol plants. *Appl. Biochem. Biotechnol.*, 120, 37, 2005.
26. Ingledew, W.M., Ethanol fuel production: Yeast processes, in: *Encyclopedia of Industrial Biotechnology*, M.C. Flickinger (Ed.), pp. 1–14, J. Wiley & Sons, Weinheim, Germany, New York, 2009.
27. Shaw, A., Lam, F.H., Hamilton, M., Consiglio, A., MacEwen, K., Brevnova, E.E., Greenhagen, E., LaTouf, W.G., South, C.R., van Dijken, H., Stephanopoulos, G., Metabolic engineering of microbial competitive advantage for industrial fermentation processes. *Science*, 353, 583, 2016.
28. Delavarrafiee, M. and Christopher, F.H., Real-world fuel use and gaseous emission rates for flex fuel vehicles operated on E85 versus gasoline. *J. Air Waste Manag. Assoc.*, 68, 235, 2017.
29. Aransiola, E.F., Ojumu, T.V., Oyekola, O.O., Madzimbamuto, T.F., Ikhu-Omoregbe, D.I.O., A review of current technology for biodiesel production: State of the art. *Biomass Bioenergy*, 61, 276, 2014.

30. Atabani, A.E., Silitonga, A.S., Badruddin, I.A., Mahlia, T.M.I., Masjuki, H.H., Mekhilef, S., A comprehensive review on biodiesel as an alternative energy resource and its characteristics. *Renew. Sustain. Energy Rev.*, 16, 2070, 2012.
31. Fargione, J., Hill, J., Tilman, D., Polasky, S., Hawthorne, P., Land clearing and the biofuel carbon debt. *Science*, 319, 1235, 2008.
32. Hosseini, S.E. and Wahid, M.A., Development of biogas combustion in combined heat and power generation. *Renew. Sustain. Energy Rev.*, 40, 868, 2014.
33. Berla, B.M., Saha, R., Immethun, C.M., Maranas, C.D., Moon, T.S., Pakrasi, H.B., Synthetic biology of cyanobacteria: Unique challenges and opportunities. *Front. Microbiol.*, 4, 246, 2013.
34. Carriquiry, M.A., Du, X., Timilsina, G.R., Second generation biofuels: Economics and policies. *Energy Policy*, 39, 4222, 2011.
35. Ail, S.S. and Dasappa, S., Biomass to liquid transportation fuel via Fischer Tropsch synthesis–Technology review and current scenario. *Renew. Sustain. Energy Rev.*, 58, 267, 2016.
36. Williams, R.H., Larson, E.D., Liu, G., Kreutz, T.G., Fischer–Tropsch fuels from coal and biomass: Strategic advantages of once-through ("polygeneration") configurations. *Energy Proc.*, 1, 4379, 2009.
37. Chaudhari, S.T., Dalai, A.K., Bakhshi, N.N., Production of Hydrogen and/or Syngas (H_2 + CO) via steam gasification of biomass-derived chars. *Energy Fuels*, 17, 1062, 2003.
38. Zhang, W., He, J., Engstrand, P., Bjorkqvist, O., Economic evaluation on bio-synthetic natural gas production integrated in a Thermomechanical pulp mill. *Energies*, 8, 12795, 2015.
39. Wongkhorsub, C. and Chindaprasert, N., A Comparison of the use of pyrolysis oils in diesel engine. *Energy Power Eng.*, 5, 350, 2013.
40. Naik, S.N., Goud, V.V., Rout, P.K., Dalai, A.K., Production of first- and second-generation biofuels: A comprehensive review. *Renew. Sustain. Energy Rev.*, 14, 578, 2010.
41. Brennan, L. and Owende, P., Biofuels from microalgae-A review of technologies for production, processing, and extractions of biofuels and co-products. *Renew. Sustain. Energy Rev.*, 14, 557, 2010.
42. Demirbas, M.F., Current technologies for biomass conversion into chemicals and fuels. *Energy Source Part A*, 28, 1181, 2006.
43. Behera, S., Singh, R., Arora, R., Sharma, N.K., Shukla, M., Kumar, S., Scope of algae as third generation biofuels. *Front. Bioeng. Biotechnol.*, 2, 90, 2015.
44. Cuellar-Bermudez, S.P., Garcia-Perez, J.S., Rittmann, B.E., ParraSaldivar, R., Photosynthetic bioenergy utilizing CO_2: An approach on flue gases utilization for third generation biofuels. *J. Clean. Prod.*, 98, 53, 2015.
45. Lu, J., Sheahan, C., Fu, P., Metabolic engineering of algae for fourth generation biofuels production. *Energy Environ. Sci.*, 4, 2451, 2011.

46. Krishnan, C., Sousa, Lda C., Jin, M., Chang, L., Dale, B.E., Balan, V., Alkali-based AFEX pre-treatment for the conversion of sugarcane bagasse and cane leaf residues to ethanol. *Biotechnol. Bioeng.* 107, 3, 441–50, 2010.
47. Chen, X., Shekiro, J., Elander, R., Tucker, M., Improved Xylan hydrolysis of corn stover by deacetylation with high solids dilute acid pre-treatment. *Ind. Eng. Chem. Res.*, 51, 1, 70–76, 2012b.
48. Noureddini, H. and Byun, J., Dilute-acid pre-treatment of distillers' grains and corn fiber. *Bioresour. Technol.*, 101, 3, 1060–1067, 2010.
49. Sindhu, R., Kuttiraja, M., Binod, P., Janu, K.U., Sukumaran, R.K., Pandey, A., Dilute acid pre-treatment and enzymatic saccharification of sugarcane tops for bioethanol production. *Bioresour. Technol.*, 102, 23, 10915–10921, 2011.
50. Speers, A.M. and Reguera, G., Consolidated bioprocessing of AFEX-pretreated corn stover to ethanol and hydrogen in a microbial electrolysis cell. *Environ. Sci. Technol.*, 46, 14, 7875–7881, 2012.
51. Jin, M., Gunawan, C., Balan, V., Lau, M.W., Dale, B.E., Simultaneous saccharification and co-fermentation (SSCF) of AFEX(TM) pretreated corn stover for ethanol production using commercial enzymes and Saccharomyces cerevisiae 424A(LNH-ST). *Bioresour. Technol.*, 110, 587–594, 2012.
52. Kobayashi, T., Kohn, B., Holmes, L., Faulkner, R., Davis, M., Maciel, G.E., Molecular-level consequences of biomass pre-treatment by dilute sulfuric acid at various temperatures. *Energy Fuels*, 25, 1790–1797, 2011.
53. Sierra-Ramírez, R., Garcia, L.A., Holtzapple, M.T., Selectivity and delignification kinetics for oxidative short-term lime pre-treatment of poplar wood, part I: Constant-pressure. *Biol. Prog.*, 27, 4, 976–985, 2011.
54. Rabelo, S.C., Carrere, H., Maciel, F.R., Costa, A.C., Production of bioethanol, methane and heat from sugarcane bagasse in a biorefinery concept. *Bioresour. Technol.*, 102, 17, 7887–7895, 2011.
55. Zhao, X., Cheng, K., Liu, D., Organosolv pre-treatment of lignocellulosic biomass for enzymatic hydrolysis. *Appl. Microbiol. Biotechnol.*, 82, 5, 815–827, 2009.
56. Arato, C., Pye, E.K., Gjennestad, G., The lignol approach to biorefining of woody biomass to produce ethanol and chemicals. *Appl. Biochem. Biotechnol.*, 121–124, 871–882, 2005.
57. Yong, J. and Soo, S., Compositional changes in industrial hemp biomass (Cannabis sativa L.) induced by electron beam irradiation pre-treatment. *Biomass Bioenergy*, 35, 7, 3267–3270, 2011.
58. Khanal, S.K., Montalbo, M., Leeuwen, J.H.V., Srinivasan, G., Grewell, D., Ultrasound enhanced glucose release from corn in ethanol plants. *Biotechnol. Bioeng.*, 98, 978–985, 2007.
59. Khan, F., Ahmad, S.R., Kronfli, E., Gamma-radiation induced changes in the physical and chemical properties of lignocelluloses. *Biomacromol.*, 7, 2303–2309, 2006.

60. Nitayavardhana, S., Rakshit, S.K., Grewell, D., Leeuwen, J.H.V., Khanal, S.K., Ultrasound pre-treatment of cassava chip slurry to enhance sugar release for subsequent ethanol production. *Biotechnol. Bioeng.*, 101, 487–496, 2008.
61. Valery, B.A., Nazim, C., Richard, S., Alex, B., David, L., Biomass pretreatment: Fundamentals toward application. *Biotechnol. Adv.*, 29, 6, 675–685, 2011.
62. Bak, J.S., Kim, M.D., Choi, I.G., Kim, K.H., Biological pre-treatment of rice straw by fermenting with Dichomitus squalens. *New Biotechnol.*, 27, 4, 424–434, 2010.
63. Singh, D. and Chen, S., The white-rot fungus Phanerochaete chrysosporium: Conditions for the production of lignin-degrading enzymes. *Appl. Microbiol. Biotechnol.*, 81, 3, 399–417, 2008.
64. Faulon, J.-L., Carlson, G.A., Patrick Hatcher, G., A three-dimensional model for lignocellulose from gymnospermous wood. *Org. Geochem.*, 21, 12, 1169–1179, 1994. doi:10.1016/0146-6380(94)90161-9
65. Ullmann, F. and Bohnet M., *Ullmann's encyclopedia of industrial chemistry*, 6th edn, Wiley-VCH; J. Wiley & Sons, Weinheim, Germany, New York, 2002.
66. Alvira, P., Tomás-Pejó, E., Ballesteros, M., Negro, M.J., Pre-treatment technologies for an efficient bioethanol production process based on enzymatic hydrolysis: A review. *Biores. Technol.*, 101, 4851–4861, 2010.
67. Kumar, P., Barrett, D.M., Delwiche, M.J., Stroeve, P., Methods for pretreatment of lignocellulosic biomass for efficient hydrolysis and biofuel production. *Ind. Eng. Chem. Res.*, 48, 8, 3713–3729, 2009.
68. Himmel, M.E., Ding, S.Y., Johnson, D.K., Adeny, W.S., Nimlos, M.R., Brady, J.W., Foust, T.D., Biomass recalcitrance: Engineering plants and enzymes for biofuels production. *Science*, 315, 804–807, 2007.
69. Digman, M.F., Shinners, K.J., Casler, M.D., Optimizing on-farm pretreatment of perennial grasses for fuel ethanol production. *Biores. Technol.*, 101, 5305–5314, 2010.
70. Wyman, C.E., Dale, B.E., Elander, R.T., Comparative sugar recovery and fermentation data following pre-treatment of poplar wood by leading technologies. *Biotechnol. Prog.*, 25, 2, 333–339, 2009.
71. Xu, J., Thomsen, M.H., Thomsen, A.B., Pre-treatment on corn stover with low concentration of formic acid. *J. Microbiol. Biotechnol.*, 19, 8, 845–850, 2009.
72. Rocha, M.V., Rodrigues, T.H., De Macedo, G.R., Gonçalves, L.R., Enzymatic hydrolysis and fermentation of pretreated cashew apple bagasse with alkali and diluted sulfuric acid for bioethanol production. *Appl. Biochem. Biotechnol.*, 155, 407–417, 2009.
73. Cara, C., Ruiz, E., Oliva, J.M., Sáe, F., Castro, E., Conversion of olive tree biomass into fermentable sugars by dilute acid pre-treatment and enzymatic saccharification. *Bioresour. Technol.*, 99, 1869–1876, 2008.
74. Huang, H.J., Lin, W.L., Ramaswamy, S., Tschirner, U., Process modeling of comprehensive integrated forest biorefinery: An integrated approach. *Appl. Biochem. Biotechnol.*, 154, 1–3, 205–216, 2009.

75. Lu, X., Yimin, Z., Irini, A., Optimization of H2SO4-catalyzed hydrothermal pre-treatment of rapeseed straw for bioconversion to ethanol: Focusing on pre-treatment at high solids content. *Bioresour. Technol.*, 100, 12, 3048–3053, 2009.
76. Cheng, G., Varanasi, P., Li, C., Liu, H., Melnichenkos, B.Y., Simmons, B.A., Kent, M.S., Sing, S., Transition of cellulose crystalline structure and surface morphology of biomass as a function of ionic liquid pre-treatment and its relation to enzymatic hydrolysis. *Biomacromol.*, 12, 4, 933–941, 2011.
77. Shi, Y., Tao, Y., Wang, Y., Zhao, J., Zhou, S., The study of the dilute acid pre-treatment technology of corn stover and rice straw. *Adv. Mater. Res.* Trans Tech Publications, Ltd., 550–553, 480–483, 2012.
78. Chen, Y., Stevens, M.A., Zhu, Y., Holmes, J., Moxley, G., Xu, H., Reducing acid in dilute acid pre-treatment and the impact on enzymatic saccharification. *J. Ind. Microbiol. Biotechnol.*, 39, 5, 691–700, 2012a.
79. Roland, D.P., Cellulose: Pores, internal surfaces, and the water interface: Textile and paper chemistry and technology. *ACS Symp. Ser.*, 49, 20, 1976.
80. Liu, C.Z., Wang, F., Stiles, A.R., Guo, C., Ionic liquids for biofuel production: Opportunities and challenges. *Appl. Energy*, 92, 406–414, 2012.
81. Wang, H., Gurau, G., Rogers, R.D., Ionic liquid processing of cellulose. *Chem. Soc Rev.*, 41, 1519–1537, Wiley-VCH, Weinheim, Germany, 2012.
82. Xu, J. and Cheng, J.J., Pre-treatment of switchgrass for sugar production with the combination of sodium hydroxide and lime. *Biores. Technol.*, 102, 4, 3861–3868, 2011.
83. Chang, V.S., Nagwani, M., Holtzapple, M.T., Lime pre-treatment of crop residues bagasse and wheat straw. *Appl. Biochem. Biotechnol.*, 74, 135–159, 1998.
84. Falls, M., Ramirez, R.S., Haltzapple, M.T., Oxidative lime pre-treatment of Dacotah switchgrass. *Appl. Biochem. Biotechnol.*, 165, 243–259, 2011.
85. Xu, J., Cheng, J.J., Sharma, R.R., Burns, J.C., Sodium hydroxide pre-treatment of switchgrass for enzymatic Saccharification improvement. *Energy Fuels*, 24, 3, 2113–2119, 2010.
86. Harmsen, P., Huijgen, W., Bermúdez, L., Bakker, R., *Literature review of physical and chemical pre-treatment processes for lignocellulosic biomass*, Wageningen UR Food & Biobased Research 1st ed., 2010.
87. Sun, R. and Tomkinson, J., Comparative study of lignins isolated by alkali and ultrasound-assisted alkali extractions from wheat straw. *Ultrason. Sonochem.*, 9, 85–93, 2002.
88. Klinke, H.B., Ahring, B.K., Schmidt, A.S., Thomsen, A.B., Characterization of degradation products form alkaline wet oxidation of wheat straw. *Biores. Technol.*, 82, 15–26, 2002.
89. Mass, R.H., Bakker, R.R., Boersma, A.R., Bisschops, I., Pels, J.R., Jong, E.D., Weusthuis, R.A., Reith, H., Pilot-scale conversion of lime-treated wheat straw into bioethanol: Quality assessment of bioethanol and valorization of side streams by anaerobic digestion and combustion. *Biotechnol. Biofuels*, 1, 14, 2008.

90. Cheng, Y.S., Zheng, Y., Dooley, C.W., Jenkins, B.M., VanderGheynst, J.S., Evaluation of high solids alkaline pre-treatment of rice straw. *Appl. Biochem. Biotechnol.*, 162, 1768–1784, 2010.
91. Kim, S. and Holtzapple, M.T., Delignification kinetics of corn stover in lime pre-treatment. *Biores. Technol.*, 97, 778–785, 2006.
92. Kim, Y., Kreke, T., Hendrickson, R., Parenti, J., Ladisch, M.R., Fractionation of cellulase and fermentation inhibitors from steam pretreated mixed hardwood. *Bioresour. Technol.*, 135, 30–38, 2013.
93. Jung, Y.H., Kim, I.J., Han, J.I., Choi, I.G., Kim, K.H., Aqueous ammonia pre-treatment of oil palm empty fruit bunches for ethanol production. *Biores. Technol.*, 102, 20, 9806–9809, 2011.
94. Du, B., Sharma, L.N., Becker, C., Chenn, S.F., Mowery, R.A., van Walsum, G.P., Chambliss, C.K., Effect of varying feedstock–Pre-treatment chemistry combinations on the formation and accumulation of potentially inhibitory degradation products in biomass hydrolysates. *Biotechnol. Bioeng.*, 107, 3, 430–440, 2010.
95. Ince, N.H., Tezcanli, G., Belen, R.K., Ultrasound as a catalyzer of aqueous reaction systems: The state of the art and environmental applications. *Appl. Catal.*, 29, 167–176, 2001.
96. Thompson, L.H. and Doraiswamy, L.K., Sonochemistry: Science and engineering, *Ind. Eng. Chem. Res.*, 38, 4, 1215–1249, 1999.
97. Margulis, M.A. and Margulis, I.M., Contemporary review on nature of sonoluminescence and sonochemical reactions. *Ultrason. Sonochem.*, 9, 1–10, 2002.

4

Present Status and Future Prospect of Butanol Fermentation

Rashmi Mishra[1], Aakansha Raj[2] and Satyajit Saurabh[3]*

[1]ABA Division, National Institute of Secondary Agriculture (ICAR-NISA), Namkum, Ranchi, Jharkhand, India
[2]Microbial and Molecular Genetics Laboratory, Department of Botany, Patna University, Patna, Bihar, India
[3]DNA Fingerprinting Laboratory, Bihar State Seed and Organic Certification Agency, Mithapur, Patna, Bihar, India

Abstract

Biofuel has become a desirable option that may be used effectively in the modern environment. In terms of energy content, moisture affinity, and blending capabilities, biobutanol outperforms bioethanol. The biomass used to make biobutanol is lignocellulosic, which is abundant and can often be found for free or at a significantly reduced cost. Microbial fermentation can effectively produce biobutanol from home, commercial, and agricultural waste. In a process known as ABE fermentation (i.e., acetone, butanol, ethanol fermentation), butanol is synthesized by bacteria that produce solvents (such as *Clostridia acetobutylicum*). Large-scale waste production occurs in many industries, and if it is not adequately handled, it can result in serious air, water, and soil contamination. Several bacteria may utilize carbohydrate-rich waste components to produce biobutanol through aerobic as well as anaerobic fermentation, including *Escherichia coli*, *Clostridium acetobutylicum*, *Bacillus subtilis*, *Clostridium beijerinkckii*, *Pseudomonas putida*, and *Saccharomyces cerevisiae*, among others. In addition to offering a perfect, environmentally acceptable clean energy source, the production of biobutanol from industrial wastes has the ability to more fully address global challenges such as pollution, global warming, the greenhouse effect, etc. Recently, the demand for renewable resources and advances in biotechnology is driving renewed interest in fermentative butanol production.

Keywords: Biofuel, butanol, biobutanol production, ABE fermentation, biorefinery

---

Corresponding author: satyajitsaurabh@gmail.com

4.1 Introduction

Globally rapid technological development has increased demand for conventional fuels like never before. Renewable energy sources are q desirable choice for guaranteeing future energy security [1] since they can lessen reliance on fossil fuels and combat climate change [2]. Although conventional fuels meet nearly 85% of the world's overall energy consumption, there are still some drawbacks to be considered. One of these disadvantages is that their sources are finite and will eventually run out, and another is that their broad use has terrible environmental effects. Due to these two drawbacks, researchers have been searching for environmentally acceptable and renewable energy sources to replace conventional fuels. During this search, energy from biomass has become one of the most alluring options [3], it has a significant and expanding impact on the world's energy system [4]. The International Energy Agency (IEA) estimates that biofuels might provide a quarter of the world's transport fuels needs by 2050. Figure 4.1 provides a schematic illustration of the production of biofuel from biomass.

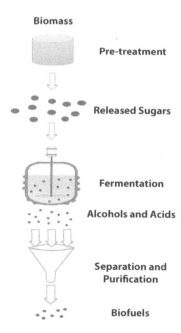

Figure 4.1 A schematic illustration of the production of biofuel from biomass.

In the realm of biofuel research today, keywords like biodiesel, bioethanol, and biobutanol are common. The latter, however, is drawing more attention from researchers and industrialists because of its several distinct advantagesdue to its superior ability to meet the demand for both an excellent sustainable biofuel and a crucial industrial solvent precursor.

Biobutanol has a number of benefits over other biofuels that make it a promising biofuel for use in the future. The energy content of biobutanol is rather high compared to other fuel substitutes. Biobutanol has less volatility and evaporative emissions than ethanol because it has a lower vapor pressure. Butanol, also referred to as butyl alcohol, can be found in four different isomeric forms: 1-butanol (n-butanol), 2-butanol, tert-butanol, and isobutanol. It is believed that 1-butanol and isobutanol's characteristics make them potential candidates for use as biofuels. Butanol is an essential precursor utilized in the manufacture of polymers and plastics [5]. It is used as a solvent in the production of vitamins, hormones, antibiotics, and other products for the textile and cosmetic industries [6, 7]. Lignocellulose biomass can also be converted into biobutanol through the fermentation of acetone butanol ethanol (ABE), just like second generational bioethanol [8].

A variety of feedstocks can be used to manufacture biobutanol domestically. Comparing biobutanol to petroleum fuels, fewer emissions are produced. By balancing the carbon dioxide emitted during biobutanol combustion with the carbon dioxide absorbed during feedstock growth, overall greenhouse gas emissions are decreased. Because biobutanol and water are insoluble, it would be possible to transport it through pipes to save money on transportation. Thus, biobutanol has emerged as a new type of renewable biofuel with superior properties over other conventional fuels like gasoline and ethanol. It has high octane number, lower volatility, low vapor pressure, flexible fuel blends, higher viscosity, and lubricity. Table 4.1 describes the fuel characteristics of gasoline, butanol, ethanol, and methanol. Due to its distinctive quality, it has continually captured the interest of scientists and industrialists around the world. It has potentials to be used commercially in several applications, such as solvent, additive, chemical intermediate, and extracting agent (Table 4.2).

4.2 Biobutanol Production

Al-Shorgani *et al.* (2012) described the production of biobutanol (also known as butyl alcohol) as the anaerobic fermentation of sugars from biomass, including corn stover [14], sugarcane bagasse (SCB) [15], inulin,

Table 4.1 Comparison of fuel characteristics of different biofuel.

Fuel	Energy density (MJ/L)	Water solubility (mL 100 mL^{-1})	Heat of vaporization (MJ/kg)	Air-to-fuel ratio	Research octane testing	Motor
Gasoline	32	<0.01	14.6	0.36	91–99	81–89
Butanol	29	9.1	11.2	0.43	96	78
Ethanol	19.6		9	0.92	130	96
Methanol	16	Miscible	6.5	1.20	136	104

Table 4.2 List of commercial applications of biobutanol.

S. no.	Commercial uses	References
1.	Used as fossil fuel extender or as biofuel	[8]
2.	Diminish environmental concerns like reduction of greenhouse gases, air, and water pollution	[9]
3.	Production of IPK, a jet fuel blend	[10]
4.	Solvents in chemical and pharmaceuticals and brake fluids, paint thinner	[11]
5.	Important substitute for commercial gasoline (petroleum)	[12]

algal biomass [16], palm oil mill effluent [13], cheese whey [17], cassava [105] starch [18], cane molasses [19], and food waste [20]. This process, sometimes referred to as ABE fermentation, yields acetone, butanol, and ethanol [21].

The industrial-scale traditional biobutanol fermentation process uses solventogenic Weizmann organism (Clostridium sp.), which rank second after ethanol fermentation by yeast and are capable of absorbing both simple and complex sugars, such as pentose, hexose, etc. [22]. Acidogenesis and solventogenesis are the two different processes that make up the metabolism of clostridia that produce ABE (Figure 4.2). Carbon and electrons are essential for the metabolism of clostridia that produce ABE [23]. The first phase, known as the acidogenesis phase, of the Emden–Meyerhof–Pranas (EMP) pathway, involves the conversion of glucose to pyruvate, glycolysis results in the production of 2 mol of ATP [24]. Carboxylic acids are produced after pyruvate is produced. These organics promotes exponential cell growth by providing energy to the strains and decreasing the pH value in the fermentation medium [23, 25]. The principal byproducts are organic acids, carbon dioxide, and hydrogen.

The second stage, known as solventogenesis, is when ABE is produced [26] and is connected to the rise in external pH brought on by acid absorption [27]. An abrupt shift in the patterns of gene expression is what separates the two periods [28]. The solventogenic fermentation process is usually constrained by a number of problems, including substrate inhibition, butanol toxicity in the medium, delayed development, and therefore decreased cell density. In addition to these limitations, the production of acetone and ethanol lowers the yield of biobutanol. In order to solve

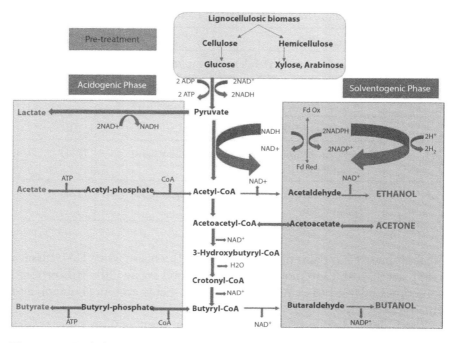

Figure 4.2 Metabolic pathways in ABE fermentation by *Clostridia*. In the yellow region, the creation of organic acids is depicted in red (acidogenesis), whereas in the blue region, the formation of solvents is depicted in blue (solventogenesis).

these problems, scientists have developed microbial strains with enhanced biobutanol output and tolerance.

4.2.1 Microbes and Biobutanol Production

Clostridia are the primary microorganism involved in butanol production. *Clostridium acetobutylicum*, *Clostridium beijerinckii*, *Clostridium saccharobutylicum*, and *Clostridium saccharoperbutylacetonicum* are the main solventogenic clostridia [100]. These four strains are all mesophilic bacteria with high levels and yields of butanol synthesis, but they have different capacities for utilizing carbon sources and 16 different ideal pH and temperature ranges [28, 29]. Among these, *C. acetobutylicum* and *C. beijerinckii* are frequently utilized in research. The first bacteria utilized for ABE fermentation was *C. acetobutylicum* [30]. This rod-shaped, Grampositive endospore-forming, obligatorily anaerobic clostridium produces butanol [23, 31, 32]. However, these bacteria also include species that produce butanol, are aerotolerant, and endospore-forming species,

C. acetobutylicum YM1 [13, 33]. *C. acetobutylicum* ATCC 824 is one of the most extensively investigated and first genome-sequenced strains of solventogenic *Clostridium*. But various solvent-forming clostridia may have distinct genomes from one another [29].

The only solvent that is produced to a level that becomes toxic to cells and significantly inhibits their growth is butanol [93]. High concentrations of butanol (>10 g/L) can increase the fluidity of cell membranes by degrading the phospholipid component, which further destabilizes the membrane and reduces its capacity to carry out its related function. Additionally, it has been reported that butanol can prevent cells from maintaining their internal pH and sugar uptake [35–37]. When butanol concentration reaches 12 to 16 g/L, cell proliferation stops [91]. The final butanol titer from batch fermentation is typically less than 12 g/L [28]. The amount of substrate that may be used for fermentation is severely constrained by butanol inhibition, which also lowers butanol concentration and productivity [28].

ABE-producing clostridia can ferment a variety of carbon sources, such as glucose, xylose, fructose, sucrose, galactose, mannose, inulin, and glycerol [15, 24, 25, 38]. The majority of these carbon sources come from a combination of free sugars and carbs, including sweet sorghum [92] and a mixture of different agricultural [39–41] and domestic [42] wastes as feedstocks for butanol production using butanol-producing clostridia. These cheap and inedible feedstocks could increase the profitability of butanol fermentation.

4.2.2 Substrate for Biobutanol Production

Currently, only 10% of the world's annual energy demand is fulfilled by the energy supply from biomass (about 50 EJ), and this trend is increasing. It is estimated that 1500 EJ of energy will be produced from biomass by 2050. There are several generations of biofuels, including biobutanol, depending on the type of biomass used. Edible crops, nonedible crops, microalgae, etc. are used to produce first-generation, second-generation, and third-generation biobutanol, respectively [43, 44]. Due to the exponential rise in demand for renewable energy, various biomasses are now being used. Biobutanol is one of the most important by-products of these biomasses. Third-generation biomass comprises bacteria and algae that can produce photosynthetic substances. These are used because they are high in oil/lipids, proteins, or carbohydrates. They do not require cultivable land or other agricultural inputs and have a high lipid content of 20% to 40% of their dry weight, so they are advantageous over the other two types of biomasses. Although this method is expected to become commonplace in the coming years [43], technical immaturity is now a major obstacle to the production of ethanol from this group of biomasses. Microalgal biomass when compared to macroalgae (as for example brown macroalgae, red macro algae, green macroalgae, mixed macroalgae) is more easily available and conveniently convertible to biobutanol since the amount of alginate present in macroalgae is quite high.

4.2.3 ABE Fermentation Process

ABE fermentation is the second-most popular process after ethanol fermentation in the fermentation industry. However, compared with other procedures, the ABE fermentation process is more desirable in terms of the diversity of goods generated because it produces three separate products as opposed to only one. Similar to other fermentation processes, ABE fermentation is divided into three categories batch, fed-batch, and continuous modes of fermentation. Other strategies employed in the continuous mode of fermentation include free cell, immobilized cell, and cell recycling [44]. The fermentation process is mostly controlled bacteria's metabolic activity. Acidogenesis and solventogenesis are two steps of the biphasic ABE fermentation process. Several *Clostridium* species share similar metabolic pathways, such as those of *C. acetobutylicum* and *C. beijerinckii* are alike. The byproducts from the metabolic pathway of Clostridium sp. during the ABE fermentation process includes utanol, acetone, ethanol, acetic acid, butyric acid, CO_2, and H_2 [45].

4.2.4 Recovery of Biobutanol from Fermentation Broth

The recovery of biobutanol from fermentation broth requires low-energy consumption. The technique uses a polymer resin as an adsorbent in an adsorption/desorption process. The desorption step is performed more efficiently and combined with the condensation and decantation steps. The production of biobutanol depends on this phase because it is the most expensive in the fermentation process.

Low butanol concentrations in fermentation broths make it difficult to recover butanol using conventional methods, which consume a lot of energy. Reverse osmosis, liquid–liquid extraction, gas stripping, and procedures based on membranes, such as perstraction, pervaporation, and distillation, are only a few of the recovery technologies that are now in use. The optimal technique has not been determined despite the existence of all these methods [46, 47].

The separation process is the most notable and significant phase in the manufacture of biofuel, particularly biobutanol. For the recovery of acetone, butanol, and ethanol from the fermentation broth, a variety of approaches have been used. This procedure is crucial because it not only increases yield by lowering the possibility of solvent toxicity toward microorganisms but also saves time and money by recovering products rapidly. Apart from the conventional distillation technique, adsorption [48–50], gas stripping [51–53], liquid–liquid extraction (LLE)n [48–50], pervaporation [54, 55], and reverse osmosis (RO) are a few of the separation processes now in use, particularly for recovering butanol from fermentation broth.

Gas stripping
Gas stripping is a well-known separation method that is basically based on the principle of gas purging. This technique selectively removes the volatile components from the broth formed during ABE fermentation [51]. Gas is sparged into the fermenter, which leads to the condensation of volatiles, and they are thus recovered from the condenser. This technique is applicable owing to the volatile nature of the products formed during ABE fermentation. Different parameters for example availability of ethanol and acetone, rate of gas recycling as well as the size of the bubble were tested for the recovery of ABE from the fermentation broth in terms of performance.

Research showed that by applying the sparger gas stripping mode, the formation of a large amount of foam in the bioreactor took place. This required addition of a huge amount of antifoam during the process. This addition of the excess quantity of antifoam leads to the reduced production of fermentation products, which is often referred to as the toxicity of microbes [51]. Gas recycle rate of 0.08 m^3 s^{-1} and constant gas stripping rate of 0.058 h^{-1} were studied to be enough for maintaining the concentration of n-butanol far below the level of toxicity during ABE fermentation process. The study further reported that bubbles less than 0.5 cm in size had no effect on the stripping rate of n-butanol during the run of ABE fermentation [51].

Liquid–liquid extraction
This method works on the principle of organic solvent contact. This technique involves the extraction of a dissolved substance from a solution mixture through a solvent, by using another solvent. Butyric acid-saturated n-decanol was used for the selective extraction of butyl alcohol from a cell-free cultivation medium. Once free from n-butanol, the culture medium

was refed into the reactor. Under optimum conditions, the concentration of n-butanol was found to be 8 g L^{-1} and the productivity rate observed was around 0.51 g $L^{-1} \cdot h^{-1}$. However, the addition of n-decanol in the fermenter resulted in the toxicity of the cells, thereby reducing the fermentation productivity rate and thus the concentration of products obtained. To avoid this situation, the least interaction should be maintained between the cell culture and n-decanol. A reduction in cell mass after the extraction procedure was also observed during this separation technique [89]. This phenomenon led to less productivity during the second run of the cycle. To overcome this, toxic decanol in combination with nontoxic oleyl alcohol was used.

Adsorption
Adsorption is another separation procedure that is based on the principle of hydrophobic adsorbents. Particles from a gaseous or liquid combination are preferentially adhered to a solid surface in this instance. Adsorption of ethanol, as well as butanol, was carried out on mesoporous carbons. The surface area was between 500 and 1300 $m^2 g^{-1}$, and it was observed that adsorption of n-butanol in comparison to ethanol was more efficient. It was also noted that the adsorption of these alcohols was directly proportional to the surface area of these mesoporous carbons [89]. This simply suggests that as the surface area is increased, the adsorption rate of ethanol as well as n-butanol increases. The basic purpose for using mesoporous carbons as adsorbents was that they were thermally and chemically stable during the entire process.

Perstraction/Membrane Extraction
Membrane extraction is a kind of liquid–liquid extraction method where a porous membrane is kept in between two phases. Membrane extraction is well known as perstraction is based on the principle of interaction between membrane and solvent. Perstraction makes use of the installation of a membrane area that separates the extracting liquid from the extractant. Membrane extraction method applied directly along with fermentor at 35°C with silicone membrane used as selective boundary produced butanol in a concentration of 8.89 g L^{-1} during the first cycle and around 10.29 g L^{-1} in the second cycle. The extractant used during this method was oleyl alcohol. Despite the recovery of efficient concentration of butanol, the only issue that comes along with this method of extraction is the removal of acetone from the fermentation broth is poor [89].

4.3 Perspectives

The growing population with the growing demand for food, feed, and fuel needs to be addressed with a sustainable solution. Consumption of non-renewable fossil fuels causes global environmental issues and rises in fuel prices. To address this, alternative renewable sustainable biofuel production has been promoted and thus biofuel production expanded. Biofuels are mostly produced from biomasses of food or feed. This may have a negative impact on the growing population in terms of shortage of food and feed. Alternatively, inedible lignocellulosic biomass, having cellulose and hemicelluloses, is being used. But, high-order crystal structure of lignocelluloses needs an expensive pre-treatment and saccharification step, for strains to utilize the lignocellulosic biomass. Presently, butanol production through ABE fermentation is facing several challenges, such as high production cost, butanol feedback inhibition, and low titer, tolerance and yield of butanol produced (Figure 4.3). Still, several studies are carried to address these problems by suggesting potential remedies for several gridlocks such as substrate, fermentation techniques, strain development, and butanol removal/recovery techniques.

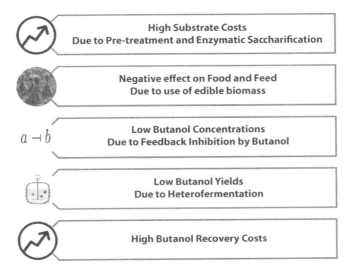

Figure 4.3 Challenges in biobutanol production.

4.3.1 Substrate

Several crops are being cultivated on arable land for biofuel production, and are termed energy crops. These crops are converted to biodiesel by transesterification and butanol through fermentation. However, this energy crop cultivation is stained with controversies being considered as a threat to the production of the agricultural crop for food security. Then biofuels are produced from agricultural and forestry residues having lignocellulosic biomass and organic components of municipal wastes are appreciable. But it has its own limitation of (i) being expensive due to transportation and pre-treatment of substrate biomass to utilize the same as feedstocks for biobutanol production, (ii) composition of lignocelluloses (depends on crop and age), (iii) lignocelluloses is recalcitrant due to its crystalline structure having high degree of polymerization, (iv) constant availability throughout the year, and (v) pre-treatment is required significantly to hydrolyze the lignocelluloses. There are several methods of pre-treatment, such as mechanical, chemical, physicochemical, and biological, which has to be utilized alone or in combination depending on the physicochemical properties of the lignocellulosic biomass for better results. A better understanding of biochemical, thermodynamic, and molecular mechanisms involved in hydrolysis might help to adopt a proficient method for pre-treatment. Although enzymatic hydrolysis is preferred, the enzyme properties, kinetics, and optimization of parameters may help to increase the enzymatic activity efficiently.

The involvement of molecular biology may help in designing a plant with low lignin, to reduce the cost of pre-treatment [56–58]. Studies are ongoing to make these substrates and the pre-treatments cost-effective in biobutanol production. The year-round constant supply could be ensured by improving agricultural practices and utilizing the alternative source of lignocellulosic biomass. The operational parameters from supply and storage could be optimized by Geographic Information System (GIS)-based dynamic approach like Integrated Biomass Supply Analysis and Logistics (IBSAL) [59]. Additionally, depending on geographical locations, climate and soil-type energy crops, like *Miscanthus giganteus*, can be cultivated on marginal fallow lands.

Since 1995, researchers at Argonne National Laboratory have been into developing a model known as Greenhouse Gases, Regulated Emissions and Energy Use in Transportation (GREET) model. The preliminary version of this model was GREET 1.0 and was released in 1996. The multidimensional spreadsheet model GREET, which is mostly focused on Microsoft, addresses well-to-wheels analytical challenges associated with renewable transport fuels, such as ethanol or butanol, as well as vehicle of this model, GREET 1.7 was released in November 2005 and was efficient enough to analyze around 100 pathways related to transportation fuel. This model is being updated on regular intervals depending on the feedstocks introduced, emerging technologies, fuels as well as vehicle systems [60].

Generally, for a specific vehicle fuel system, GREET is utilized to calculate

- Energy consumption
- Emissions of equivalent greenhouse gases
- Emissions of pollutants such as carbon monoxide, nitrogen oxides, particulate matters less than 10μm, sulfur oxides and particulate matters less than 2.5 μm.

Genetic manipulation of the substrate generally, only makes sense in the fourth generation of production, since the first generation already concerns food crops, the second generation trying to reuse/recycle, the lignocellulosic or other such waste/residues while the third generation uses algae as feedstock and further fodder for future lines of production. On the scale of sustainability, this generation thus, aims for "carbon-negative" biofuel generation. As with any generation, economic feasibility is dependent on the biomass productivity of the feedstock. Furthermore, studies are carried to utilize algae for butanol production. As, the microalgae is a photosynthetic organism, grow easily in large quantities at faster rate, and can be utilized by *Clostridium* and *E. coli* to produce butanol through ABE fermentation. Notably, it is grown without affecting the crop fields as a solution for the controversies associated with the threat to the food security.

4.3.2 Alleviate Carbon Catabolite Repression

During fermentation, sugar catabolism is inhibited by glucose. This phenomenon is known as carbon catabolite repression (CCR). The CCR is reported in bacteria and yeast, including *Clostridia* [61]. To reduce the CCRs, the utilization of feasible biomass and genetic manipulation is recommended to utilize other sugars like xylose and arabinose in presence of glucose. The genes associated with the expression of sugar metabolism-related transcription factors, transporters, and metabolic enzymes are studied in *C. acetobutylicum* [62]. The butanol production has been increased by disrupting the carbon catabolite protein-encoding gene (ccpA) [63]. Protein, the CcpA, acts as a metabolic repressor for many sugars in presence of glucose. Even, the genome of *E. coli* has been engineered to alleviate carbon catabolite repression [66].

4.3.3 Fermentation Improvement

The fermentation technologies using batch culture, fed-batch culture, and continuous culture are being improvised to enhance butanol production. Along with the fermentation technologies, studies are also focused on the integrated butanol recovery techniques to avoid feedback inhibition by high titers of butanol. The high titer of butanol is also a challenge as it inhibits cell growth and butanol production, too. *In situ* separation and recovery is highly recommended to address such issues [65]. This may enhance productivity by enhancing sugar conversion, better cell survival and retention, and avoiding feedback inhibition by high titer butanol in the fermentation broth. The overall butanol concentration is reported to increase by approximately 250% in the integrated fermenters compared to the nonintegrated fermenter [66].

The batch culture is the simplest of the three fermentation processes. In this, several parameters (such as pH, C/N ratio, CO, H_2, and electron carriers) have been studied. Low pH, moderate C/N ratio, increased CO, increased H_2, and electron carriers like methyl viologen (MV) and methylene blue (MB) are recommended for improved yield. Still, batch culture has limitations like inhibition at high levels. The fed-batch culture has been utilized to overcome the disadvantages of batch cultures and improved butanol production. Different methods have been utilized to avoid inhibition by recovering butanol. The continuous culture is advantageous to batch culture and fed-batch culture in terms of operational stability, as the butanol concentration in continuous culture should be lower [67]. The immobilization or cell recycle technique coupled with continuous bioreactors shows beneficial results in terms of efficiency of butanol production [68].

A significant number of researches have been carried to employ different fermentation strategies. This requires an insightful knowledge to regulate the metabolic flux and redirecting the carbon flux to the production of butanol. It may aid to increase the butanol production by suppressing the production of unwanted solvents, like ethanol and acetone. The advancement in ABE fermentation is optimizing the variables, such as the dilution rate, the carbohydrate feed concentration, the cell retention factor, and the recovery of product. The production of butanol could be optimized by utilizing mathematical modeling and simulation analysis tool like Resonance Surface Methodology (RSM) to set up experiments with different factors, and analyzing the experimental data recorded by the analysis of variance (ANOVA) and other related statistical tools.

4.3.4 Strain Development

The development of DNA ligases and restriction endonucleases in the 1970s signaled the beginning of genetic engineering as a transgenic technology. This develops into a promising tool for improving traits in virtually all living things, including microbes, plants, and animals. However, because to worries about biosafety, biodiversity loss, and ecological imbalances, the use of this technology is still controversial on a number of platforms [69, 70]. Other secure and innovative approaches are required to address these problems in order to get broad acceptance across many platforms. The advancement of various innovative technologies over the past two decades has drawn attention as the preferred tool for genetic modification to control genes at the transcriptional and posttranscriptional levels. The generation of strains through genetic engineering is being done to increase butanol production efficiency and solvent toxicity tolerance. Several microbes have also been genetically altered to produce biobutanol in addition to clostridium.

Clostridium
By using conventional mutagenesis and screening, efforts are made to generate mutants with increased butanol tolerance and yield [34, 91]. Dong reported two mutants of *C. acetobutylicum*, Rh9 with a butanol tolerance of 19 g/L and EA2018 with an enhanced butanol ratio of 70% (vs. 60%) among solvents [71]. With the improved butanol tolerance, however, these strains did not give better butanol productions. Recently, a mutant strain *C. acetobutylicum* JB200 with high butanol production derived from *C. acetobutylicum* ATCC 55025 was obtained by spontaneous mutation in a fibrous-bed bioreactor (FBB) which can produce up to ~25 g/L butanol [72]. Through the use of contemporary recombinant DNA technology, numerous metabolically engineered strains have been produced that contain the appropriate genes. *C. acetobutylicum* ATCC 824 mutant strains have been modified to produce more butanol/solvents [73, 74]. For example, Jiang knocked out the acetoacetate decarboxylase gene (adc) and the mutant had an increased butanol/acetone ratio [75]. Jiang, for instance, altered the mutant's butanol/acetone ratio by knocking out the acetoacetate decarboxylase gene (adc) [75]. Metabolic engineering allowed for the generation of butanol in nonnative hosts in addition to native solventogenic clostridia. Different strategies have been employed for improving butanol tolerance in *Clostridium acetobutylicum*; e.g., by deletion of a gene by *astE* (succinylglutamatedesuccinylase), knockout of gene *Cac*-3319 (histidine kinase production), and overexpression ofgenes *entC* (isochorismate

synthase), *FeoA* (small iron transport protein), and *groESL* (heat shock proteins) [76, 77].

From *C. tyrobutyricum*, Yu got another strain that had been successfully developed. Yu overexpressed aldehyde/alcohol dehydrogenase 2 (adhE2), a protein that transforms butyryl-CoA into butanol, in the butyric acid-producing bacteria *C. tyrobutyricum* ATCC 25755 to produce butanol [80]. Besides, higher butanol titer was observed by knocking out the acetate kinase (ack) gene in the host bacterium [78]. This engineered *C. tyrobutyricum* (Δack-adhE2) can produce 10 to 16 g/L butanol depending on the carbon source used in the fermentation [78].

Escherichia coli
Escherichia coli is a microbe that has undergone in-depth research and has been engineered to produce butanol. The process of producing butanol using recombinant *E. coli* that expresses genes related to the clostridial butanol biosynthetic pathway, such as thl, crt, hbd, bcd-etfAB, and adhE, was initially described by Atsumi et al. (2008) [79]. The butanol-producing recombinant *E. coli* had a low titer (less than 1 g/L). The E. coli strain was also modified by Dellomonaco to generate butanol at a significantly higher 14 g/L concentration [80]. Moreover, transgenic *E. coli* was driven by outside pressures to manufacture butanol at a high rate of 30 g/L, according to Shen [81].

Cyanobacteria
Carbon dioxide, which is the only available carbon for plants, can in addition be utilized for chemicals and also for biofuel production [82]. Synthetic pathway for the production of different alcohols such 2,3-butandiol, ethanol and n-butanol was constructed in the PCC7942 strain of cyanobacteria *Synechococcus elongates*. The production of n-butanol was observed to be 14.5 mg/L. The strain and the plasmid utilized during this production process were EL14 and NSIT.d-ter and NSII atoB, hbd, crt, and adhE2. The NADH mediated production process which involves NADP-dependent Adh from *E. coli* in addition to Bldh from C. beijerinckii in EL22 strain of *Synechococcus* showed around 29.9 mg/L of n-butanol output. The low productivity was because of the toxicity caused during the process [82]. Due to the low productivity of n-butanol, butandiol was targeted owing to its less toxic nature as well as its similarity with the pathway of cyanobacteria. Production of butandiol through this process was observed to be 2.38 g/L which is a significant amount when exogenous pathway of cyanobacteria is considered [90].

Thermoanaero bacterium saccharolyticum

Thermoanaero bacterium is a Gram-positive bacterium that is well known and can be genetically constructed for the synthesis of biohydrogen, ethanol, as well as biobutanol [6, 7]. The clusters of genes used were obtained from *Thermoanaero bacterium thermosaccharolyticum* (hbd, crt, bcd, eftA, eftB) and *C. acetobutylicum* (DSM571 and adhE2). The synthetic pathway for the butanol production involving *C. acetobutylicum* was seen to be 0.89mg/L from 10g/L of xylose [90].

Klebsiella pneumoniae

A well-known rod-shaped, facultatively anaerobic, gram negative bacterium is called *Klebsiella*. This bacterium is genetically engineered for the production and synthesis of n-butanol, sec-butanol, butandiol, propylene glycol, ethanol, and hydrogen [23]. The bacterium was genetically constructed for the production of n-butanol and sec-butanol from crude glycerol as the only carbon source. For the synthesis of n-butanol, modification of CoA-dependent pathway as well as 2,2-keto acid pathway was performed. This was done by carrying out the active expression of two genes ter-bdhB-bdhA and kvid, respectively, for the above mentioned two pathways. The butanol titer and butanol production obtained from CoA-dependent pathway was estimated to be 15.03 mg/L and 27.79 mg butanol/g-cell and that from 2,2-keto acid pathway was around 28.7 mg/L and 51.58 mg butanol/g-cell [90].

Geobacillus thermoglucosidasius

Geobacillus is a gram-positive, anaerobic bacterium that is known to survive between 40°C to 70°C. This bacterium was genetically engineered for ethanol and isobutanol production [83, 84]. The construction process involved two genes. The first one being acetohydoxy acid synthase gene from *B. subtilis* and the other one being 2-ketoisovalerate dehydrogenase gene from *L. lactis*. In addition to these, a lactate hydrogenase gene was inserted at the promoter region which was derived from *Geobacillus thermodenitrificans*. The amount of isobutanol obtained through this process was around 3.3g/L where glucose was used as a major substrate, at a temperature of 50°C [84].

Pyrococcusfuriosus

Pyrococcus is a flagellated, heterophillic, cocci-shaped archaebacteria whose metabolic products are carbon dioxide and hydrogen. Genetic modification of *Pyrococcusfuriosus* was done and butanol production was observed at higher temperatures. Genes responsible for the production of n-butanol and sec-butanol were isolated from *Thermoanaerobacter tengeongensis* and *Spirochaete* as well as from *Thermoanaerobacter* sp X514 and were established in *Pyrococcusfuriosus*. The amount of n-butanol and sec-butanol after 2 days from this genetically engineered microorganism was observed to be 70 and 15 mg/L respectively at 60°C [85].

Yeast

Saccharomyces is common yeast which is potentially used in fermentation process. It is preferably used in beverage industry as well as for alcohol production. Genetic manipulation of *Saccharomyces cerevisiae* was done for obtaining was done for obtaining n-butanol, sec-butanol, and isobutanol. Deletion of genes $\Delta adh1$, $\Delta ilv2$ of YSG52 strain was brought about for the production of n-butanol using glucose. The amount of n-butanol recovered was 242.8 mg/L. n-butanol recovered using glycine, followed Ehlich pathway and was estimated around 92 mg/L. This pathway involved the conversion of glycine to glycoxylate and finally to butanol having β-ethylmalate and α-ketovalerate as intermediate products [86].

4.3.5 Butanol Recovery

Separating butanol from the fermentation broth and purifying it are the first two steps in butanol recovery. The separation is being employed by several processes, having comparable advantages and disadvantages (mentioned in Table 4.3). Low butanol concentration in fermentation broth makes the process energy intensive and expensive. A multistage recovery process having combination of different recovery processes can be utilized for efficient butanol recovery with other important intermediates [87]. Employing an alternative extraction process with controlled laminar flow operation in microreactor system is attractive [88]. The highest butanol titre reported is 30 g/L [81]. It was achieved by utilizing *in situ* gas stripping recovery technique and intermittent linear glucose feeding. Although other recovery techniques are integrated with the fermentation process for butanol separation, the integrated recovery techniques can not only efficiently recover butanol but also release the product inhibition effect and therefore boost butanol production.

Table 4.3 Advantages and drawbacks of some butanol recovery methods.

Method	Basic principle	Major advantages	Major drawbacks
Gas stripping	Gas purging	Simple, less clogging	Incomplete removal
Liquid-liquid extraction	Organic solvent contact	High capacity	Expensive
Membrane evaporation	Selective diffusion	Smaller membrane needs	Clogging occurs
Perstraction	Membrane and solvent	High selectivity	Large membrane area
Adsorption	Hydrophobic adsorbents	High uptake capacity	Desorption is tedious

4.4 Conclusion

The rapid increase in depletion of fossil fuels and environmental health provoked researchers to explore sustainable strategy for the production of biofuel. Nonrenewable energy sources including coal, natural gas, and oil predominantly supply the world's energy needs. However, the lack of fossil fuels and rising costs of them, together with the severe environmental challenges, prompted biologists to find a microbial solution to these problems. The focus of research is shifting from other biofuels like bioethanol and biodiesel to biobutanol because it has the astonishing potential to replace gasoline as the only fuel in internal combustion engines without requiring any modifications and performs better than bioethanol across a number of parameters.

Selecting the most suitable strains with improved tolerance through screening is the most desired way for efficient butanol production. Additionally, a microbe could be designed through genetic manipulations and genome editing to have excellent strain for producing biobutanol more efficiently with better tolerance to butanol in a bioreactor. On the other hand, genetic manipulations and genome editing could be used for designed biomass for establishing an efficient fermentation process. These strategies is required for improving butanol productivity and yield economically. Furthermore, the possible way outs to address the challenges associated with biobutanol production could be utilization of microbial cell factories, the optimization of fermentation parameters with recovery techniques, and genetic manipulations leading to central carbon metabolism for alternate biosynthetic pathways, i.e., metabolic engineering.

References

1. Sharma, B., Ingalls, R.G., Jones, C.L., Khanchi, A., Biomass supply chain design and analysis: Basis, overview, modeling, challenges, and future. *Renewable Sustain. Energy Rev., 24*, 608–627, 2013.
2. Cherubini, F. and Strømman, A.H., Life cycle assessment of bioenergy systems: State of the art and future challenges. *Bioresour. Technol., 102*, 2, 437–451, 2011.
3. Srirangan, K., Akawi, L., Moo-Young, M., Chou, C.P., Towards sustainable production of clean energy carriers from biomass resources. *Appl. Energy, 100*, 172–186, 2012.
4. Reid, W.V. and Ali, M.K., The future of bioenergy. *Global Change Biol., 26*, 1, 274–286, 2020.
5. Ibrahim, M.F., Ramli, N., Bahrin, E.K., Abd-Aziz, S., Cellulosic biobutanol by Clostridia: Challenges and improvements. *Renewable Sustain. Energy Rev., 79*, 1241–1254, 2017.
6. Zheng, J., Tashiro, Y., Wang, Q., Sonomoto, K., Recent advances to improve fermentative butanol production: Genetic engineering and fermentation technology. *J. Biosci. Bioeng., 119*, 1, 1–9, 2015.
7. Harris, O., Wilbur, S., George, J., Eisenmann, C., *Toxicological profile for 2-butoxyethanol and 2-butoxyethanol acetate*, p. 296, US Department of Health and Human Services, Agency for Toxic Substances and Disease Registry, Atlanta, GA. USA, 1998.
8. Durre, P., Biobutanol: An attractive biofuel. *Biotechnol. J., 2*, 1525–34, 2007.
9. Kumar, M. and Gayen, K., Developments in biobutanol production: New insights. *Appl. Energy, 1*, 88, 6, 1999–2012, 2011.
10. Wang, W.-C., Tao, L., *Bio-jet fuel conversion technologies. Renew. Sustain. Energy Rev., 53*, 801–822, 2016.
11. Liu, H., Wang, G., Zhang, J., The promising fuel-biobutanol, in: *Liquid, Gaseous and Solid Biofuels - Conversion Techniques*, F. Zhen (Ed.), IntechOpen, 2013.
12. Devi, P.B., Joseph, D.R., Gokulnath, R., Manigandan, S., Gunasekar, P., Anand, T.P., Vimal, M.R., The effect of TiO2 on engine emissions for gas

turbine engine fueled with jatropha, butanol, soya and rapeseed oil. *Int. J. Turbo Jet-Engines*, 37, 1, 85–94, 2020.
13. Al-Shorgani, N.K.N., Kalil, M.S., Ali, E., Yusoff, W.M.W., Hamid, A.A., Enhancement of biobutanol production by butyric acid addition using Clostridium saccharoperbutylacetonicum N1-4 (ATCC 13564). *Biotechnology*, 11, 6, 326–332, 2012.
14. Gao, X., Zhao, H., Zhang, G., He, K., Jin, Y., Genome shuffling of Clostridium acetobutylicum CICC 8012 for improved production of acetone–butanol–ethanol (ABE). *Curr. Microbiol.*, 65, 2, 128–132, 2012.
15. Lu, C., Yu, L., Varghese, S., Yu, M., Yang, S.T., Enhanced robustness in acetone-butanol-ethanol fermentation with engineered Clostridium beijerinckii overexpressing adhE2 and ctfAB. *Bioresour. Technol.*, 243, 1000–1008, 2017.
16. Ellis, J.T., Hengge, N.N., Sims, R.C., Miller, C.D., Acetone, butanol, and ethanol production from wastewater algae. *Bioresour. Technol.*, 111, 491–495, 2012.
17. Becerra, M., Cerdán, M.E., González-Siso, M.I., Biobutanol from cheese whey. *Microb. Cell Fact.*, 14, 1, 1–15, 2015.
18. Li, S., Guo, Y., Lu, F., Huang, J., Pang, Z., High-level butanol production from cassava starch by a newly isolated Clostridium acetobutylicum. *Appl. Biochem. Biotechnol.*, 177, 4, 831–841, 2015.
19. Li, H.G., Luo, W., Gu, Q.Y., Wang, Q., Hu, W.J., Yu, X.B., Acetone, butanol, and ethanol production from cane molasses using Clostridium beijerinckii mutant obtained by combined low-energy ion beam implantation and N-methyl-N-nitro-N-nitrosoguanidine induction. *Bioresour. Technol.*, 137, 254–260, 2013.
20. Arancon, R.A.D., Lin, C.S.K., Chan, K.M., Kwan, T.H., Luque, R., Advances on waste valorization: New horizons for a more sustainable society, in: *Energy Science & Engineering*, vol. 1, pp. 53–71, Apple Academic Press, 2017.
21. Gao, M., Tashiro, Y., Yoshida, T., Zheng, J., Wang, Q., Sakai, K., Sonomoto, K., Metabolic analysis of butanol production from acetate in Clostridium saccharoperbutylacetonicum N1-4 using 13 C tracer experiments. *Rsc Adv.*, 5, 11, 8486–8495, 2015.
22. Lütke-Eversloh, T. and Bahl, H., Metabolic engineering of Clostridium acetobutylicum: Recent advances to improve butanol production. *Curr. Opin. Biotechnol.*, 22, 5, 634–647, 2011.
23. Tashiro, Y., Yoshida, T., Noguchi, T., Sonomoto, K., Recent advances and future prospects for increased butanol production by acetone-butanol-ethanol fermentation. *Eng. Life Sci.*, 13, 5, 432–445, 2013.
24. Ezeji, T., Milne, C., Price, N.D., Blaschek, H.P., Achievements and perspectives to overcome the poor solvent resistance in acetone and butanol-producing microorganisms. *Appl. Microbiol. Biotechnol.*, 85, 6, 1697–1712, 2010.
25. Visioli, L.J., Enzweiler, H., Kuhn, R.C., Schwaab, M., Mazutti, M.A., Recent advances on biobutanol production. *Sustain. Chem. Process.*, 2, 1, 1–9, 2014.

26. Ndaba, B., Chiyanzu, I., Marx, S., n-Butanol derived from biochemical and chemical routes: A review. *Biotechnol. Rep.*, *8*, 1–9, 2015.
27. Lee, S.Y., Park, J.H., Jang, S.H., Nielsen, L.K., Kim, J., Jung, K.S., Fermentative butanol production by Clostridia. *Biotechnol. Bioeng.*, *101*, 2, 209–228, 2008.
28. García, V., Päkkilä, J., Ojamo, H., Muurinen, E., Keiski, R.L., Challenges in biobutanol production: How to improve the efficiency? *Renewable Sustain. Energy Rev.*, *15*, 2, 964–980, 2011.
29. Yang, S.T. and Lu, C., Extraction-fermentation hybrid (extractive fermentation, in: *Separation and Purification Technologies in Biorefineries*, pp. 409–437, 2013.
30. Cheng, C.L., Che, P.Y., Chen, B.Y., Lee, W.J., Chien, L.J., Chang, J.S., High yield bio-butanol production by solvent-producing bacterial microflora. *Bioresour. Technol.*, *113*, 58–64, 2012.
31. Karimi, K., Tabatabaei, M., SárváriHorváth, I., Kumar, R., Recent trends in acetone, butanol, and ethanol (ABE) production. *Biofuel Res. J.*, *2*, 4, 301–308, 2015.
32. Lütke-Eversloh, T. and Bahl, H., Metabolic engineering of Clostridium acetobutylicum: Recent advances to improve butanol production. *Curr. Opin. Biotechnol.*, *22*, 5, 634–647, 2011.
33. Guo, T., He, A.Y., Du, T.F., Zhu, D.W., Liang, D.F., Jiang, M., Wei, P., Ouyang, P.K., Butanol production from hemicellulosichydrolysate of corn fiber by a Clostridium beijerinckii mutant with high inhibitor-tolerance. *Bioresour. Technol.*, *135*, 379–385, 2013.
34. Jones, D.T. and Woods, D., Acetone-butanol fermentation revisited. *Microbiol. Rev.*, *50*, 4, 484–524, 1986.
35. Bowles, L.K. and Ellefson, W., Effects of butanol on Clostridium acetobutylicum. *Appl. Environ. Microbiol.*, *50*, 5, 1165–1170, 1985.
36. Gottwald, M. and Gottschalk, G., The internal pH of Clostridium acetobutylicum and its effect on the shift from acid to solvent formation. *Arch. Microbiol.*, *143*, 1, 42–46, 1985.
37. Ounine, K., Petitdemange, H., Raval, G., Gay, R., Regulation and butanol inhibition of D-xylose and D-glucose uptake in Clostridium acetobutylicum. *Appl. Environ. Microbiol.*, *49*, 4, 874–878, 1985.
38. Mao, S., Luo, Y., Zhang, T., Li, J., Bao, G., Zhu, Y., Chen, Z., Zhang, Y., Li, Y., Ma, Y., Proteome reference map and comparative proteomic analysis between a wild type Clostridium acetobutylicum DSM 1731 and its mutant with enhanced butanol tolerance and butanol yield. *J. Proteome Res.*, *9*, 6, 3046–3061, 2010.
39. Khedkar, M.A., Nimbalkar, P.R., Gaikwad, S.G., Chavan, P.V., Bankar, S.B., Sustainable biobutanol production from pineapple waste by using Clostridium acetobutylicum B 527: Drying kinetics study. *Bioresour. Technol.*, *225*, 359–366, 2017.

40. Nimbalkar, P.R., Khedkar, M.A., Chavan, P.V., Bankar, S.B., Biobutanol production using pea pod waste as substrate: Impact of drying on saccharification and fermentation. *Renewable Energy*, *117*, 520–529, 2018.
41. Harde, S.M., Jadhav, S.B., Bankar, S.B., Ojamo, H., Granström, T., Singhal, R.S., Survase, S.A., Acetone-butanol-ethanol (ABE) fermentation using the root hydrolysate after extraction of forskolin from Coleus forskohlii. *Renewable Energy*, *86*, 594–601, 2016.
42. Claassen, P.A., Budde, M.A., López-Contreras, A.M., Acetone, butanol and ethanol production from domestic organic waste by solventogenic clostridia. *J. Mol. Microbiol. Biotechnol.*, *2*, 1, 39–44, 2000.
43. Procentese, A., Raganati, F., Olivieri, G., Russo, M.E., Rehmann, L., Marzocchella, A., Low-energy biomass pre-treatment with deep eutectic solvents for bio-butanol production. *Bioresour. Technol.*, *243*, 464–473, 2017.
44. Kumar, M. and Gayen, K., Developments in biobutanol production: New insights. *Appl. Energy*, *88*, 6, 1999–2012, 2011.
45. Zheng, Y.N., Li, L.Z., Xian, M., Ma, Y.J., Yang, J.M., Xu, X., He, D.Z., Problems with the microbial production of butanol. *J. Ind. Microbiol. Biotechnol.*, *36*, 9, 1127–1138, 2009.
46. Cheng, C.L., Che, P.Y., Chen, B.Y., Lee, W.J., Chien, L.J., Chang, J.S., High yield bio-butanol production by solvent-producing bacterial microflora. *Bioresour. Technol.*, *113*, 58–64, 2012.
47. Abdehagh, N., Tezel, F.H., Thibault, J., Separation techniques in butanol production: Challenges and developments. *Biomass Bioenergy*, *60*, 222–246, 2014.
48. Lin, X., Wu, J., Fan, J., Qian, W., Zhou, X., Qian, C., Jin, X., Wang, L., Bai, J., Ying, H., Adsorption of butanol from aqueous solution onto a new type of macroporous adsorption resin: Studies of adsorption isotherms and kinetics simulation. *J. Chem. Technol. Biotechnol.*, *87*, 7, 924–931, 2012.
49. Nielsen, D.R., Leonard, E., Yoon, S.H., Tseng, H.C., Yuan, C., Prather, K.L.J., Engineering alternative butanol production platforms in heterologous bacteria. *Metab. Eng.*, *11*, 4-5, 262–273, 2009.
50. Qureshi, N., Hughes, S., Maddox, I.S., Cotta, M.A., Energy-efficient recovery of butanol from model solutions and fermentation broth by adsorption. *Bioprocess. Biosyst. Eng.*, *27*, 4, 215–222, 2005.
51. Ezeji, T.C., Qureshi, N., Blaschek, H.P., Production of acetone, butanol and ethanol by Clostridium beijerinckii BA101 and *in situ* recovery by gas stripping. *World J. Microbiol. Biotechnol.*, *19*, 6, 595–603, 2003.
52. Lu, C., Zhao, J., Yang, S.T., Wei, D., Fed-batch fermentation for n-butanol production from cassava bagasse hydrolysate in a fibrous bed bioreactor with continuous gas stripping. *Bioresour. Technol.*, *104*, 380–387, 2012.
53. Xue, C., Zhao, J., Liu, F., Lu, C., Yang, S.T., Bai, F.W., Two-stage *in situ* gas stripping for enhanced butanol fermentation and energy-saving product recovery. *Bioresour. Technol.*, *135*, 396–402, 2013.

54. Dong, Z., Liu, G., Liu, S., Liu, Z., Jin, W., High performance ceramic hollow fiber supported PDMS composite pervaporation membrane for bio-butanol recovery. *J. Membr. Sci.*, *450*, 38–47, 2014.
55. Li, S.Y., Srivastava, R., Parnas, R.S., Separation of 1-butanol by pervaporation using a novel tri-layer PDMS composite membrane. *J. Membr. Sci.*, *363*, 1-2, 287–294, 2010.
56. Green, E.M., Fermentative production of butanol—The industrial perspective. *Curr. Opin. Biotechnol.*, *22*, 3, 337–343, 2011.
57. Saurabh, S., Vidyarthi, A.S., Prasad, D., RNA interference: Concept to reality in crop improvement. *Planta*, *239*, 3, 543–564, 2014.
58. Saurabh, S., Genome editing: Revolutionizing the crop improvement. *Plant Mol. Biol. Rep.*, *39*, 4, 1–21, 2021.
59. Sokhansanj, S., Kumar, A., Turhollow, A.F., Development and implementation of integrated biomass supply analysis and logistics model (IBSAL). *Biomass Bioenergy*, *30*, 10, 838–847, 2006.
60. Brinkman, N., Wang, M., Weber, T., Darlington, T., *GM Study: Well-to-wheels analysis of advanced fuel/vehicle systems – A North American study of energy use, greenhouse gas emissions, and criteria pollutant emissions*, EERE Publication and Product Library, Washington, DC USA, 2005, May, http://www.transportation.anl.gov/pdfs/TA/339.pdf.
61. Tracy, B.P., Jones, S.W., Fast, A.G., Indurthi, D.C., Papoutsakis, E.T., Clostridia: The importance of their exceptional substrate and metabolite diversity for biofuel and biorefinery applications. *Curr. Opin. Biotechnol.*, *23*, 3, 364–381, 2012.
62. Servinsky, M.D., Kiel, J.T., Dupuy, N.F., Sund, C.J., Transcriptional analysis of differential carbohydrate utilization by Clostridium acetobutylicum. *Microbiology*, *156*, 11, 3478–3491, 2010.
63. Ren, C., Gu, Y., Hu, S., Wu, Y., Wang, P., Yang, Y., Yang, C., Yang, S., Jiang, W., Identification and inactivation of pleiotropic regulator CcpA to eliminate glucose repression of xylose utilization in Clostridium acetobutylicum. *Metab. Eng.*, *12*, 5, 446–454, 2010.
64. Wang, X., Goh, E.B., Beller, H.R., Engineering E. coli for simultaneous glucose—Xylose utilization during methyl ketone production. *Microb. Cell Fact.*, *17*, 1–11, 2018. https://doi.org/10.1186/s12934-018-0862-6.
65. De Brabander, P., Uitterhaegen, E., Verhoeven, E., Vander Cruyssen, C., De Winter, K., Soetaert, W., In Situ product recovery of bio-based industrial platform chemicals: A guideline to solvent selection. *Fermentation.*, *7*, 1, 26, 2021. https://doi.org/10.3390/fermentation7010026.
66. Azimi, H., Tezel, H., Thibault, J., Optimization of the *in situ* recovery of butanol from ABE fermentation broth via membrane pervaporation. *Chem. Eng. Res. Des.*, *150*, 49–64, 2019.

67. Tashiro, Y., Yoshida, T., Noguchi, T., Sonomoto, K., Recent advances and future prospects for increased butanol production by acetone-butanol-ethanol fermentation. *Eng. Life Sci.*, 13, 5, 432–445, 2013.
68. Ezeji, T.C., Qureshi, N., Blaschek, H.P., Bioproduction of butanol from biomass: From genes to bioreactors. *Curr. Opin. Biotechnol.*, 18, 3, 220–227, 2007.
69. Singh, A., Kiran, S., Saurabh, S., Kumari, S., Rhizosphere engineering for crop improvement, in: *Rhizosphere Engineering*, pp. 417–444, Academic Press, 2022.
70. Stephens, J. and Barakate, A., Gene editing technologies–ZFNs, TALENs, and CRISPR/Cas9, in: *Encyclopedia of Applied Plant Sciences, 2nd*, B Thomas, B.G. Murray, D.J. Murphy, pp. 157–161, Academic, Cambridge, MA, 2017.
71. Yu, S., Jiang, P., Dong, Y., Zhang, P., Zhang, Y., Zhang, W., Hydrothermal synthesis of nanosized sulfated zirconia as an efficient and reusable catalyst for esterification of acetic acid with n-butanol. *Bull. Korean Chem. Soc*, 33, 2, 524–528, Academic, Cambridge, MA, 2012.
72. Zhao, Y.L., Dolat, A., Steinberger, Y., Wang, X., Osman, A., Xie, G.H., Biomass yield and changes in chemical composition of sweet sorghum cultivars grown for biofuel. *Field Crops Res.*, 111, 1-2, 55–64, 2009.
73. Huang, H.J., Ramaswamy, S., Liu, Y., Separation and purification of biobutanol during bioconversion of biomass. *Sep. Purif. Technol.*, 132, 513–540, 2014.
74. Papoutsakis, E.T., Engineering solventogenic clostridia. *Curr. Opin. Biotechnol.*, 19, 5, 420–429, 2008.
75. Jang, Y.S., Lee, J.Y., Lee, J., Park, J.H., Im, J.A., Eom, M.H., Lee, J., Lee, S.H., Song, H., Cho, J.H., Seung, D.Y., Enhanced butanol production obtained by reinforcing the direct butanol-forming route in Clostridium acetobutylicum. *MBio*, 3, 5, e00314-12, 2012.
76. Wu, P., Wang, G., Wang, G., Børresen, B.T., Liu, H., Zhang, J., Butanol production under microaerobic conditions with a symbiotic system of Clostridium acetobutylicum and Bacillus cereus. *Microb. Cell Fact.*, 15, 1, 1–11, 2016.
77. Tomas, C.A., Beamish, J., Papoutsakis, E.T., Transcriptional analysis of butanol stress and tolerance in *Clostridium acetobutylicum*. *J. Bacteriol.*, 186, 7, 2006–18, 2004 Apr 1.
78. Park, H.M., Malaviya, A., Yu-Sin, J.A.N.G., Park, J.H., Lee, S.Y., Step towards the realization of economic viability for biofuel industry: Competitive production of biobutanol by glycerol fermentation. *Korean Society of Biological Engineering Conference*, 196, 196–196, 2011.
79. Atsumi, S. and Liao, J.C., Directed evolution of Methanococcus-jannaschiicitramalate synthase for biosynthesis of 1-propanol and 1-butanol by Escherichia coli. *Appl. Environ. Microbiol.*, 74, 24, 7802–7808, 2008.

80. Dellomonaco, C., Clomburg, J.M., Miller, E.N., Gonzalez, R., Engineered reversal of the β-oxidation cycle for the synthesis of fuels and chemicals. *Nature*, 476, 7360, 355–359, 2011.
81. Shen, C., Lan, E., Dekishima, Y., Baez, A., Cho, K., Liao, J., Driving forces enable high titer anaerobic 1-butanol synthesis in Escherichia coli. *Appl. Environ. Microb.*, 77, 2904–2915, 2011.
82. Angermayr, S.A., Hellingwerf, K.J., Lindblad, P., de Mattos, M.J.T., Energy biotechnology with cyanobacteria. *Curr. Opin. Biotechnol.*, 20, 3, 257–263, 2009.
83. Cripps, R.E., Eley, K., Leak, D.J., Rudd, B., Taylor, M., Todd, M., Boakes, S., Martin, S., Atkinson, T., Metabolic engineering of Geobacillusthermoglucosidasius for high yield ethanol production. *Metab. Eng.*, 11, 6, 398–408, 2009.
84. Lin, P.P., Rabe, K.S., Takasumi, J.L., Kadisch, M., Arnold, F.H., Liao, J.C., Isobutanol production at elevated temperatures in thermophilicGeobacillusthermoglucosidasius. *Metab. Eng.*, 24, 1–8, 2014.
85. Keller, M.W., Lipscomb, G.L., Loder, A.J., Schut, G.J., Kelly, R.M., Adams, M.W., A hybrid synthetic pathway for butanol production by a hyperthermophilic microbe. *Metab. Eng.*, 27, 101–106, 2015.
86. Lian, J., Si, T., Nair, N.U., Zhao, H., Design and construction of acetyl-CoA overproducing Saccharomyces cerevisiae strains. *Metab. Eng.*, 24, 139–149, 2014.
87. Morone, A. and Pandey, R.A., Lignocellulosic biobutanol production: Gridlocks and potential remedies. *Renewable Sustain. Energy Rev.*, 37, 21–35, 2014.
88. Jovanovic, J., Rebrov, E.V., Nijhuis, T.A., Kreutzer, M.T., Hessel, V., Schouten, J.C., Liquid–liquid flow in a capillary microreactor: Hydrodynamic flow patterns and extraction performance. *Ind. Eng. Chem. Res.*, 51, 2, 1015–1026, 2012.
89. Kujawska, A., Kujawski, J., Bryjak, M., Kujawski, W., ABE fermentation products recovery methods—A review. *Renewable Sustain. Energy Rev.*, 48, 648–661, 2015.
90. Rao, A., Sathiavelu, A., Mythili, S., Genetic engineering in biobutanol production and tolerance. *Braz. Arch. Biol. Technol.*, 59, e16150612, 2016.
91. Sakuragi, H., *Studies on applications of Clostridium species for biorefinery*, vol. 33, Kyoto University, 2014.
92. Rochon, E., Ferrari, D.M., Lareo, C., Integrated ABE fermentation-gas stripping process for enhanced butanol production from sugarcane-sweet sorghum juices. *Biomass and Bioenergy*, 2098, 153–160, 2017.
93. Berezina, O.V., Zakharova, N.V., Brandt, A., Yarotsky, S.V., Schwarz, W.H., Zverlov, V.V., Reconstructing the clostridial n-butanol metabolic pathway in Lactobacillus brevis. *Appl. Microbiol. Biotechnol.*, 87, 2, 635–646, 2010.

94. Tomas, C.A., Welker, N.E., Papoutsakis, E.T., Overexpression of groESL in Clostridium acetobutylicum results in increased solvent production and tolerance, prolonged metabolism, and changes in the cell's transcriptional program. *Appl. Environ. Microbiol.*, 69, 8, 4951–4965, 2003.
95. Li, S.B., Qian, Y., Liang, Z.W., Guo, Y., Zhao, M.M., Pang, Z.W., Enhanced butanol production from cassava with Clostridium acetobutylicum by genome shuffling. *World J. Microbiol. Biotechnol.*, 32, 4, 1–10, 2016.

5

Strategies of Strain Improvement for Butanol Fermentation

Shreya, Nikita Bhati and Arun Kumar Sharma*

Department of Bioscience and Biotechnology, Banasthali Vidyapith, Rajasthan, India

Abstract

Biobutanol is an economic, attractive, and sustainable fuel because petroleum oil leads to costly fuel because of more greenhouse gases and depletion of oil reserves in the atmosphere. Restricted accessibility of carbonaceous conventional fossil fuels is linked with growing pollution due to overexploitation has increased the demand expedition for renewable fuels. The main challenges in the production of butanol are product inhibition, cost-effective feedstock availability, and low butanol titer. These obstacles are overcome by using various metabolic engineering techniques, and genetic engineering and incorporating capable continuous fermentation processes with well product recovery strategies. The production of biobutanol can be either via a chemical route by petroleum or via a biotechnological route by microorganisms like *Clostridia*. Presently, more interest has been developed in the catalytic coupling of bioethanol into butanol over different hetero systems. This catalytic process has high potential and is a growing step in overcoming the demerits of bioethanol as a sustainable transportation fuel. *Clostridium beijerinckii* is a potential strain for the production of biobutanol from cellulosic feedstocks. However, the production of high biobutanol is being emphasized through genetic engineering of Clostridia and non-Clostridia organisms (such as *Bacillus subtilis, Saccharomyces cerevisiae, Escherichia coli,* and *Pseudomonas putida*) in both conditions anaerobic and aerobic fermentation. Numerous applications of novel genome sequencing of Clostridial organism of higher butanol production enhance the genetic engineering scope for biobutanol production. This chapter, more emphasized suitable bacterial strain selection, cheaper biomass availability for butanol production, genetic engineering tools for different microorganisms, and process development attempts.

Corresponding author: arun.k.sharma84@gmail.com

Keywords: Biofuel, acetone–butanol–ethanol fermentation, *Clostridium* spp., bioethanol, biomass, development, mutagenesis

5.1 Introduction

The rising population and existing lifestyle alterations have rapidly increased energy consumption globally. This also enhances the rate of exhaustion of conventional energy sources and enhanced environmental pollution [1, 2]. In this era, continuous exhaustion of fuel is a major concern, and retrieving them will turn out to be enormously hard in the future [3, 4]. With the rising concern about the harmful effects of environmental pollution and global warming, more awareness has been gained for developing renewable and green energy sources. Among various existing biofuel resources like biodiesel, bioalcohols, bioelectricity, and biogases, the bioalcohols are potentially great to reduce greenhouse gas emissions, versatile chemical feedstock, decrease fossil fuel dependency, and also consumed as a transportation fuel. Particularly biobutanol, it is derived from biomass fermentation and it is less corrosive and explosive compared to bioethanol. More significantly, bioethanol can be used in a few vehicles without any alteration in their engines [5, 6].

Biobutanol was first industrially synthesized during 1912 to 1919 by acetone–butanol–ethanol (ABE) fermentation of cereal grains and molasses via *Clostridium acetobutylicum* (Jones and Woods, 1986). Afterward, *Clostridium saccharobutylicum*, and *Clostridium saccharoperbutylacetonicum* were also reported as potential strains which have a high potential for the production of biobutanol with maximum yield [7]. Currently, butanol is used in industries for the production of coating, lacquers, brake fluids, detergents, and plasticizers. Now approximately, every year production of butanol is around 10 to 12 billion pounds [8, 9].

Butanol has less vapor pressure, is less hygroscopic, has higher energy content, and is highly tolerant to water contamination in gasoline blends compared with methanol and ethanol, which exhibits its usage in feedstocks for various chemical products and energy. Butanol has gained increased awareness due to its usage as a substitute fuel. The production of butanol traditionally resulted from nonrenewable petroleum, a declining resource whose utilization led to important global environmental concerns [10].

Particularly, petroleum fuel consumption for energy generation increases the carbon dioxide, nitrous oxide, and methane in the atmosphere [11]. Therefore, renewable fuel is required by the world as a substitute for fossil fuel.

In these circumstances, biofuel production has the possible potential to recompense for the present requirement of petroleum fuels [12, 13].

Different approaches for biobutanol production have been reported, for example, batch, fed-batch, and continuous methods. Instead of various disadvantages like contamination and high capital cost, the continuous approach has a more economical propensity over the other two approaches [14–16]. The various advantages of continuous processes are reinoculation time, reduction in sterilization, reduction in butanol inhibition, and superior productivity. The wide issues in the fermentation of butanol are endorsed by low production, high product recovery cost, and selection of sustainable biomass. Genetic engineering and gene manipulation also contribute an important role in tool development to enhance the potential of microbial strains assisting the yield of fermentation. Present investigations focus highly on the examination of the cost-effective lignocellulosic materials as feedstocks, competent recovery methods, and new approaches to compensating product inhibition [17–19].

The waste agricultural residue material provides a cost-effective solution for biofuel production. Over 40 million tonnes of nonedible plants are produced each year, much of which is wasted. Therefore, these agricultural residues present a feasible economic solution for biofuel production.

Today, Potent and abundant sources of biomass comprise switchgrass, flax shives, wheat straw, poplar, hemp hurds, salix, food processing waste, and organic waste from nonedible plant parts. Nowadays, the opportunities that are exhibited from the conversion of this woody biomass to biofuels must be of primary concern [20].

However, additional chemical and physical processing is required to convert lignocellulosic biomass into fuels directly. The standard approaches for this conversion are steam explosion pre-treatment and dilute acid pre-treatment. With a high probability of biofuel conversion from lignocellulose, enzymatic hydrolysis is measured as an additional environmentally friendly preference. However, great amounts of costly cellulases are necessary during hydrolysis, which significantly enhances the saccharification cost [21].

Significant experimental efforts have been made to genetically modify or develop butanol-producing cultures that can endure maximized levels of acetone–butanol concentration in the range of 20 to 30 g/L. In recent reports, it has been reported that *Escherichia coli* developed for butane production with a 14- to 16-g/L range of concentration. The developed microbial strain cannot tolerate > 15 g/L butanol because of its much more toxic level than isobutanol. Similarly, more than 25 g/L butanol concentration is toxic to *C. beijerinckii* BA101 [22, 23].

5.2 Background

Historically, acetone–butanol–ethanol fermentation was mostly emphasized for the synthesis of acetone which was the solvent for smokeless nitrocellulose explosive preparation. For the time being, butanol was highly used in quick-drying lacquer and in rubber production as a solvent to provide a finishing car body. Butanol also serves as a manufacturer for various production of amino resins, butyl acetate, glycol ethers, butyl amines, acryl esters, and butylacetate [24]. A few more applications are gasoline additive (anti-icing), humectants for cellulose nitrate, additive in cleaners and polishes, mobile phase in a thin layer, and paper chromatography [25].

Presently, additional interest has been paid to biobutanol production after its compatibility as a biofuel has been confirmed. In 2005, the use of biobutanol had a major impact when David Ramey wrestled with butanol in an unmodified vehicle in the United States [24].

5.3 Microorganism

Clostridia genus holds a huge range of bacteria that produces biobutanol such as *Clostridium aurantibutyricum, C. acetobutylicum, Clostridium saccharoacetobutylicum, C. beijerinckii, Clostridium pasteurianum, Clostridium tetanomorphum, Clostridium sporogenes,* and *Clostridium cadaveris.* Significantly, *C. beijerinckii, C. saccharoacetobutylicum, C. acetobutylicum,* and *C. saccaroperbutylacetonicum* demonstrated high activity for butanol synthesis with maximum yields [26, 27].

An obligate anaerobic bacteria *Clostridium acetobutylicum*, which is spore-forming and gram-posi tive was primarily utilized microbial strain for acetone–butanol–ethanol fermentation at a commercial scale. Carbohydrates like mannose, starch, fructose, glucose, lactose, dextrins, and sucrose are fully consumed, whereas arabinose, insulin, mannitol, galactose, glycerol, raffinose, and xylose are not fully consumed by *Clostridial* sp., but trehalose, melibiose and rhamnose are cannot be fermented [28, 29]. Coculture of one or two anaerobic and different aerobic strains is an important technique for escaping the rigorous anaerobic conditions to maximize butanol production. Culturing of two aerobic and anaerobic strains such as *Clostridium butylicum* TISTR 1032 with *Bacillus subtilis* WD161 exhibited 6.5 fold increase in production of butanol (7.4 g/L ABE) from cassava starch still exclusive of exhausting anaerobic maintenance. Coculturing is a different technique where one anaerobic clostridial strain liberates intermediates for fermentation of butanol which were further consumed by another anaerobic strain for butanol production [30].

The main disadvantage in the fermentation of butanol by parent strains is led to inhibition of product at the maximum level of butanol concentration, accumulation of acid during fermentation, and microorganisms' inability for industrial-level production [31]. *Clostridia* sp. when frequently subcultured, decreases its butanol production activity because of an increase in butanol hydrophobicity that enhances the fluidity of the cell membrane and alters the pH gradient of transmembrane, and capability of glucose uptake, all these parameters in turn unfavorably affect the productivity of butanol [32, 33].

5.4 ABE Fermentation

ABE fermentation process is classified into two stages: acidogenic phase and solventogenic phase. The acidogeneic phase is connected to butyric acid and acetic acid production with carbon dioxide and hydrogen as the key products. Whereas, solventogenic phase is associated with pH decline and further sporulation, which led to the form of acetone, 1-butanol, and carbon dioxide from formed acids with carbohydrates [34, 35].

During the fermentation process, first, phosphorylation of glucose into glucose-6-phosphate, followed by pyruvate conversion through Embden–Meyerhoff pathway (EMP). During glycolysis, the production of pyruvate is conducted by pyruvate ferredoxin oxidoreductase in coenzyme presence. This process resulted in CO_2, reduced ferredoxin, and acetyl-CoA. Further conversion of acetyl-CoA to acetate by phosphate acetyltransferase and acetate kinase. However, butyryl-CoA conversion into Butyrate is performed in the presence of butyl kinase and phosphate butyltransferase. Medium pH declines due to acetic acid and butyric acid accumulation as soon as starting of solventogeneic phase occurs. Reassimilation of acidogenic phase byproducts occurs during the solventogenic phase. Conversion of acetyl-CoA into acetoacetyl–CoA via thiolase which undergoes further reduction into 3-hydroxybutyryl-CoA by hydroxybutyryl-CoA dehydrogenase. Dehydration of 3-hydroxybutyryl-CoA to crotonyl–CoA performed by crotonase, further reduced by butyryl-CoA dehydrogenase and NADH in butyryl–CoA. Further by the use of enzymes butanol dehydrogenase, butyraldehyde dehydrogenase, and NADH, the butyryl-CoA is converted to butanol and butyraldehyde [36].

5.4.1 The Obstacle in ABE Fermentation from *Clostridium* sp.

Among clostridia, *C. acetobutylicum, C. saccharobutylicum, C. saccharoperbutylacetonicum,* and *C. beijerinckii* are the main producers of solvent. This solventogenic process is linked with the sporulation process which is begun by a few transcription factors, CoA transferase, and acetoacetate decarboxylase resulting in the production of solvent in *C. acetobutylicum*. The independent spore regulation and activation of an acid affected by formic acid presence in clostridial cells have been recognized as an inhibitory factor for acetone–butanol–ethanol# fermentation [37]. The two-phase metabolism continued to the solvent formation which hurdles the continuous fermentation process. The by-products formation (acetate, acetone, and ethanol) and butanol production results in the purification of specific products which is expensive. In *C. acetobutylicum*, loss of megaplasmid pSOL1 occurred which is also a limiting factor, and this loss is called strain degeneracy. During degeneracy, cells lose their solvent production capability and become asporogenous [38]. Butanol toxicity and tolerance is also the main obstacle to butanol fuel production on an economic scale. Clostridium has a low tolerance of butanol up to 1% to 2% [39].

5.5 Selection of Biomass for the Production of Butanol

For fermentation, the feedstock is mainly a liable factor based on its economic viability [40–42]. Based on feedstocks utilization, biofuels were divided into two categories, i.e, first- and second-generation biofuels. Raw materials used as feedstocks in first-generation biofuel were cereal grains and sugarcane, whereas in second-generation biofuel used were lignocellulosic sources (forest and agriculture wastes). Therefore, it should be illustrated that the first-generation used raw materials were food competitive whereas, the raw materials used for second-generation biofuels were nonedible biomass. Presently, more emphasis is paid to second-generation biofuels because of the abundant availability of cost-effective raw materials [42–44].

The utility of various starch (defibrated-sweet-potato slurry, liquefied corn, cassava, extruded corn, and sago) and (whey permeate) lactose were reported as carbon and energy sources for microbes in acetone–butanol–ethanol fermentation (Table 5.1) [45]. Reporters also postulated that liquefied corn starch would be cheaper raw material in biobutanol production after deletion of $Na_2S_2O_5$ during the raw material pre-treatment and by

Table 5.1 Various raw materials used in acetone–butanol–ethanol fermentation [46].

Raw materials	Bacterial strain	Composition	Gelatinization required
Domestic organic waste	C. acetobutylocum	13% lignin, 59% sugars, 17% ash	Yes
Barley straw	C. beijerinckii	7% lignin, 42% cellulose, 11% ash 28% hemicellulose,	Yes
Corn stover	C. beijerinckii	6% Ash, 26% hemicellulose, 23% lignin, 38% cellulose	Yes
Defibrated-sweet-potato-slurry	C. acetobutylocum	Starch	No
Liquefied corn starch	C. beijerinckii	45% moisture, 39% starch	Yes
Switchgrass	C. beijerinckii	19% lignin, 29% hemicellulose, 37% Cellulose,	Yes
Whey permeate	C. acetobutylocum	0.36% fat, 5% lactose, 0.86% protein	No
Extruded corn	C. acetobutylocum	3.8% corn oil, 11.2% fiber, 61% starch, 8.0% protein	Yes
Cassava	Coculture of C. butylicum and B. Subtilis	0.2% ash, 2.4% fiber, 2.7% protein, 70% Starch,	No
Sago	C. saccharobutylicum	Small amounts of nitrogenous matters and minerals, 86% starch	Yes

gas stripping. Separately starch substrates, the lactose including substrate suitability such as cheese whey were examined using *C. acetobutylicum* AS 1.224 and *C. acetobutylicum* DSM 792 in batch fermentation. Results reported that cheese whey had a superior ability to synthetic media or lactose solution [46].

5.6 Processes Improvement

Simultaneous wheat straw hydrolysis with regards to simple sugar and cultivation of butanol is a significant substitute for the utilization of exclusive glucose in the fermentation processes of butanol.

Qureshi *et al.* [47] reported both stages of cultivation trials to determine the batch process results for butanol production from different polysaccharide sources by using various pre-treatment combinations such as fermentation with pretreated wheat straw, anaerobic hydrolysis, and fermentation of wheat straw, cofermentation, and hydrolysis of sugar-supplemented straw, separate fermentation and hydrolysis of wheat straw without sediments removal, simultaneous fermentation and hydrolysis of wheat straw with agitation by gas stripping. Different monosaccharides, such as arabinose, glucose, mannose, xylose, and galactose, have been reported as hydrolysates. Therefore, additional sugar enhances the fed-batch fermentation productivity [48].

Many factors led to the effect of the fermentation of butanol such as salt concentration inhibition, low water activity, nutrient deficiency, O2 diffusion, substrate inhibition, accumulation of macromolecules, and the presence of dead cells when nutrients are added to the fermenter. The main fermentation inhibitors of lignocellulosic material are furfural, salts, acetic, coumaric acids, phenolic compounds, and glucuronic acid which are present in the lignocellulosic hydrolysates [49, 50].

The competency of fed-batch and batch fermentation processes is not only harmed by inhibitors present in fermentation but also by time-consuming and additional steps such as reinoculation or bioreactor sterilization. To avoid these demerits, continuous fermentation is used. For continuous fermentation, the common steps are cell recycling in the free cell fermentation process, cell-free systems, immobilized cell systems, and movement of a cell in fermentation broth freely, due to air-lift and mechanical agitation. It holds nutrients and microbial cells in suspension and assists in mass transfer promotion. The fermentation activity using immobilized cells is more advantageous because of high fermentation productivity, less downstream processing cost, less recovery cost, and stable fillings on continuous processing. The immobilization method removes the lag phase and validates proficient continuous fermentation without performing frequent inoculation. This process also helps to enhance the catalytic effects of biocatalysts and the stability of the cell. The immobilization process may provide shelter against shear forces to cells and enhances genetic stability [51].

5.7 Strain Improvement

The major confront of commercialization of acetone–butanol–ethanol fermentation depicts on lesser butanol acceptance capability of the microorganism because it leads to damage to the cell membrane. To conquer this task, mutagenesis strategies were practiced to alter the highly butanol liberating strains [53, 54]. To enhance the butanol yield the genetic development of strains has been done. Therefore, research has emphasized metabolic engineering techniques, the use of cheaper carbon sources for butanol production, bacterial strain selection, techniques of biobutanol recovery, and the developmental process (Figure 5.1) [55].

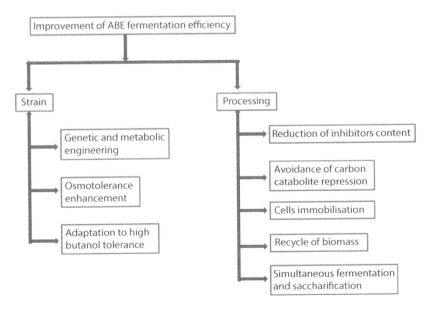

Figure 5.1 Improvement strategies for biobutanol production.

5.7.1 Mutagenesis

Mutagenesis is a process of mutation development in DNA molecules. Various mutations in DNA can occur like chromosomal rearrangement or changes in DNA sequence. During the processes of mitosis or replication due to error, these mutations can occur. Natural selection is significant for producing genetic variations (Figure 5.2). Mutations took place when the environment is exposed to genotoxin, a chemical that alters DNA structure. Mutagenesis can cause irreversible effects which may cause various effects on the organism and also on the level population. There are more than thousands of potent genotoxic and mutagenic agents which came in contact with the organism. Genotoxicity can lead to cause various DNA damage types [115]. Mutations are conventional methods for the improvement of strain for industrially valuable microorganisms. This is an efficient and effective method. No special equipment is required. It involves potential mutagenic strain selection and insertion of mutations. Mutation can be spontaneous or induced. Mutagenesis alters the nitrogenous sequences of DNA (frameshift, transversion, transition mutations, etc). Parekh *et al.* reported the usage of mutagenic agents for the improvement of strain to enhance its productivity level [116].

5.7.1.1 Spontaneous Mutations

Mutations can be occurred through endogenously or as a result of mutagen exposure. One of its types is spontaneous which occur due to error in DNA replication like, pairing occurs between G and T, not with C. This leads to a mismatch. Various enzymes that correct this mismatch are known as

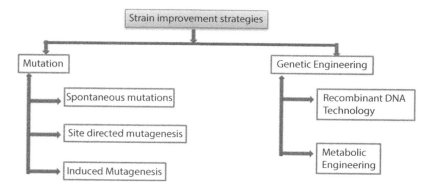

Figure 5.2 Overview of different strain improvement strategies.

mismatch repair enzymes. However, this error may not be resolved and mutations may occur. The breakage of strands and abasic sites can also lead to formation spontaneously, such as due to inherent instability in the chemical bonds and also because of an increase in thermal energy generated by cell metabolism. This spontaneous damage of DNA involves amino group loss due to chemical structure rearrangements within the bases. Endogenous mutation occurs as a result of DNA damage occurred due to ROSs formation through routine oxidative metabolism [119].

5.7.1.2 Induced Mutation

The process of random mutagenesis includes chemical and physical mutagenesis for the enhancement of strain. Mutagens can be either chemical or physical. Physical mutagens include gamma rays, UV rays, and X-rays. On the other hand, chemical mutagenesis includes HNO_2, ethidium bromides (EtBr), N-methyl-N-nitro-N-nitrosoguanidine (NMG), and ethyl methyl sulphonate (EMS) [118].

Even though naturally occurring, exposure to chemicals can accelerate the development of spontaneous mutations. Such as, the faster a cell divides, the more likely it is to be a spontaneous mutation. Few chemicals enhance the cell division rate in several tissues, which increases the spontaneous mutation occurrence probability. Additionally, DNA repair inhibition through cadmium, arsenic, and additional metals may increase spontaneous mutations due to endogenous DNA damage and less removal of mismatches. In addition, the subjection of genotoxic materials causes a mutation in repair genes or mismatch repair, resulting in a reduced repair rate. Among the various physical agents, UV exposure treatment is broadly utilized in industries because it is very significant and effective and also has no requirement for other types of equipment. The maximum absorption of DNA is at 260 nm, however, most UV radiations used exist in the range of 200 to 300 nm [117].

Strain improvement through mutation
A random mutagenesis process was performed to alter the DNA sequences of genes that are accountable for acetone–butanol–ethanol production. The SA-1 mutant strain was successfully enterprise in 1983 by continuous improvement of n-butanol diluted from *C. acetobutylicum* ATCC 824 in 1983. Mutagenic strains were maintained to maximize the titer of fermented butanol and minimize resistance to acetone (121% higher). More additional merits of mutant strains include maximum utilization of

carbohydrates and maximum activity of a-amylase endorsing viable utilization of waste lignocellulosic sources as a feedstock [48, 56].

One more mutated strain (MEMS-7) was engineered with higher potential in molasses from *C. acetobutylicum*. This strain was attained when the parent organism was treated with UV exposure, with N-methyl-N-nitro-N-nitrosoguanidine, and ethyl methane sulphonate. Reporters examined that the mutant MEMS-7 strain was found to be more effective with a 20% butanol higher yield than the parent strain (Table 5.2). The highest worthy successful gaining was *C. beijerinckii* BA101 strain with maximum butanol producing ability (19–20 g/L) [57]. One more emphasis was focused on the mutagenesis technique to depletion of expensive and tedious maintenance of anaerobic fermentation conditions. A successful *E. coli* mutant was developed for 3-methyl-1-butanol (9.5 g/L) synthesis and was mutagenized by treatment of 4-aza-D, L-leucine which is a structural analogue of L-leucine [58].

The primary process response of *C. acetobutylicum* for butanol production is weakly understood. Wang *et al.* [59] reported that the TCA cycle by *C. acetobutylicum* is promoted and glycolysis may be inhibited. The significant factors exhibiting the metabolic response toward butanol stress of *Clostridium* spp. are deliberately altered in fatty acids and lipid compositions of bacterial cells to osmoregulator concentration and intracellular metabolism. Wang *et al.* [59] also suggested that the cells of *C. acetobutylicum* alter their levels of branched-chain amino acids and long acyl chain saturated fatty acids to balance their cell membrane integrity and fluidity under butanol stress conditions. Enhanced levels of glycine, tyrosine, aspartate, phenylamine, alanine, threonine, glutamate, and tryptophan may also be accountable for enhancing butanol tolerance of *C. acetobutylicum*. These observed outcomes pointed to the maximum chances of butanologenic strains synthesis having maximum butanol tolerance.

5.7.2 Strain Improvement Through Genetic Engineering

A genetic modification strategy was performed after recognizing the acetone and butanol constructing genes (Table 5.3). Before the genetic engineering technique, the manipulation of target genes was provided by metabolic engineering and it came into existence that microbes were not capable of producing a considerable amount of fermented products. The TargeTron technology was performed to deplete the acetone consequences in acetone–butanol–ethanol fermentation to the removal of acetoacetate decarboxylase gene (adc), this gene is highly important for the production of acetone in commercial strain EA 2018 for hyper butanol

Table 5.2 Different mutagenesis experiments in organisms producing butanol.

Parent organism	Mutants	Mutagens	Observations	Reference
C. acetobutylicum PTCC-23	MEMS-7	ethyl methane sulphonate, N-methyl-N-nitro-N-nitrosoguanidine, and UV exposure	20% yield enhanced	[110]
E. coli JCL16	AL-1, AL-2	(a structural analogue to L-leucine) 4-aza-D,L-leucine	The yield obtained a maximum of 33% of the theoretical (0.33 g/g)	[111]
C. beijerinckii NCIMB 8052 or C. acetobutylicum ATCC 824	BA101, BA-105	N-methyl-N0-nitro-N-nitrosoguanidine collectively with selective improvement on the glucoseanalog 2-deoxyglucose.	Starch degrading activity improved by 82% (BA-101), 25% (BA-105), and increased to a last concentration of almost 2% of butanol	[112, 113]

Table 5.3 Various genes are concerned in the genetic engineering of clostridia for acetone–butanol–ethanol fermentation [62–64].

Genes	Approach of genetic modification	Encoding genes	Host	Products	Outcomes
crt, hbd, thiL, adhe1 (adhe) and bcd-etfB-etfA	Insertion	b-hydroxybutyryl-CoA dehydrogenase, butyryl-CoA dehydrogenase, butyr aldehydedehydrogenase, Acetyl-CoA acetyltransferase, butanol dehydrogenase and crotonase	E. coli	Butanol	All genes expression were detected except bcd-etfBetfA
bcd-etfB-etfA, crt, thiL, hbd, and adhe2 (deletion of ldhA, fnr, frdBC, pflB, adhE, from E. coli)	Deletion and insertion	b-hydroxybutyryl-CoA dehydrogenase, butyraldehydedehydrogenase, butanol dehydrogenase and crotonase, butyryl-CoA dehydrogenase, acetyl-CoA acetyltransferase,	E. coli	Butanol	All genes expression were detected except bcd-etfBetfA
adc	Deletion	Acetoacetate decaboxylase	C. acetobutylicum	Acetone	Yield improved 57–70.8%

(Continued)

Table 5.3 Various genes are concerned in the genetic engineering of clostridia for acetone–butanol–ethanol fermentation [62–64]. (*Continued*)

Genes	Approach of genetic modification	Encoding genes	Host	Products	Outcomes
ctfA, bdhB, aldA, hbdA, bdhA, bcd-etfB-etfA, crt A, adh-1, adhA, aad andadcA	Insertion	Butyryl-CoA dehydrogenase, aldehyde dehydrogenase, crotonase, electrontransport protein subunits A and B, 3-hydroxybutyryl-CoA dehydrogenase, alcohol dehydrogenase, and acetoacetate decaboxylase, CoA-transferase (subunits A and B)	*E. coli*	Acetone, and ethanol, and butanol	Expression of all genes except adhA was detected
thiL, crt, hbd, bcd-etfB-etfA, adhe2, and adhe1	Insertion	Crotonase, b-hydroxybutyryl-CoA dehydrogenas, butyraldehydedehydrogenas, Acetyl-CoA acetyltransferase, butanol dehydrogenase, and butyryl-Coadehydrogenase	*E. coli, P. puttida,* and *B. subtilis*	Butanol	The pathway was successfully constructed in three organisms and the butanol titer was highest in *E. coli*.

production [60]. This results in the enhancement of the butanol ratio by 10.05%, whereas the production of acetone declined to 0.21 g/L. Presently, complete genome sequencing of microorganisms has been done. In the future, genome sequencing of additional high butanol producing bacteria will present the genetic engineering scope for the butanol production enhancement [61].

The metabolic engineering tool firstly needs a metabolic analysis system and kinetics analysis system of its intracellularly performed enzymatic reactions. The chosen microorganism is further used for environmental modifications and genetic modifications. Both enzymatic profile and protein content should be altered.

5.7.2.1 Recombinant DNA Technology

One additional strategy of the genetic engineering technique is Recombinant DNA Technology (RDT) used in making the microbial strain to produce solvent. Some trials have been made with C. acetobutylicum ATCC 824 as a host. due to the degeneration process of solvent liberating genes (aad, ctfA, adc, and ctfB) after subculturing, the strain cannot be able to produce butanol and acetone. These four genes were incorporated in Plasmid pSOLI, afterward, Plasmid was incorporated in degenerated mutants for expressing suitable enzymes concerned in acetone and butanol production. Results exhibited that acetone and butanol production using modifies organisms was stopped because of a lack of plasmid pSOLI destruction. Similarly, aad more expression for butanol production in non-solventogenic and non-sporulating C. acetobutylicum M5 strain using Plasmid pSOLI was observed [38]. Hence, Clostridial strains have high genetic complexity and lack appropriate genetic techniques all the attempts to use it as a host were unsuccessful. Therefore, this problem confident investigators in experimenting with other host organisms for butanol-producing genes.

Another investigator isolated *C. saccharobutylicum* having maximum hemicellulosic activity, and inserted its genes in *E. coli*, encoding butyryl-CoA dehydrogenase (bcd), crotonase, 3-hydroxybutyryl-CoA dehydrogenase, acetoacetate decarboxylase, CoA-transferase, aldehyde dehydrogenase, and alcohol dehydrogenase. Approximately all the genes showed expression in *Lactobacillus brevis* as the host bacterium. In *S. cerevisiae*, a successful bcd gene expression was achieved but with less improvement in the production of butanol.

Some heterologous organisms were used as a host to incorporate genes that are responsible for butanol production of Clostridia [50]. According to this context, butanol synthesis genes of *C. acetobutylicum* ATCC 824 were incorporated into *E. coli*. During this experiment, genes hbd, bcd-etfB-etfA, thiL, adhE2, adhE1, and crt, which denote b-hydroxybutyryl-CoA dehydrogenase (HBD), butyryl-CoA dehydrogenase (BCD), acetyl-CoA acetyltransferase (THL), butanol dehydrogenase (BDH), butyraldehyde dehydrogenase (BYDH), 3-hydroxybutyryl-CoA dehydratase (CRT) enzymes. Results exhibited that the genes responsible for butanol production were achievable in aerobic conditions in *E. coli*. However, concentration of 1-butanol around 1.5% is tolerable by *E. coli* similar to *C. acetobutylicum* (a native strain) [65, 66].

5.7.2.1.1 Escherichia coli

E. coli has been examined as a commercially industrial compatible host for biofuel and chemicals production. *E. coli* can tolerate up to 1% to 1.5% of butanol and is capable of fermenting xylose and glucose [66] (Table 5.4). Atsumi *et al.* [65] genetically modified *E. coli* with *Clostridium* during the butanol pathway which resulted in enhanced production of butanol level (13 mg/L). Further modifications in recombinant *E. coli*, ligation of

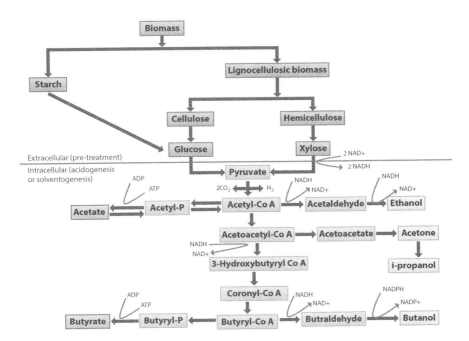

Figure 5.3 Metabolic pathway of acetone–butanol–ethanol fermentation of *Clostridia*.

Table 5.4 Butanol production from recombinant strains [64, 78, 80, 100].

Organisms	O2 tolerance	Cellular type	n-butanol production (via acetyl-co)	n-butanol tolerance (growth) (w/v)	n-butanol tolerance (viability) (w/v)
Escherichia coli	Facultative anaerobe	Prokaryote	30 g/L	1.5%	~2%
Pseudomonas putida	Facultative aerobe	Prokaryote	122 mg/L	0.75%	6%
Synechoccus elongatus	Facultative anaerobe	Prokaryote	29.9 mg/L	ND	ND
Clostridium acetobutylicum	Obligate anaerobe	Prokaryote	19 g/L	1.5%	~2%
Bacillus subtilis	Obligate aerobe	Prokaryote	24 mg/L	1.25%	5%
Saccharomyces cerevisiae	Facultative anaerobe	Eukaryote	2.5 mg/L	2%	ND

Clostridium adhE2 gene and atoD gene exhibited enhanced production of butanol up to 6.2 g/L [67]. NADH acts as inhibiting factor in the synthesis of butanol, during glucose fermentation. Therefore, to increase the availability of NADH coupling was done with the pentose phosphate pathway. Metabolism of glucose through the pentose phosphate pathway produces high reducing power per gram mole of glucose than glycolysis [68]. So if glycolytic flux is diverted toward the pentose phosphate pathway return increases NADH intracellularly. Isomerization of glucose-6-phosphate is done by phosphoglucose isomerase, and inactivation of this enzyme constructs the primary route of glucose catabolism through the pentose phosphate pathway [69]. Further, pdh upregulation and FDH addition have been projected to enhance the availability of NADH favoring the production of n-butanol (Figure 5.3). With all these experimental techniques, E. coli was made capable of producing 6.1 g/L of n-butanol with 0.31 g/g yield of glucose [67, 70].

5.7.2.1.2 *Saccharomyces cerevisiae*
Saccharomyces cerevisiae is broadly utilized for the production of bioethanol at an industrial scale and is well known and considered the best host for the production of biobutanol. *Saccharomyces cerevisiae* can tolerate butanol up to 2% [71]. *Saccharomyces cerevisiae* was genetically altered with *Clostridium* butanol pathway and it was capable to produce butanol up to 2.5 mg/L [72]. This genetically modified *Saccharomyces cerevisiae* was further altered with trans-enoyl-CoA reductase combined with *Clostridium* butanol pathway to obtain maximum butanol production and exhibited 14.1 mg/L butanol [73].

5.7.2.1.3 *Pseudomonas putida*
Pseudomonas putida can highly tolerate solvents due to pumps present in the cell, which helps in detoxification [74]. *P. putida* can easily grow between 0.75% and 1% of butanol. Ruhl *et al.* [75] reported a *P. putida* recombinant strain has a tolerance of 6% butanol. The maximum tolerance of butanol production was examined up to 120 mg/L in *P. putida* [76].

5.7.2.1.4 Cyanobacteria
Cyanobacteria are strictly anaerobic and were genetically engineered for the production of butanol [77]. A mutant sp. strain PCC7942 of *Synechococcus* was engineered with ter gene from *Treponema denticola*, atoB gene from E. coli, and adhE2, cert, hbd from *C. acetobutylicum* and was capable of

butanol production approximately 30 mg/L. By enhancing the supply of ATP butanol production can be improved because it increases butanol tolerance. This strategy was applied in *Cyanobacteria* sp and exhibited the enhanced production of biofuel [78, 79].

5.7.2.1.5 Lactobacillus

Lactobacillus has a high tolerance ability of solvent and has up to 30% butanol tolerance. Berezina *et al.* [80] reported that *Lactobacillus brevis* ATCC367 produces butanol up to 300 mg/L when engineered with *Clostridium acetobutylicum* butanol pathway.

5.7.3 Genetic Engineering in *Clostridial* sp. for Improved Butanol Tolerance and Its Production

Genetic modification in *Clostridial* sp. has been done either to enhance butanol tolerance or butanol yield. The genetic modification strategies involve regulation of sporulating genes, deletions of competing pathway genes, or overexpression of a few butanol-producing genes by targeted or random mutagenesis. Further experiments were conducted to identify the genetic behavior of Clostridial cells during stress conditions to butanol [81]. Both physical and chemical mutagenesis of *Clostridium acetobutylicum* CICC 8012 was performed to enhance its butanol tolerance. The mutant strain F2-GA was obtained after NTG (Nitrosoguanidine)or UV treatment followed by exchanging of the genome by protoplast fusion produced 16.5 g/L ABE with 10.46 g/L butanol v/s 22.21 g/L ABE with 14.15 g/L butanol by its wild type strain [82].

When *Clostridium acetobutylicum* PJC4BK was treated through random mutagenesis by NTG (N-methyl-N-nitoso-N-nitroguanidine) resulted in a BKM19 mutant which produced 32.5 g/L ABE with 17.6 g/L butanol which was approximately 31% more than parent strain producing 13.9 g/L ABE with 7.6 g/L butanol [83]. After various repetitive cycles of chemical mutagenesis by NTG treatment of *Clostridium acetobutylicum* EA 2018 mutant exhibited deletion of 26 genes and insertion of 46 genes in addition to a lesser level of acid-forming genes expression and high adhe gene expression. More butanol production was exhibited by mutant *Clostridium acetobutylicum* EA 2018 up to 14 g/L in comparison to wild strain *Clostridium acetobutylicum* ATCC 824, i.e., 9 g/L [84]. Further through genome shuffling *Clostridium acetobutylicum* mutant strain GS4-3 was obtained capable of 20.1 g/L of butanol and 32.6 g/L of ABE [84].

In a few *Clostridial* sp. targeted mutagenesis was performed with the aim of deletion of some novel genes, spoOA (sporulating transcription factor), or opposing pathways which led to overexpression of butanol-producing genes or deficiency toward the synthesis of butanol. The spoOA being the sporulation key regulator has always been unspecified to be aiding in solventogenesis [85]. Xu *et al.* [86] constructed *Clostridium acetobutylicum* ATCC 55025 a mutant strain through deletion of single base incac3319 gene which further led to histidine kinase gene knockout concerned in SpoOA activation and this mutant JB200, produced butanol 45% more [86]. Afterward, it was examined that SpoOA gene knockout of *Clostridium pasteurianum* ATCC 6013 by NTG treatment resulted in butanol production (11.7 g/L) which was 80% higher than the wild strain [87].

In *Clostridium acetobutylicum*, SMB_G1518 protein deletion exhibited increased tolerance of butanol and cell growth by 70 % at 1% (v/v) butanol as compared to the wild-type strain, therefore it is signified that these zinc finger proteins are the negative regulator of tolerance [88]. Competing pathways deletion, i.e., phosphotransbutyrylase and acetate kinase knockout and alcohol dehydrogenase (adhe2) gene from *Clostridium acetobutylicum* ATCC824 overexpressed in *Clostridium tyrobutyricum* ATCC 25755 strain exhibited in higher production of butyryl Co-A which results in 16 g/L butanol with no production of acetone by a mutant [89]. In metabolism and stress tolerance, Glutathione exhibits an important role [90].

5.8 Production of Butanol From Bioethanol Through Chemical Processes

The standard strategy for butanol production from bioethanol occurs in basic various three steps. Firstly, dehydration of commercial products should be done to achieve an anhydrous state (100%). It is further oxidized in aluminum isopropanolate in the acetone presence to form acetaldehyde. A fractional distillation process was used for the separation of extremely volatile acetaldehyde. Further, in the second step, a strongly alkaline environment was used to condense acetaldehyde into crotonaldehyde. In the last step, crotonaldehyde was treated with isopropanol to convert it into butanol in aluminum isopropanolate and titanium presence.

This multistep process can be simplified by a borrowed hydrogen called a Gerber reaction [65]. This reaction sequence is very efficient for broad range alcohol conversion but inefficient for ethanol. Acetaldehyde condensation led to a combination of polymeric materials and higher molecular products however ethanol substrate dehydrogenation is more difficult. A multistage approach, involving the base-catalyzed aldol condensation of acetaldehyde, consecutive oxidation of bioethanol affording acetaldehyde into crotonaldehyde, and lastly the partially hydrogenated intermediate products with crotonaldehyde hydrogenation, reported with 1-butanol with 67% selectivity and 12% conversion [91, 92, 99].

In addition to product inhibition, several other factors affect butanol fermentation. However, calcium hydroxide is utilized for the successful removal of inhibitors from the hydrolysate. The reporter examined that corn stover pre-treatment with alkaline peroxide, fortunately, enhanced the yield of butanol from wheat straw enzymatic hydrolysis using *C. beijerinckii*. The high process costs linked with the toxicity of the product, few feedstocks, and less concentration of product are some of the challenges faced in butanol production. Resolving the challenges concerned with converting lignocellulosic material to butanol and main approaches improvement could contribute to producing more attractive biobutanol commercially [93].

A process named manganese-catalyzed Guerbet-type is involved in continuing selectivity of butanol transformation from ethanol. This process involves 79% butanol conversion of 12.6% ethanol in presence of a distinct manganese pincer complex at 160°C [94, 95]. This first ever report exhibits the effectiveness of non-noble–metal catalysts for the advancement of ethanol into advanced ethanol. A comprehensive catalog of homogenous and heterogeneous catalysts utilized in the valorization process of butanol from ethanol and summarized part of projected reaction mechanisms for these processes [96]. The different catalytic processes operating under the liquid phase are of useful interest in various chemical industries due to the low consumption of gas and energy requirements. Despite all the processes, bioethanol transformation to butanol in the fluid stage is still a challenge [97, 98].

5.9 Advances in Genetically Engineered Microbes can Produce Biobutanol

Bulk production of butanol is highly challenging to manage fermentation conditions under anaerobic criteria and this restricts the various commercial and economic purposes of *Clostridium* strains. In anaerobic conditions, the formation of biomass is less due to less gain of ATP. Subculturing, Butanol toxicity, and continuous culture conditions cause *Clostridium* spp. to decrease their high butanol production ability. However, because of these demerits experimentation have been shifted toward other microorganisms which have the potential for butanol production after genetic modifications.

In the butanol production process, cloning of valuable genes and expression for the viable production of proteins in a few microbial strains which are tolerant to butanol like *Saccharomyces cerevisiae, Bacillus subtilis,* and *E. coli*. Various genetically modified microbial strains which have the potential to produce butanol are *Klebsiella, Candida, Enterococcus, Rhodococcus, Pichia,* and *Zymomonas* [101]. The E. Coli mutant strain was reported to have a butanol production ability of 552 mg/L after transferring its native producer to a butanol biosynthetic pathway [102]. This report also demonstrated that tolerating ability of E. coli up to 1.5% butanol. Shen *et al.* [103] reported the production of 1-butanol and 1-propanol by using native biosynthetic pathways of microorganisms. During this investigation the E. coli strain used was capable of converting glucose into 2-ketobutyrate, further converted into 1-butanol and 1-propanol using significant enzymes incorporated from *S. cerevisiae* and *Lactococcus lactis*.

The enhanced production of butanol and ethanol was obtained through depletion of genes concerned with side pathways competing, which led to 1:1 of propanol and butanol with a final production titer of 2 g/L [104]. In further investigation, Atsumi *et al.* [105] substitute the threonine pathway with a keto-acid pathway for 2-ketobutyrate production from *Methanococcus jannaschii*. The production of n-butanol does not depend on the activities of enzymes. The formation of a product also requires adequate acetyl-CoA, NADH, reducing power, and carbon precursor [105].

5.10 Economics of Biobutanol Fermentation

During the biobutanol fermentation, the substrates have high energy than the products because it's an exothermic process. Based on product ratio and energy combustions, the energy yield and theoretical mass of acetone–butanol–ethanol fermentation are 94% and 37% respectively obtained in fermentation [106]. While investigating, it was examined that ABE fermentation could not achieve a 100% yield of economic feasibility. Therefore, the cost of a substrate, considerable to 60% approximately of the total cost of production, plays vital importance in fermentation economics. Based on the recent scenario and market needs, none of the sugar and starch-containing crops can formulate economically feasible fermentation. However, industrial wastes and agricultural waste materials may be appropriate for acetone–butanol–ethanol fermentation economically. Ethanol distilleries are more productive economically when cheaper raw materials are utilized. Consequently, economically ABE fermentation is highly susceptible to the substrate, not the yield, but it was examined that if yield becomes less than 25% then fermentation of butanol will not be sufficient [107]. The continuous fermentation process is more economical compared to batch fermentation processes because the batch requires extra valves, piping, and sterilization of equipment. Market value is sufficient enough toward product consumption of acetone–butanol–ethanol fermentation. It is assumed that after the adaption of biobutanol as a fuel, the requirement will be high. Byproducts of fermentation involving ethanol, CO_2, H_2, and acetone may also help in the production of butanol economically.

The economic possibility of acetone–butanol–ethanol fermentation can be materialized. The first step requires to strain tolerance toward butanol which improves productivity and fermentation time. Approximately 40 to 60 hours are essential for biobutanol production commercially. Second step, depletion in recovery cost and product inhibition could be eliminated by selecting the product removal process online [108]. However, product recovery and raw materials used are essential factors for butanol fermentation. It is a significant challenge to mount feasible recovery processes and select cost-effective substrates for economical ABE fermentation [109].

5.11 Applications of Butanol

In addition to the probable role as biofuel engine, butanol is a significant bulk chemical having a wide industrial usage. Around half of the worldwide production is used in the form of methacrylate esters and butyl acrylate is used in the production of elastomers, flocculants, adhesives, superabsorbents, plastics, nitrocellulose lacquers, and latex surface coatings. Other significant butanol-derived compounds are plasticizers, butyl acetate, and butyl glycol ether. The complex of lesser applicability is amino resins and butyl amines. The derived compounds and butanol are brilliant diluents in brake fluid formulations, and hydraulic and paint thinners. It is moreover, utilized as a solvent in the perfume industry and for the manufacturing of hormones, antibiotics, and vitamins. Other applications include the manufacturing of detergents, nail care products, deicing fluids, cosmetics like eye makeup, safety glass, and hygienic products. It is also used in the food and flavor industries [114].

5.12 Butanol Advantages

Biobutanol is easily added to conventional gasoline due to its low vapor pressure. Presently, the biobutanol can be merged at higher concentrations with the existing fuels up to 11.5% (v/v) in US gasoline and 10% (v/v) in European gasoline. There are maximum potential chances in the future to enhance the maximum permissible use of gasoline by up to 16% volume. In recent years, it is expected that there will be an increased response to alternative fuels to reduce carbon (C) emissions from burning fossil fuels. However, the success of alternative fuels depends on several criteria. The main issue affecting the competitiveness of biobutanol and petrochemicals is the cost of substrate and product recovery, and therefore an important target of research, both the properties of butanol and bioethanol.

5.13 Conclusion

Butanol has several advantages over bioethanol as a sustainable energy option and is particularly promising as a liquid transport fuel. The studies summarized in this chapter represent the immense potential of various processes and methods for improving both traditional chemical and bioengineering pathways for the large-scale production of butanol. Numerous genetic techniques can be modified to improve the properties of various butanol-producing bacteria. The best solution is coupling in a continuous process (one-step process) of bioethanol into butanol. These strategies are implemented for butanol fermentation modification, strain improvement, chemical catalysis, and recovery processing in successful industrial parameters that the production of butanol is probable to become a cost-effective process in the future. Researchers are also focusing on butanol production through aerobic conditions using various genetically modified microorganisms such as *S. cerevisiae*, *E. coli*, etc. Through optimized commercial use of waste and appropriate organisms to achieve economic progress. This will be a fascinating milestone for the attention of commercial, research, and government organizations to future support the operation of extraction technology and innovative fermentation.

References

1. Gullison, R.E., Frumhoff, P.C., Canadell, J.G., Field, C.B., Nepstad, D.C., Hayhoe, K., Avissar, R., Curran, L.M., Friedlingstein, P., Jones, C.D., Nobre, C., Tropical forests and climate policy. *Science*, 316, 985–986, 2007.

2. Ramanjaneyulu, G. and Reddy, B.R., Emerging trends of microorganism in the production of alternative energy, in: *Recent Developments in Applied Microbiology and Biochemistry*, pp. 275–305, 2019.
3. Lin, L., Cunshan, Z., Vittayapadung, S., Xiangqian, S., Mingdong, D., Opportunities and challenges for biodiesel fuel. *Appl. Energy*, 88, 1020–1031, 2010.
4. Martien, J.I. and Amador-Noguez, D., Recent applications of metabolomics to advance microbial biofuel production. *Curr. Opin. Biotechnol.*, 43, 118–126, 2017.
5. Durre, P., New insights an novel developments in clostridial acetone/butanol/isopropane fermentation. *Appl. Microbiol. Biotechnol.*, 49, 639–648, 1998.
6. Fan, Y.X., Zhang, J.Z., Zhang, Q., Ma, X.Q., Ma, Q., Liu, Z.Y., Lu., M., Qiao., K., Chapter One - Biofuel and chemical production from carbon one industry flux gas by acetogenic bacteria. *Adv. Appl. Microbiol.*, 116, 1–34, 2021.
7. Keis, S., Shaheen, R., Jones, T.D., Emended descriptions of *Clostridium acetobutylicum* and *Clostridium beijerinckii*, and descriptions of *Clostridium saccharoperbutylacetonicum* sp. nov. and *Clostridium saccharobutylicum* sp.nov. *Int. J. Syst. Evol. Microbiol.*, 51, 2095–2103, 2001.
8. Qureshi, N. and Blaschek, H.P., ABE production from corn: A recent economic evaluation. *J. Ind. Microbiol. Biotechnol.*, 27, 292e297, 2001.
9. Luo, H., Yang, R., Zhao, Y., Wang, Z., Liu, Z., Huang, M., Zeng, Q., Recent advances and strategies in process and strain engineering for the production of butyric acid by microbial fermentation. *Bioresour. Technol.*, 253, 343–354, 2018.
10. Cooksley, C.M., Zhang, Y., Wang, H., Redl, S., Winzer, K., Minton, N.P., Targeted mutagenesis of the *Clostridium acetobutylicum* acetone-butanol-ethanol fermentation pathway. *Metab. Eng.*, 14, 630–641, 2012.
11. Antoni, D., Zverlov, V.V., Schwarz, W.H., Biofuels from microbes. *Appl. Microbiol. Biotechnol.*, 77, 23–35, 2007.
12. Huang, H., Liu, H., Gan, Y.R., Genetic modification of critical enzymes and involved genes in butanol biosynthesis from biomass. *Biotechnol. Adv.*, 28, 651–657, 2010.
13. Ma, H., Oxley, L., Gibson, J., Li, W., A survey of China's renewable energy economy. *Renew. Sustain. Energy Rev.*, 14, 438–445, 2010.
14. Maddox, I.S., Qureshi, N., Thomson, K.R., Production of acetone–butanol–ethanol from concentrated substrates using *Clostridium acetobutylicum* in an integrated fermentation-product removal process. *Process Biochem.*, 30, 209–215, 1995.
15. Qureshi, N. and Blaschek, H.P., Recovery of butanol from fermentation broth by gas stripping. *Renew. Energy*, 22, 557–564, 2001.
16. Zhang, J.X., Zou, D., Singh, S., Cheng, G., Recent developments in ionic liquid pre-treatment of lignocellulosic biomass for enhanced bioconversion. *Sustain. Energy Fuels*, 5, 1655–1667, 2021.

17. Qureshi, N. and Maddox, I.S., Reduction of butanol inhibition by perstraction: Utilization of concentrated lactose/whey permeate by *Clostridium acetobutylicum* to enhance butanol fermentation economics. *Food Bioprod. Process*, 83, 43–52, 2005.
18. Assobhei, O., Kanouni, A.E., Ismaili, M., Loutfi, M., Petitdemange, H., Effect of acetic and butyric acids on the stability of solvent and spore formation by *Clostridium acetobutylicum* ATCC 824 during repeated subculturing. *J. Ferment. Bioeng.*, 85, 209–212, 1998.
19. Bram, S., Ruyck, J.D., Lavric, D., Using biomass: A system perturbation analysis. *Appl. Energy*, 86, 194–201, 2009.
20. Stabnikova, O., Wang, J.Y., Ivanov, V., Value-added biotechnological products from organic wastes, in: *Environmental Biotechnology*, pp. 343–394, Humana Press, Totowa, NJ, USA, 2010.
21. Sanderson, K., Lignocellulose: A chewy problem. *Nature*, 474, S12–S14, 2011.
22. Atsumi, S., Cann, A.F., Connor, M.R., Shen, C.R., Smith, K.M., Brynildsen, M.P. et al., Metabolic engineering of *Escherichia coli* for 1-butanol production. *Metab. Eng.*, 10, 305–311, 2008.
23. Qureshi, N. and Blaschek, H.P., Butanol production using *Clostridium beijerinckii* BA101 hyper-butanol producing mutant strain and recovery by pervaporation. *Appl. Biochem. Biotechnol.*, 84, 225–236, 2000.
24. Durre, P., Biobutanol: An attractive biofuel. *Biotechnol. J.*, 2, 1525–1534, 2007.
25. Survase, S.A., Nimbalkar, P., Jurgens, G., Granström, T., Chavan, P., Bankar, S.B., Efficient strategy to alleviate the inhibitory effect of lignin-derived compounds for enhanced butanol production. *ACS Sustain. Chem. Eng.*, 9, 1172–1179, 2021.
26. Huang, H., Liu, H., Gan, Y.R., Genetic modification of critical enzymes and involved genes in butanol biosynthesis from biomass. *Biotechnol. Adv.*, 28, 651–657, 2010.
27. Narueworanon, P., Laopaiboon, L., Phukoetphim, N., Laopaiboon, P., Impacts of initial sugar, nitrogen and calcium carbonate on butanol fermentation from sugarcane molasses by *Clostridium beijerinckii*. *Energies*, 13, 1–19, 2020.
28. Kumar, M. and Gayen, K., Developments in biobutanol production: New insights. *Appl. Energy*, 88, 1999–2012, 2011.
29. Lin, D.S., Yen, H.W., Kao, W.C., Cheng, C.L., Chen, W.M., Huang, C.C., Chang, J.S., Bio-butanol production from glycerol with *Clostridium pasteurianum* CH4: The effects of butyrate addition and *in situ* butanol removal via membrane distillation. *Biotechnol. Biofuels*, 8, 168, 2015.
30. Tran, H.T.M., Cheirsilp, B., Hodgson, B., Umsakul, K., Potential use of *Bacillus subtilis* in a co-culture with *Clostridium butylicum* for acetone-butanol-ethanol production from cassava starch. *Biochem. Eng. J.*, 48, 260–267, 2010.

31. Garcia, V., Pakkila, J., Ojamo, H., Muurinen, E., Keiski, R.L., Challenges in biobutanol production: How to improve the efficiency. *Renew. Sust. Energy Rev.*, 15, 964–980, 2011.
32. Huffer, S., Clark, M.E., Ning, J.C., Blanch, H.W., Clark, D.S., Role of alcohols in growth, lipid composition, and membrane fluidity of yeasts, bacteria and archaea. *Appl. Environ. Microbiol.*, 77, 6400–6408, 2011.
33. Haldar, D. and Purkait, M.K., Lignocellulosic conversion into value-added products: A review. *Process Biochem.*, 89, 110–133, 2020.
34. Patakova, P., Linhova, M., Rychtera, M., Paulova, L., Melzoch, K., Novel and neglected issues of acetone–Butanol–Ethanol (ABE) fermentation by clostridia: *Clostridium* metabolic diversity, tools for process mapping and continuous fermentation systems. *Biotechnol. Adv.*, 31, 58–67, 2013.
35. Etteh, C.C., Ibiyeye, A.O., Jelani, F.B., Rasheed, A.A., Ette, O.J., Victor, I., Production of biobutanol using Clostridia *Spp* through novel ABE continuous fermentation of selected waste streams and industrial by-products. *Sci. Afr.*, 12, e00744, 2021.
36. Ranjan, A. and Moholkar, V.S., Biobutanol: Science, engineering, and economics. *Int. J. Energy Res.*, 36, 277–323, 2012.
37. Wang, S., Zhang, Y., Dong, H., Mao, S., Zhu, Y., Wang, R., Li, Y., Formic acid triggers the "acid crash" of acetone-butanol-ethanol fermentation by *Clostridium acetobutylicum*. *Appl. Environ. Microbiol.*, 77, 1674–1680, 2011.
38. Sillers, R., Chow, A., Tracy, B., Papoutsakis, E.T., Metabolic engineering of the nonsporulating, non-solventogenic *Clostridium acetobutylicum* strain M5 to produce butanol without acetone demonstrate the robustness of the acidformation pathways and the importance of the electron balance. *Metab. Eng.*, 10, 321–332, 2008.
39. Berezina, O.V., Brandt, A., Yarotsky, S., Schwarz, W.H., Zverlov, V.V., Isolation of a new butanol-producing *Clostridium* strain: High level of hemicellulosic acidity and structure of solventogenesis genes of a new *Clostridium saccharobutylicum* isolate. *Syst. Appl. Microbiol.*, 32, 449–459, 2009.
40. Lenz, T.G. and Morelra, A.R., Economic evaluation of the acetone–Butanol fermentation. *Ind. Eng. Chem. Prod. Res. Dev.*, 19, 478–483, 1980.
41. Gapes, J.R., The economics of acetone–Butanol fermentation: Theoretical and market considerations. *J. Microbiol. Biotechnol.*, 2, 27–32, 2000.
42. Qureshi, N. and Blaschek, H.P., ABE production from corn: A recent economic evaluation. *J. Ind. Microbiol. Biotechnol.*, 27, 292–297, 2001.
43. Naik, S.N., Goud, V.V., Rout, P.K., Dalai, A.K., Production of first and second generation biofuels: A comprehensive review. *Renew Sustain. Energy Rev.*, 14, 578–597, 2010.
44. Gressel, J., Transgenics are imperative for biofuel crops. *Plant Sci.*, 174, 246–263, 2008.
45. Ezeji, T., Qureshi, N., Blaschek, H.P., Production of acetone–Butanol–Ethanol (ABE) in a continuous flow bioreactor using degermed corn and *Clostridium beijerinckii*. *Process Biochem.*, 42, 34–39, 2007.

46. Foda, M.I., Dong, H., Li, Y., Study the suitability of cheese whey for biobutanol production by *Clostridia*. *J. Am. Sci.*, 6, 39–46, 2010.
47. Qureshi, N., Saha, B.C., Hector, R.E., Hughesb, S.R., Cottaa, M.A., Butanol production from wheat straw by simultaneous saccharification and fermentation using *Clostridium beijerinckii*, Part I-batch fermentation. *Biomass Bioenergy*, 32, 168–175, 2008.
48. Zhou, Z., Luo, Y., Peng, S., Zhang, Q., Li, Z., Li, H., Enhancement of butanol production in a newly selected strain through accelerating phase shift by different phases C/N ratio regulation from puerariae slag hydrolysate. *Biotechnol. Bioprocess Eng.*, 26, 256–264, 2021.
49. Ezeji, T.C., Qureshi, N., Blaschek, H.P., Bioproduction of butanol from biomass: From genes to bioreactors. *Curr. Opin. Biotechnol.*, 18, 220–227, 2007.
50. Zheng, Y.N., Li, L.Z., Xian, M., Ma, Y.J., Yang, J.M., Xu, X., He, D.Z., Problems with the microbial production of butanol. *J. Ind. Microbiol. Biotechnol.*, 36, 1127–1138, 2009.
51. Dolejs, I., Rebroš, M., Rosenberg, M., Immobilisation of *Clostridium* spp. for production of solvents and organic acids. *Chem. Pap.*, 68, 1–14, 2014.
52. Mishra, A. and Ghosh., S., Bioethanol production from various lignocellulosic feedstocks by a novel "fractional hydrolysis" technique with different inorganic acids and co-culture fermentation. *Fuel*, 236, 544–553, 2019.
53. Huang, H., Liu, H., Gan, Y.R., Genetic modification of critical enzymes and involved genes in butanol biosynthesis from biomass. *Biotechnol. Adv.*, 28, 5, 651–657, 2010.
54. Ezeji, T.C., Qureshi, N., Blaschek, H.P., Acetone butanol ethanol (ABE) production from concentrated substrate: Reduction in substrate inhibition by fed-batch technique and product inhibition by gas stripping. *Appl. Microbiol. Biotechnol.*, 63, 653–658, 2004.
55. Du, G., Zhu, C., Xu, M., Wang, L., Yang, S.T., Xue, C., Energy-efficient butanol production by *Clostridium acetobutylicum* with histidine kinase knockouts to improve strain tolerance and process robustness. *Green Chem.*, 23, 2155–2168, 2021.
56. Lin, Y.L. and Blaschek, H.P., Butanol production by a butanol-tolerant strain of *Clostridium acetobutyricum* in extruded corn broth. *Appl. Environ. Microbiol.*, 45, 966–973, 1983.
57. Qureshi, N., Saha, B.C., Hector, R.E., Dien, B., Hughes, S., Liu., S. *et al.*, Production of butanol (a biofuel) from agricultural residues: Part II – Use of corn stover and switchgrass hydrolysates. *Biomass Bioenergy*, 34, 566–571, 2010.
58. Connor, M.R., Cann, A.F., Lio, J.C., 3-Mithyl-1-butanol production in *Escherichia coli*: Random mutagenesis and two-phase fermentation. *Appl. Microbiol. Biotechnol.*, 86, 1155–1164, 2010.

59. Wang, Y.F., Tian, J., Ji, Z.H., Song, M.Y., Li, H., Intracellular metabolic changes of *Clostridium acetobutylicum* and promotion to butanol tolerance during biobutanol fermentation. *Int. J. Biochem. Cell Biol.*, 78, 297–306, 2016.
60. Davis, G. and Kayser, K.J., *Chromosomal mutagenesis*, Humana Press Inc., Totowa, New Jersey, 2008.
61. Papoutsakis, E.T., Engineering solventogenic *Clostridia*. *Curr. Opin. Biotechnol.*, 19, 420–429, 2008.
62. Jiang, Y., Xu, C., Dong, F., Yang, Y., Jiang, W., Yang, S., Disruption of the acetoacetate decarboxylase gene insolvent-producing *Clostridium acetobutylicum* increases the butanol ratio. *Metab. Eng.*, 11, 284–291, 2009.
63. Inui, M., Suda, M., Kimura, S., Yasuda, K., Suzuki, H., Toda, H. *et al.*, Expression of *Clostridium acetobutylicum* butanol synthetic genes in *Escherichia coli*. *Appl. Microbiol. Biotechnol.*, 77, 1305–1316, 2008.
64. Nielsen, D.R., Leonard, E., Yoon, S.H., Tseng, H.C., Yuan, C., Prather, K.L.J., Engineering alternative butanol production platforms in heterologous bacteria. *Metab. Eng.*, 11, 262–273, 2009.
65. Atsumi, S., Cann, A.F., Connor, M.R., Shen, C.R., Smith, K.M., Brynildsen, M.P. *et al.*, Metabolic engineering of *Escherichia coli* for 1-butanol production. *Metab. Eng*, 10, 305–311, 2008.
66. Inui, M., Suda, M., Kimura, S., Yasuda, K., Suzuki, H., Toda, H. *et al.*, Expression of *Clostridium acetobutylicum* butanol synthetic genes in *Escherichia coli*. *Appl. Microbiol. Biotechnol.*, 77, 1305–1316, 2008.
65. Veibel, S. and Nielsen, J.I., On the mechanism of the Guerbet reaction. *Tetrahedron*, 23, 1723–1733, 1967.
66. Knoshaug, E.P. and Zhang, M., Butanol tolerance in a selection of microorganisms. *Appl. Biochem. Biotechnol.*, 153, 13–20, 2009.
67. Saini, M., Li, S.Y., Wang, Z.W., Chiang, C.J., Chao, Y.P., Systematic engineering of the central metabolism in Escherichia coli for effective production of n-butanol. *Biotechnol. Biofuels*, 9, 69, 2016.
68. Huerta-Beristain, G., Cabrera-Ruiz, R., Hernandez-Chavez, G., Bolivar, F., Gosset, G., Martinez, A., Metabolic engineering and adaptive evolution of *Escherichia coli* KO11 for ethanol production through the Entner–Doudoroff and the pentose phosphate pathways. *J. Chem. Technol. Biotechnol.*, 92, 990–996, 2017.
69. Kabir, M.M. and Shimizu, K., Gene expression patterns for metabolic pathway in pgi knockout *Escherichia coli* with and without phb genes based on RTPCR. *J. Biotechnol.*, 105, 11–31, 2003.
70. Koppolu, V. and Vasigala, V.K., Role of *Escherichia coli* in biofuel production. *Microbiol. Insights*, 9, 29, 2016.
71. Fischer, C.R., Klein-Marcuschamer, D., Stephanopoulos, G., Selection and optimization of microbial hosts for biofuels production. *Metab. Eng.*, 10, 295–304, 2008.

72. Steen, E.J., Chan, R., Prasad, N., Myers, S., Petzold, C.J., Redding, A., Ouellet, M., Keasling, J.D., Metabolic engineering of *Saccharomyces cerevisiae* for the production of n-butanol. *Microb. Cell Fact.*, 7, 36-1-8, 2008.
73. Sakuragi, H., Morisaka, H., Kuroda, K., Ueda, M., Enhanced butanol production by eukaryotic *Saccharomyces cerevisiae* engineered to contain an improved pathway. *Biosci. Biotechnol. Biochem.*, 79, 314– 320, 2015.
74. Molina-Santiago, C., Daddaoua, A., Fillet, S., Duque, E., Ramos, J.L., Interspecies signalling: *Pseudomonas putida* efflux pump TtgGHI is activated by indole to increase antibiotic resistance. *Environ. Microbiol.*, 16, 1267–1281, 2014.
75. Rühl, J., Schmid, A., Blank, L.M., Selected *Pseudomonas putida* strains able to grow in the presence of high butanol concentrations. *Appl. Environ. Microbiol.*, 75, 4653–4656, 2009.
76. Nielsen, D.R., Leonard, E., Yoon, S.H., Tseng, H.C., Yuan, C., Prather, K.L.J., Engineering alternative butanol production platforms in heterologous bacteria. *Metab. Eng.*, 11, 262–273, 2009.
77. Lan, E.I. and Liao, J.C., Metabolic engineering of cyanobacteria for 1-butanol production from carbon dioxide. *Metab. Eng.*, 13, 353–363, 2011.
78. Erdrich, P., Knoop, H., Steuer, R., Klamt, S., Cyanobacterial biofuels: New insights and strain design strategies revealed by computational modeling. *Microb. Cell Fact.*, 13, 128, 2014.
79. Hara, K.Y. and Kondo, A., ATP regulation in bioproduction. *Microbial Cell factories*, 14, 198, 2015.
80. Berezina, O.V., Zakharova, N.V., Brandt, A., Yarotsky, S.V., Schwarz, W.H., Zverlov, V.V., Reconstructing the clostridial n-butanol metabolic pathway in *Lactobacillus brevis*. *Appl. Microbiol. Biotechnol.*, 87, 635–646, 2010.
81. Xu, M., Zhao, J., Yu, L., Tang, I.C., Xue, C., Yang, S.T., Engineering *Clostridium acetobutylicum* with a histidine kinase knockout for enhanced n-butanol tolerance and production. *Appl. Microbiol. Biotechnol.*, 99, 1011–1022, 2015.
82. Gao, X., Zhao, H., Zhang, G., He, K., Jin, Y., Genome shuffling of *Clostridium acetobutylicum* CICC 8012 for improved production of acetone–butanol–ethanol (ABE). *Curr. Microbiol.*, 65, 128–132, 2012.
83. Jang, Y.S., Malaviya, A., Lee, S.Y., Acetone–butanol–ethanol production with high productivity using *Clostridium acetobutylicum* BKM19. *Biotechnol. Bioeng.*, 110, 1646–1653, 2013.
84. Li, H.G., Zhang, Q.H., Yu, X.B., Wei, L., Wang, Q., Enhancement of butanol production in *Clostridium acetobutylicum* SE25 through accelerating phase shift by different phases pH regulation from cassava flour. *Bioresour. Technol.*, 201, 148– 155, 2016.
85. Hu, S., Zheng, H., Gu, Y., Zhao, J., Zhang, W., Yang, Y., Jiang, W., Comparative genomic and transcriptomic analysis revealed genetic characteristics related to solvent formation and xylose utilization in *Clostridium acetobutylicum* EA 2018. *BMC Genomics*, 12, 93, 2011.

86. Xu, M., Zhao, J., Yu, L., Tang, I.C., Xue, C., Yang, S.T., Engineering *Clostridium acetobutylicum* with a histidine kinase knockout for enhanced n-butanol tolerance and production. *Appl. Microbiol. Biotechnol.*, 99, 1011–1022, 2015.
87. Sandoval, N.R., Venkataramanan, K.P., Groth, T.S., Papoutsakis, E.T., Whole-genome sequence of an evolved *Clostridium pasteurianumstrain* reveals Spo0A deficiency responsible for increased butanol production and superior growth. *Biotechnol. Biofuels*, 8, 227, 2015.
88. Jia, K., Zhang, Y., Li, Y., Identification and characterization of two functionally unknown genes involved in butanol tolerance of *Clostridium acetobutylicum*. *PLoS One*, 7, e38815, 2012.
89. Yu, M., Zhang, Y., Tang, I.C., Yang, S.T., Metabolic engineering of *Clostridium tyrobutyricum* for n-butanol production. *Metab. Eng.*, 13, 373– 382, 2011.
90. Zhu, L., Dong, H., Zhang, Y., Li, Y., Engineering the robustness of *Clostridium acetobutylicum* by introducing glutathione biosynthetic capability. *Metab. Eng.*, 13, 426–434, 2011.
91. Carlini, C., Di Girolamo, M., Macinai, A., Marchionna, M., Noviello, M., Galletti, A.M.R., Sbrana, G., Selective synthesis of isobutanol by means of the Guerbet reaction Part 2. Reaction of methanol/ethanol and methanol/ethanol/n-propanol mixtures over copper based/MeONa catalytic systems. *J. Mol. Catal. A Chem.*, 200, 137–146, 2003.
92. Ogo, S., Onda, A., Yanagisawa, K., Selective synthesis of 1-butanol from ethanol over strontium phosphate hydroxyapatite catalysts. *Appl. Catal.A*, 402, 188–195, 2011.
93. Kumar, M. and Gayen, K., Developments in biobutanol production: New insights. *Appl. Energy*, 88, v1999–2012, 2011.
94. Fu, S., Shao, Z., Wang, Y., Liu, Q., Manganese-catalyzed upgrading of ethanol into 1-butanol. *J. Am. Chem. Soc*, 129, 11941–11948, 2017.
95. Kulkarni, N.V., Brennessel, W.W., Jones, W.D., Catalytic upgrading of Ethanol to n Butanol via manganese-mediated guerbet reaction. *ACS Catal.*, 8, 997–1002, 2018.
96. Wu, X., Fang, G., Tong, Y., Jiang, D., Liang, Z., Leng, W., Liu, L., Tu, P., Wang, H., Ni, J. et al., Catalytic upgrading of ethanol to n-butanol: Progress in catalyst development. *Chem. Sustain. Chem.*, 11, 71–85, 2018.
97. Wiebus, E. and Cornils, B., Water-soluble catalysts improve hydroformulation of olefins. *Hydrocarb. Process.*, 75, 63–66, 1996.
98. Michaels, W., Zhang, H., Luyben, W.L., Baltrusaitis, J., Design of separation section in an ethanol-to-butanol process. *Biomass Bioenergy*, 109, 231–238, 2018.
99. Van Noorden, R., Chemical treatment could cut cost of biofuel. *Nature*, 34, 566–571, 2014.
100. Atsumi, S., Chann, A.F., Connor, M.R., Shen, C.R., Smith, K.M., Brynildsen, M.P., Chou, J.Y., Hanai, T., Liao, J.C., Metabolic engineering of *Escherichia coli* for 1-butanol production. *Metab. Eng.*, 10, 305– 3011, 2008.

101. Li, J., Zhao, J.B., Zhao, M., Yang, Y.L., Jiang, W.H., Yang, S., Screening and characterization of butanol-tolerant microorganism. *Appl. Microbiol.*, 50, 373–379, 2010.
102. Kharkwal, S., Karimi, I.A., Chang, M.W., Lee, D.Y., Strain improvement and process development for Biobutanol production. *Recent Pat. Biotechnol.*, 3, 202–210, 2009.
103. Shen, C.R., Lan, E.I., Dekishima, Y., Baez, A., Cho, K.M., Liao, J.C., Driving forces enable high-titer anaerobic 1-butanol synthesis in *Escherichia coli*. *Appl. Environ. Microbiol.*, 77, 2905–2915, 2011.
104. Steen, E.J., Chan, R., Prasad, N., Myers, S., Petzold, C.J., Redding, A., Keasling, J.D., Metabolic engineering of *Saccharomyces cerevisiae* for the production of n-butanol. *Microb. Cell Fact.*, 7, 36, 2008.
105. Atsumi, S., Cann, A.F., Connor, M.R., Shen, C.R., Smith, K.M., Brynildsen, M.P., Liao, J.C., Metabolic engineering of *Escherichia coli* for 1-butanol production. *Metab. Eng.*, 10, 305–311, 2008.
106. Jones, D.T., Shirley, M., Wu, X., Keis, S., Bacteriophage infections in the industrial acetone butanol (AB) fermentation process. *J. Mol. Microbiol. Biotechnol.*, 2, 21–26, 2000.
107. Yazdani, S.S. and Gonzalez, R., Anaerobic fermentation of glycerol: A path to economic viability for the biofuels industry. *Curr. Opin. Biotechnol.*, 18, 213–219, 2007.
108. Zheng, Y.N., Li, L.Z., Xian, M., Ma, Y.J., Yang, J.M., Xu, X. *et al.*, Problems with the microbial production of butanol. *J. Ind. Microbiol. Biotechnol.*, 36, 1127–1138, 2009.
109. Gapes, J.R., The economics of acetone–Butanol fermentation: Theoretical and market considerations. *J. Microbiol. Biotechnol.*, 2, 27–32, 2000.
110. Syed, Q., Nadeem, M., Nelofer, R., Enhanced butanol production by mutant strains of *Clostridium acetobutylicum* in molasses medium. *Turk J. Biochem.*, 33, 25–30, 2008.
111. Connor, M.R., Cann, A.F., Lio, J.C., 3-Mithyl-1-butanol production in *Escherichia coli*: Random mutagenesis and two-phase fermentation. *Appl. Microbiol. Biotechnol.*, 86, 1155–1164, 2010.
112. Annous, B.A. and Blaschek, H.P., Isolation and characterization of *Clostridium acetobutylicum* mutants with enhanced amylolytic activity. *Appl. Environ. Microbiol.*, 57, 2544–2548, 1991.
113. Formanek, J., Mackie, R., Blaschek, H.P., Enhanced butanol production by *Clostridium beijerinkii* BA101 grown in semidefined P2 medium containing 6 percent maltodextrin or glucose. *Appl. Environ. Microbiol.*, 63, 2306–2310, 1997.
114. Lee, J., Lee, S.M., Kim, B.W., Butanol fermentation using wood powder hydrolysates produced by a modified supercritical water treatment. *New Biotechnol.*, 25, Suppl. 1, S230, 2009.
115. Theodorakis, C.W., *Mutagenesis*, pp. 2475–2484, Southern Illinois University Edwardsville, Edwardsville, IL, USA, 2008.

116. Parekh, S., Vinci, V., Strobel, R.J., Improvement of microbial strains and fermentation processes. *Appl. Microbiol. Biotechnol.*, 54, 287–301, 2000.
117. Mala, J.G.S., Kamini, N.R., Puvanakrishnan, R., Strain improvement of *Aspergillus niger* for enhanced lipase production. *J. Gen. Appl. Microbiol.*, 47, 181–186, 2001.
118. Waites, M.J., Morgan, N.L., Rockey, J.S., Higton, G., *Industrial microbiology-an introduction*, First Edition, pp. 1–302, Wiley Blackwell Publishing, 2002.
119. Sharma, A.K., Sharma, V., Saxena, J., A review on strain improvement of fungi for enhanced lipase production. *J. Biol. Nat.*, 6, 173–180, 2016.

6
Process Integration and Intensification of Biobutanol Production

Moumita Bishai

Department of Botany, Gurudas College, Kolkata, India

Abstract

Biofuels derived from biomass have attracted more attention due to the increased demand for fossil fuels, which cause serious environmental problems. Biobutanol is thought to be a capable alternative biofuel compared to other biofuels. Along with its excellent properties as a transport fuel, it is in high demand as a chemical substitute for various industries. The main problems for the production in addition to regeneration of biobutanol occur because of the constrain of the product and its recovery in a small concentration in the fermented mixture. Therefore, intensification and integration of technologies have been introduced to address these problems and encounters via providing numerous intensification facilities and methods to unify biobutanol production and purification in a more feasible manner. Therefore, this paper describes a myriad of intensification approaches based on the available literature and also examines the various approaches that take sought to increase biobutanol production from renewable resources.

Keywords: Biobutanol, process intensification, process integration, biofuel, fermentation

6.1 Introduction

Energy is a critical requirement for humanity's survival on this planet. By 2040, India's energy consumption is anticipated to skyrocket. Natural gas, coal and oil extracted from the ground are examples of renewable energy sources [1]. Increasing global warming, high oil prices, excessive

---
Email: onlymoumita06@gmail.com

greenhouse gas emissions and adverse environmental conditions around the world are forcing researchers to find ways to produce biofuels and other renewable resources from biological resources. Preference for alternative biofuels that would have particular categories of biomass that would have significant accessibility in the field [2]. Biobutanol has proven to be the most efficient of all biofuels.

The production of biobutanol by fermentation fits well with the concept of biorefining. Due to some superior properties, it has proven to be a probable and propitious replacement for existing biofuels and organic solvents [3]. The main biobutanol production technology relies on the fermentation of *Clostridium acetobutylicum* to produce acetone–butanol–ethanol (ABE) [4]. In ABE fermentation, the product is in an acetone–butanol–ethanol ratio of 30:60:10 [5]. Using various separation methods, the energy required to recover biobutanol from the fermentation aliquot was found to be between 14.7 and 79.05 MJ/kg biobutanol [6]. Overall, the cost of recovering the product from the fermentation broth contributes significantly to the overall cost of production [6]. They then compared the valuation cycles of biobutanol with those of ethanol and gasoline, concluding that biobutanol was better tolerated than ethanol and gasoline. Biobutanol production is hampered by product inhibition and expensive recovery costs for isolating biobutanol from fermentation slurry.

Process intensification (PI) is the development of manufacturing techniques that can significantly improve a process. In 2022, Lopez-Guajardo *et al.* emphasized that process intensification can be achieved by integrating a series of processes such as operation, function and phenomenon through platform technology. Such a noble system promises a wide range of innovations. This can condense the dimensions of the device while allowing multiple operations to be performed in a single device or improve the performance of specific operations. It can be used to produce specific products more sustainably by improving existing processes or developing new alternatives [7].

Process integration, on the other hand, is a key approach to utilize the available energy in a more improved way in a prearranged set of methods so that energy demands can be met in part by using process streams instead of utilities. The integration of various biological processes into a solo process is known as integrated bioprocessing [8]. This strategy has attracted considerable curiosity in the biobutanol production from biomass due to the combination of many processing phases, such as pre-treatment, saccharification, fermentation followed by recovery in one operation and can achieve high biobutanol at the end of the titer, productivity and yield process. Therefore, it not only decreases working expenses involving materials, apparatus, energy, labor and time; but it also turns out to be a broad technology that attracts investors for industrial production.

Thus, the current chapter has systematically discussed basic aspects of biobutanol production and recovery through process intensification and integration, as well as its application in production processes, emphasizing various equipment and methods. Furthermore, problems associated with these issues were deciphered, and future strategies for process improvement were investigated.

6.2 Biobutanol

Due to the increase in global energy consumption, the focus has been on producing advanced carbon sources in a sustainable way for future generations. Due to the scarcity of fossil fuel resources and their volatile costs, as well as environmental upheavals, alternative renewable fuels are reappearing, which will lead to zero greenhouse gas emissions in the long term. Global worries about energy scarcity, greenhouse gas emissions, and the wish for energy freedom are fuelling the excitement and speed with which biofuels are being researched, developed, designed, scaled, and marketed. The global biofuels industry is expected to hit USD 102.162 billion by 2020, with an annual growth rate of 4.02% to reach $134.589 billion by 2027 [9].

Since traditional biofuels such as bioethanol and biodiesel cannot meet the demand for biofuels, another biofuel with better or similar properties has been studied, such as biobutanol [10, 11].

Biobutanol with the molecular formula C_4H_9OH, is considered one of the best substitutes for petroleum fuels owing to its excellent properties. They have low melting and boiling points and are nonflammable.

They have low vapor pressure and therefore low volatility. They produce no waste such as carbon monoxide, SOx or NOx [12]. Their energy of 29.2 MJ/dm^3 is comparable to 32 MJ/dm^3 of gasoline and 5% more than ethanol [13]. Specifically, use of biobutanol in existing machinery schemes without any modification, or combined with other fuels such as hydrogen, gasoline, ethanol, methanol and isopropanol [14, 15].

Because of its superior properties, biobutanol has found widespread acceptability in a variety of industries, including food, pharmaceutical, and textile industries, as well as chemical and paint industries, detergents, cosmetics, and polymer industries [16].

6.3 Biobutanol Production and Recovery

Both chemical and biochemical synthesis are used to create biobutanol. Ethanol is the feedstock used in the chemical method of producing biobutanol. For the production of biobutanol from ethanol, dehydrogenation, aldol condensation, and hydrogenation are carried out sequentially [17]. Initially, dehydrogenation converts ethanol to acetaldehydes by removing hydrogen. Dehydrogenation resulted in the construction of an extensive series of chemicals and petrochemicals. This is followed by the aldol condensation reaction, which converts acetaldehydes into hydroxyketone or hydroxyaldehyde. Carbon–carbon bonds are formed when carbonyl compounds react with enolate or enol compounds. The final step is hydrogenation, which produces biobutanol. Although direct ethanol production from biobutanol has many advantages, the handling of a large number of chemicals may cause environmental problems [17].

Several steps are involved in the biochemical synthesis of biobutanol. *Clostridia* species have been shown to ferment carbohydrates anaerobically [18]. Acetone biobutanol Ethanol (ABE) fermentation is the primary method for producing biobutanol. Butyryl CoA reacts with butyraldehyde to form biobutanol. Acidogenesis is the process by which organic acids are formed in fermented broth with a pH greater than 5. When formulated organic acids are transformed into solvents such as acetone, biobutanol and ethanol, the process is known as solventogenesis [19, 20]. Alcohol dehydrogenase, CoA-acylating aldehyde dehydrogenase, butyryl-CoA dehydrogenase, pyruvate dehydrogenase, 2-ketoacid decarboxylase, and crotonase are the key enzymes [21].

Clostridium saccharobutylicum, *Clostridium acetobutylicum*, *Clostridium thermocellum*, and *Clostridium tyrobutyricum* all make biobutanol naturally. *C. beijerinckii* and *C. saccharoperbutylacetonicum* [4]. *Clostridium acetobutylicum* is a significant species in the creation of biobutanol. Important organisms include *Escherichia coli* [22], Cyanobacteria [23], *Klebsiella pneumoniae* [24], *Saccharomyces cerevisiae* [25], and *Pseudomonas putida* [21].

Another constraint to biobutanol production is the selection of substrates. The nitrogen and carbon content of the substrates could influence their selection. Agricultural waste, including fruit and oil waste; lignocellulosic wastes, wood-based wastes, biodegradable waste from municipal areas, crude glycerol, waste obtained from water, some microorganisms such as microalgae and other microbes, and so on, are major substrates [5, 15, 26].

Following substrate selection, they were subjected to upstream processing such as pre-treatment, detoxification, hydrolysis, and so on [27]. This is the first step in preparing the substrate for the fermentation process.

Following upstream processing, fermentation processes such as ABE fermentation, enzymatic immobilization, synchronized saccharification and fermentation, and so on are used [28]. Figure 6.1 summarises the entire biobutanol production process.

Despite the fact that numerous advanced fermentation and recovery processes have been developed for biobutanol production, there are still some issues for sustainable biobutanol production that must be addressed for effective industrial commercialization to compensate for energy demand. Nearly the tasks entangled in the manufacture of biobutanol are as follows [26, 15]:

- Inadequate availability of renewable biomasses to diminish the price of biobutanol production by increasing its economic feasibility.
- Lower biobutanol yield in fermented broth due to high product toxicity and solvent separation costs.
- Less efficient operation of bioreactors.

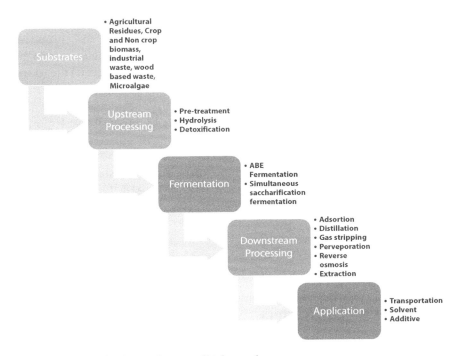

Figure 6.1 Route for the production of biobutanol.

- Production processing must still be optimized to reduce overall product costs.
- The cost of separating/recovering biobutanol from fermentation broth must be borne by those interested in the recovery process.

The above said particulars could be fixed by the strategies of integration and intensification of a process.

6.4 Process Intensification

The biggest problem in the world today is the shortage of energy. A process intensification approach can help solve the energy crisis. Numerous definitions of various writers have suggested process intensification [3, 26, 29, 30]. Recent advances in intensification of process introduced innovative tactics to processing biofuels [30]. The foremost concerns for improving manufacture of biobutanol are the growth of microorganisms to increase biobutanol output without noteworthy toxicity; bioreactor design for global productivity; innovative fermentation techniques; and the development of easy separation of biobutanol from soup of fermentation.

ABE fermentation produces biobutanol, a promising and attractive alternative to traditional fuels [31]. Yield, margin and recovery can be used to quantify the total functioning of the ABE fermentation. Due to the high toxicity of biobutanol to microbes, product inhibition occurs during processing. During batch fermentation, these microbes convert biomass into biobutanol, which is restricted to an extreme biobutanol absorption of 13 g/L [31]. Thus, characteristic fermentation by ABE process are administered at low substrate concentrations, large bioreactor and process fluids, high energy demand along with high downstream processing load. Hence, costly recovering process with low concentration ABE fermented products, creating large amounts of wastewater [31].

Biobutanol process improvement includes gas-stripping, membrane separation, such as reverse osmosis, liquid–liquid extraction, and pervaporation in combination with distillation [6, 32, 33]. There are several other process intensification procedures to accomplish biobutanol production.

6.4.1 PI Using Bioreactors

The conversion and productivity in batch reactors are typically near to the ground because of lag phase during reaction, inhibition of product, harvesting time, disinfecting, and refilling of the reactor. A continuous mode of operation with simultaneous product removal could be an option to address the problem [34]. Later, in 2022, Gurunathan and Sahadevan investigated whether a stirred tank bioreactor (STBR) using the *Clostridium* GBL1082 strain could be used for biobutanol production with a output of 0.16 g/L h [35]. They then used the same strain in an oscillatory baffled bioreactor (OBBR). When compared to STBR, efficiency (0.22 g/L h) was 38% higher. Finally, it was concluded that batch OBBR produced a greater amount of biobutanol than batch STBR. As a result, OBBR was proposed as a process intensifying bioreactor for the biobutanol production and its recovery. Later, engineers created a continuous biofilm fixed bed reactor to maximise biobutanol fermentation output. Immobilization of *Clostridium acetobutylicum* was used. The biobutanol output was 4.43 g/L h, and the selectivity was 88 wt% [35].

Another bioreactor, the fibrous bed bioreactor (FBB), with immobilized cells, was introduced for biobutanol fermentations. The FBB's superior performance was primarily due to the maximum living cell concentration preserved inside the bioreactor de to the exceptional immobilization mechanism of cells on the porous fibrous matrix [34]. Blockage in reactor and channelling effects are also common. Another advantage of the FBB is the elimination of nonviable cells immediately from the fermentation mixture [36].

6.4.2 PI Using Membranes

Cells are restricted in a membrane bioreactor, and from the side of the permeate, withdrawn of solvent is performed. Membrane separation processes are important in the chemical process industries for separation and could be improved separation methods. Membrane separation techniques include ultrafiltration, nanofiltration, microfiltration, pervaporation, reverse osmosis, gas permeation, and others. Among these, pervaporation could be combined with the process of fermentation for evading inhibition of product and therefore increasing the yield [1]. Biobutanol removal from ABE fermented mixture and aquatic phase was relatively simple using an intensifying approach based on pervaporation. For the recovery of biobutanol from the aquatic phase, the pervaporation membrane was made of polydimethylsiloxane/ceramic composite membrane. Azimi separated biobutanol from aqueous phase in 2017 using poly (ether block amide) (PEBA) based membrane and PERVAP 4060 flat-sheet commercial pervaporation membrane and polydimethylsiloxane (PDMS). The composite of ceramic PDMS membrane performed sound and consistently [37].

The pores of the supported membrane are impregnated with normal temperature ionic liquids in the membranes of ionic liquid. For the biobutanol separation, researchers used ionic liquids such as trihexyl(tetradecyl) phosphonium dicynamide, trioctylmethylammonium bis(trifluoromethylsulfonyl)imiden and trihexyl (tetradecyl) phosphonium bis(trifluoromethylsulfonyl)imide supported in membrane [38]. Azimi conducted a pervaporative separation of biobutanol from fermented broth using an ionic liquid supported membrane of tetrapropylammonium tetracyanoborate and polydimethylsiloxane in 2018 [39]. In this case, the result was 2.34 g/L h. For biobutanol pervaporative separation, he similarly used a silicalite-poly(dimethylsiloxane) nanocomposite membrane [39].

A feed with an adequate number of components is kept in thermal pervaporation (TPV), in direct interaction with the selective membrane's exterior surface, and the permeate is derived by condensation at atmospheric pressure [40]. Thermal pervaporation with phase-separated porous condensers is another approach to improve biobutanol separation using ABE fermentations operated at lower temperatures to accomplish the anticipated driving energy [41], in which contamination of the fermentation culture is eliminated with the increase in temperature destroying most microorganisms in the process. Borisov and others in (2018) then combined thermal osmotic evaporation with a porous condenser for phase separation, which was considered an innovative technique for the removal of biobutanol from fermentation mixtures [41]. Stable isolation of biobutanol has been observed using this technique. Thereafter, use pervaporation and purge gas to increase solvent mobility. For the separation of biobutanol, polytetrafluoroethylene hollow fibers using the ionic liquid 1-butyl-3-methylimidazolium hexafluorophosphate have been considered [42].

Another frequently used membrane-dependent process for water purification is reverse osmosis (RO) [43]. A semipermeable membrane is used in RO to separate permeate (pure water) from feed (a solution containing salts and retained compounds). Liu *et al.* patented in 2013 for the separation of biobutanol from aqueous solutions, describing reverse osmosis tests at 25 degrees Celsius with six organic model compounds: ethanol, biobutanol, oxalic acid, butyric acid, acetic acid along with lactic acid. Biobutanol showed a 99% refusal rate [43].

Perstraction is a membrane separation that syndicates the concepts of pervaporation and liquid–liquid extraction [44]. The toxic nature of liquid–liquid extraction can be altered by providing a film between the fermented aliquot along with the extractant. The main advantage of pressure extraction is that it keeps the microorganisms out of the solvent. The flow control step in this process is the diffusion of biobutanol through the membrane.

Another method of separating ABEs from aqueous solutions using polyvinylidene fluoride membranes is air gap membrane distillation. Biobutanol has been successfully isolated from ABE solutions [45]. Based on experimental data, a mathematical study was performed on the parting of biobutanol from aqueous phase with the help of polypropylene membranes during membrane distillation [45].

6.4.3 PI Using Distillation

Distillation is an old method of separation based on differences in component volatility [6]. The main separation technique used to recover biobutanol from fermentation broth is distillation. At 101.3 kPa and 55.5% by weight of biobutanol, the biobutanol–water mixture forms an azeotrope. The main problem with the use of distillation is the reduced solubility of biobutanol in aqueous medium, since the maximum solubility of biobutanol in aquatic condition is 7.7% by weight. Under azeotropic conditions, the upper product contained 79.9 wt % biobutanol, while the lower product contained 7.7 wt% biobutanol [46]. The fermenter is operated at ambient pressure during flash fermentation, and the ABE mixture is continuously poured in a loop that is closed to a void room (6.4–6.8 kPa) where the biobutanol is extracted by heating and evaporation. The greatest biobutanol concentration generated was 30 to 37 g/L, with a total energy of 17.0 MJ/kg [46].

Gas stripping fermentation is a substitute to conventional fermentation in which fermentation and separation of biobutanol are carried out simultaneously by stripping under inert gas. This is one of the simplest and finest methods of recover biobutanol from fermentation broth. During this process, an inert gas is usually introduced into the fermenter by infusion, where the biobutanol is stripped from the gas and then removed by condensation. The biobutanol concentration in the recycle stream was higher than in the fermentation aliquot [32]. It is simpler and less expensive than other separation methods. Additionally, gas stripping does not remove nutrients and intermediates from the fermentation reaction, and since biobutanol is continuously removed, it reduces the toxic or inhibitory effects of biobutanol. CO_2 and H_2 fermentation gases are used as extraction gases to remove biobutanol immediately. A temperature of 35 degrees Celsius and 3.21 L/h eliminating gas, the ABE clearance rate was 0.3 g/L h. When contrasted to the batch reactor, the ABE production jumped from 0.07 to 0.17 g/L h, and the amount of lactose expanded ten times [32].

Raftery and Karim (2017) [47] investigated the financial feasibility of enormous scale continuous manufacturing of biobutanol in a multifeed digester alongside *in situ* extraction of gas employing *Clostridium acetobutylicum*. The upgraded bioreactor's working capability was contrasted to the results of an experimental fermentation with no degassing for fiscal practicality. There are five feeds consisting of hybrid poplar, cellulosic biomass, sorghum, bagasse, corn stalks along with switchgrass, four fiber expansion pre-treatment methods of ammonia, hydroxide sodium, dilute sulfuric acid and liquid hot water used in various bioreactors. The results are promising, showing that process improvement practices in the bioprocess industry have the potential to progress towards new, if not, economically unfeasible methods [47].

Patrescu *et al.* (2018) conducted another study in which they united separation in an FBB along with energy-efficient ABE fermentation with two phase stripping of gas for the production and purification of biobutanol. In this reactor, a maximum ABE of 73.3 g/L was obtained using 48.5 g/L biobutanol in the first stripping step, with a 199% increase in condensate concentration. ABE encountered 48.5 g/L biobutanol and 671.1 g/L ABE encountered 515.3 g/L biobutanol in the in the second stage [16].

6.4.4 PI Using Liquid–Liquid Extraction

Separation in liquid–liquid extraction is achieved by differences in partition coefficients. This is another traditional yet highly efficient performance with the foremost benefit of working mainly in ambient conditions [48]. Since biobutanol is more soluble than aqueous phase, in many organic solvents, the extraction can be easily performed by extractant collection. Oleyl alcohol and decyl alcohol and are the most extensively used extractants in the separation of biobutanol [49].

One of the foremost problems with liquid–liquid extraction is the toxicity of the extractant to microorganisms. To recover biobutanol, a cascade of four-stage mixer settlers for liquid–liquid extraction and a membrane bioreactor were used [50]. Isolation of biobutanol from cell-free media using n-decyl alcohol saturated with butyric acid. Concentration and yield of biobutanol increased to 8 g/L h and 0.51 g/L h, respectively. Kurkijarvi *et al.* (2016) studied nonbiocompatible solvents (1-heptanol, 1-octanol, and 1-decanol) used for the ABE fermentation broth separation [51]. First, the biobutanol was removed with a nonbiocompatible solvent, then drops of the nonbiocompatible solvent were detached from the recycle stream. Among other things, liquid–liquid extraction offers high throughput and selectivity for the separation of biobutanol.

Biobutanol can be extracted easily and selectively using extractive fermentation, which requires less energy. The extractant should have a high extraction capacity, efficiency along with distribution coefficient, be biocompatible with microorganisms, be easy to phase separate, and so on [35]. Because of its high distribution coefficient in comparison to acetone extraction, oleyl alcohol is a popular extractant for biobutanol separation. Tributyrin, another extractant, can extract both biobutanol and acetone [14]. There has been a lot of literature presented on biobutanol production improvements in extraction modes in fed-batch, batch and continuous cultures [14]. They then used fermentation using extractive process to yield and eradicate biobutanol using an oleyl alcohol tributyrin mixture, which resulted in high distribution coefficients for both biobutanol with acetone.

Conventional solvents made from organic matter have a high diffusion coefficient, separation capability, and potency for biobutanol along with other compounds, but they are mainly detrimental, insecure, and dangerous. Due to their low vapor pressure, ionic liquids were cast-off as novel ecologically favorable solvents [52]. It has been widely used in the separation of biobutanol. For manipulating the properties of ionic liquids, various combinations of anions and cations can be used. Ionic liquids were used straitly in bioreactors for instantaneous recovery with other products [53]. The advantage of ionic liquids is that there is no loss of solvent, which saves money. The ionic liquid 1-octyl-3-methylimidazolium bis[(trifluoromethyl)sulfonyl] amide has been used to remove biobutanol from water with a removal efficiency greater than 74% of the preliminary biobutanol [53]. The ionic liquid 1-hexyl-3-methylimidazolium tris(pentafluoroethyl)trifluorophosphate has been shown to be vigorous in the biobutanol retrieval from the aquatic stage [54].

6.4.5 PI Using Adsorption

Adsorption is a tangible or molecular process in which particulates of a solution or gaseous vapor preferentially adhere to a rigid substrate. Recovery of biobutanol from fermented broth was carried out using adsorption. Nevertheless, considering the tiny amount of the adsorption material adsorption is challenging to execute on a commercial scale. A wide range of materials was selected as adsorbents for biobutanol removal. Commonly investigated adsorbents for biobutanol removal include zeolites, activated carbons, and polymers [55].

Silicate materials are currently extensively employed as potential adsorption agents to earn biobutanol extraction. Researchers in 2015 investigated the capability of aquaphobic polymer matrices as adsorbents to recover and purify biobutanol [56]. The mechanism by which sorbents separate n-alcohols is mainly described as hydrophobic interactions. To remove biobutanol, various poly(styrene-co-DVB), poly (butylene phthalate) and poly(methacrylate) resins have been used.

Shetty *et al.* (2018) Removal of biobutanol using new grafted calixarenes as adsorbents [57]. These sorbents are made up of hydrophobic calixarenes that have holes that are covalently attached to silica supports (both porous and hydrophilic). Mesoporous carbon was employed for biobutanol and ethanol adhesion, and it was found that biobutanol retention was higher than ethanol retention. To particularly attach n-butanol, a macroporous adsorption material (KA-I) was also used [58]. A MEL6 zeolite-like material in a membrane with a dimension of 222.24 mg/g was used for biobutanol adsorption at 30°C [55]. High silica zeolites (CBV811C300, CBV901 and CBV28014) with surface areas of 400, 620, and 700 m^2/g were used for the discerning absorption of biobutanol from ABE fermentation broths [26]. Adsorption was noted in the following order: n-butanol, acetone, and ethanol. A macroporous aquaphobic poly(styrene-copolydivinylbenzene) resin was studied in extended bed adsorption for the adsorption removal of biobutanol from ABE cultured broths. After 38.5 hours, the total liquid and biobutanol amounts were 40.7 and 27.2 g/L, separately, with a biobutanol retrieval rate of 81% [59].

Clostridium acetobutylicum was immobilized as a biofilm on a fiber-type substrate for ABE fermentation in a biofilm reactor with fixed-bed adsorption. Cellular tolerance improved considerably as biobutanol productivity grew. When compared to traditional batch fermentation, the solvent concentration and output ranged between 96.5 and 130.7 g/L and 1.0 to 1.5 g/L h rose by four to six times and three to five times, correspondingly. Aside from a high percentage and output, fixed-bed adsorption biofilm reactors use less energy and liquid than traditional batch fermentation [60].

6.5 Process Integration

The use of integrating several cellular systems into a single process is referred to as integrated processing. This strategy has glowed attention in the biobutanol production process from biomass because it combines several processing phases, such as pre-treatment, saccharification, fermentation, and separation in one operation, achieving higher titer of biobutanol, output, and profit at the end of the progression. Thus, processing integration lowers working expenses in terms of materials, personnel, machinery, energy, and time. Although this scaling method is more appealing for industrial scale due to the cheap initial cost of setting, more effort should be put into study to address the problems connected with it [61].

The *in situ* recovery procedure is well described as a current method for inducing biobutanol fermentation. Furthermore, *in situ* recovery reduces the obligation solvent on goods when gathering products from the fermentation technique. The current knowledge was validated in all of the fermentation conditions that demonstrated significantly increased biobutanol output. To avoid the toxic nature of biobutanol cells, in situ adsorption was achieved using a dynamic column constructed as an external column of a fed-batch bioreactor, adsorbing the immobilized microbial culture of (53.8–92.2) g/L biobutanol and keeping 8.9 g/L biobutanol in the fermentation broth. This method generated 230% more biobutanol than fermentation alone, which yielded only 16.6 g/L of biobutanol [62]. The fed-batch process with in situ recovery is far more efficient than the batch process with *in situ* recovery because it tackles two major flaws in ABE fermentation:

1. Minimal level of sugar (100 g per batch/L); and
2. Impossibility of producing an end product owing to fluid build-up in the process of fermentation slurry.

The rate of carbon intake was discovered to influence biobutanol synthesis in a continuous system. Accordingly, expanding the substrate's utilization rate (by reducing the harmful concentration of biobutanol and possibly enhancing cellular number) can enhance the skill of continuous ABE fermentation [63]. Direct *in situ* repair increased cell density, leading to higher glucose consumption rates and a considerably higher biobutanol output rate of up to 11 g/L h. To better the overall process, appropriate restorative performances should be evaluated in an unpredictable manner. Kujawska *et al.* (2015) investigated various separation methods in ABE fermentation for the generation of biobutanol [6].

Synchronous digestion and brewing also known as SSF is a new technique that includes the occurrence of scavenging and metabolism in the same vessel on the exact same moment. This technique lowers the spectrum of preparatory stages needed throughout biobutanolmicrobial fermentation whereas increasing total substrates transformation and production to biobutanol [28, 64]. Consequently, the use of the SSF technique reduces the amount of time as well as energy required for enzymatic dehydration. Valles *et al.* showed in 2021 that saccharification done according to circumstances akin to *Clostridium* ABE fermentations possessed no noticeable impact upon saccharification effectiveness, enabling both methods to be satisfactorily united [28]. Additionally, as soon as contrasted with the hydrolytic fermentation by itself (SHF) techniques, the technique known as SSF produces greater outputs as well as effectiveness through preserving low amounts of glucose, that prevents cellulase inhibition by glucose and thus improves the absorption of carbohydrates coming from sucrose subunits. The stacked substance construction for SSF is a major issue with the SSF process. Certain lignin-based biomass preparatory steps produce toxic chemicals following saccharification, which can interfere with brewing effectiveness [65]. As a consequence, an appropriate SSF pre-treatment method should be investigated. The substances that lack the need for processing, such as kraft pulp mill refuse, sago pulp leftovers, or starch-based the plant matter, are preferred [66, 11]. Due to restricted substrate loading, SSF may also experience low sugar concentrations in the system. Prior to fermentation, the reclaimed sugars can be concentrated using the SHF method. In 2021, Valles *et al.*, 5% of preprocessed oil palm fruit bunches gave 30 g/L of fermentable sugars and the ABE fermented sugar concentration was low; however, despite the identical starting sugar amount and output, SSF produced higher biobutanol titers, profits, and output than SHF [28].

Pretreated straw made from wheat with an average sugar concentration of 25.6 g/L yielded SSF-like results [67]. SSF production and efficacy were enhanced by the inclusion of temporary sugars. The SSF technique, like other *in situ* fermentations, can be improved by adding in situ regeneration into the system. When compared to SSF alone, a 20% rise in total solvent was found, and a 60% increase after several modification experiments [67]. Although in situ SSF repair has not received much attention, it is theoretically possible. Beyond the traditional SSF technique, there's aren't any fed-batch files or uninterrupted SSF, so such possibilities deserve to be explored for potential expansion. Generally speaking, the technique is difficult in the context of feeding the pasteurized concrete substrate into the reaction vessel, and evaluating the procedure's success which is complicated. These obstacles, nevertheless could be surmounted alongside the proper biological processing techniques as well as process infrastructure.

Technology of Consolidated bioprocessing (CBP) is a way of simultaneously generating enzymes, then saccharification followed by fermentation in the identical reactor [68]. As a consequence, the microorganism used in the present procedure essentially are accomplished with capacity of producing enzyme for saccharification form biomass followed by fermentation process for of sugar into desired goods. The applied approach was planned towards decreasing the stages of the pathways required to convert bioresource into butanol. The strategy of direct fermentation (DF), remained widely reported designed for the conversion of starchy substances into butanol, where cellular parts produce amylase towards conversion of starchy part into sugar followed by its consumption for the production of biobutanol, entirely it occurs in the identical reactor and at a fixed time. Quite a lot of research investigations were conducted regarding the DF of starch-based feedstock towards biobutanol from *Clostridia* since these Gram-positive bacteria have the capability of producing a collection of amylases. *C. acetobutylicum* of strain P262 used the double fermentation technique to produce amylase and glucoamylase by using sago, maize, potato, and cassava as raw substrate and then converting the starchy product into biobutanol [69]. *Clostridia* could produce amylase at the same time as during the fermentation of lag phase, within 4 hours, and transform carbohydrate to glucose for the metabolic process of cells. As a consequence, regarding biobutanol fermentation efficiency, the DF cassava starch process beats the SHF tapioca granule hydrolysate process [70]. With the addition of carbonate of calcium as an amylase activator to improve substrate utilization, the direct synthesis of the flour of cassava via *Clostridium* sp. isolate BOH_3 produced 17.8 g/L biobutanol, with a profit of 0.30 g/g [71]. *C. acetobutylicum* PW12 mutation ART18 produced an additional 31 percent biobutanol than the wild type during the DF process of tapioca starch. [71]. Regardless of knowing that this DF technique simultaneously produces amylase, hydrolyzes, and metabolizes ABE, all of those fermentations were finished in 4872 hours of the time it took to ferment. CBP compared to lignocellulosic waste is more difficult to transform onto biobutanol than CBP coming from starch-based feedstock. Regardless of the reality that multiple cellulose alleles were discovered as well as a complete cellulosome complex transcribed within the *Clostridial* genome, little cellulase translation or activity in brewing has been detected [72]. This malfunction might have been produced through the suppression of an organism's gene encoding a critical and function-limiting protein, a significant abnormality in this type of protein, or not being able to move and build the cellulosome in the correct cellular location [71]. A complicated cellulase enzyme combination is also required for full breakdown of cellulosic material into monomeric unit of sugar. This is more challenging process due to lignocellulosic material must be pretreated before saccharification, and most methods generated inhibitory chemicals during either by pre-treatment or by saccharification. The lignocellulose is much complicated material. As a consequence, the processes of cellulose based biomass by *Clostridia* looks to be unachievable without combine culture or else genomic manufacturing. Numerous efforts were made to use the DF method for transformation of biomass containing cellulose to biobutanol. *C. acetobutylicum* having strain ATCC 824 was found to have cellulase activity when cultured on the xylose. When a glycoside hydrolase gene was incorporated into *C. beijerinckii* it could not grow on microcrystalline cellulose, but then it produced considerably more amount of solvent as soon as cultivated than the strain which is of wild type [73]. When the culture of *Clostridium thermocellum* and *C. saccharoperbutylacetonicum* N1-4 was carried out, it produced biobutanol of 5.8 g/L, whereas *C. acetobutylicum* strain ATCC 824 is able to generate biobutanol individually of 3.73 g/L using CMC and filter paper. Such brewing operations took 12 days to complete. Genetically engineered *Clostridium cellulolyticum* had been able to produce biobutanol of 0.66 g/L, besides it also addresses several obstacles for future improvement [73]. Increasing the rate of biobutanol synthesis by the speeding up of hydrolysis of cellulase produced by *Clostridia*. Ibrahim *et al*. (2018) discovered that mixing *C. beijerinckii* and *C. thermocellum* resulted in 14.3 g/L of butanol after 7200 mins with 45 °C. 61.67 FPU/mL of cellulase was produced, that was greater than market level cellulase. Such investigational findings using lignocellulosic biomass as substrate are too excellent to be true and should be repeated by other scholars for further development [73].

Pervaporation alone may not be adequate for biobutanol extraction, so distillation may be needed to couple. A combined separation of pervaporation distillation was used to separate biobutanol to achieve high purity (99.9% wt.%) biobutanol. Van Wyk *et al*. (2018) used pervaporation distillation to intensify the ABE separation process in order to decrease downstream processing energy costs. Biobutanol was separated from a model solution of ABE using a PDMS membrane at different operating temperatures, with the best results achieved at 40°C [1].

In a process where vapor stripping vapor permeation was incorporated, a feed is introduced from the uppermost part of a vapor column which is stripping and content from the feed is stripped with upstairs vapor augmented with solvent promoting vapor liquid stabilities. The vapor of the solvent is squeezed in addition to channelled into vapor infusion membrane segment through aqueous selection membrane that excludes flush [74]. The scholar used a vapor stripping vapor permeation method to recover and dehydrate biobutanol and fermentation by ABE using diluted solution of water. 10.4 MJ/kg of energy was needed for straightforward stripping from 1.3 wt.% biobutanol. The addition of a vapor compressor and membrane device in sequence with the striping, as well as the permeate recuperating raised biobutanol content from 25 to 95% wt. Furthermore, electricity savings rose by 25%.

Xue *et al*. (2016) studied the dynamics of column on the interaction of adsorption, drying, desorption process intended for the extraction and separation of biobutanol from water using pellets of silica [75]. In the initial stage, biobutanol obtained from the water segment showed adsorption on silica crystals in a column till overload, then drying of the column was done using air at a moderate condition of 50°C to 70°C to eliminate the liquid. The full adsorption column is heated to 130°C to 150°C. for air desorption, followed by condensation to reclaim the biobutanol. This process's energy demand (about 3.4 MJ/kg) remains noticeably diminished as compared to the biobutanol's energy content (36 MJ/kg) [75].

A solitary azeotropic split wall distillation column substitutes three condensation columns when combined with heat pumping, improving efficiency and performance while using less energy [16]. Such samples split wall distillation with vapor firmness was used to remove biobutanol. Patrascu et al. in (2018) explored an intensifying method for biobutanol extraction from fermentation broth using a heat pump-assisted azeotropic split wall column. This procedure uses 58% less energy (2.7 MJ/kg) than the traditional method (6.3 MJ/kg) for biobutanol separation [16].

Several variables can affect solvent output in fermentation, as well as the acetone–butanol ratio. It was found that the culture media when supplemented with reducing agents, showed various results such as it might causes acetone–butanol ratio to be altered. Reducing substances can be biological or inanimate. Organic reducing agents comprises formic acid, glycerol, methylene blue whereas artificial reducing agents include sodium bisulphite and hydrosulphite [41]. It was recommended that glycerol being the most striking reducing agent being a high source of feedstock. Glycerol reduces the biobutanol output in *Clostridia* strain fermentation in the cereal produce flour in addition to beet molasses. This rises in overall ABE content from 60% to 80% increased ABE fermentation efficiency. Amino acids are an effective nitrogen supply and an essential precursor in many biochemical production routes. Under microbial stress, the different intracellular functions have been principally controlled by amino acids. As a consequence, the amino acids addition is helpful with an aim at raising the percentage of solvent and microbial movements. The inclusion of aspartic acid family amino acids regulated and improved biobutanol synthesis via fermentation with *C. acetobutylicum* [3]. Furthermore, the toxicity of biobutanol was greatly decreased in the increased biobutanol fermentation was performed using exogenic compounds having amino and carboxylic functional group [3]. *Clostridium acetobutylicum* NRRL B-527 uses 54.15 to 61.0 g L of glucose which produces 12.43 to 6 0.10 g L of biobutanol in batch fermentation whereas and 20.82 to 60.33 g L in fed batch combined with liquid-liquid extraction [3].

Loaded can be used in a variety of methods in fermentation using ABE. In the continuous, semibatch and batch fermentation, the feedstock is fed to the fermentor. Periodic reactive loading is most attractive for continuous fermentation and simultaneous product elimination because it improves fermentation performance by keeping ABE concentrations very low less than the obstruction threshold [41].

Ultrasonic radiation, also known as sonication, induces strong microconvection in addition to excess of shear in the matrix. Shear-induced morphological alterations influence microbial cell function. Borah *et al.* (2019) tried to enhance the ABE fermentation process by using sonication (or ultrasonic irradiation) [76]. They used the noxious plant as a raw material to conduct ABE fermentation with *Clostridium acetobutylicum* strain MTCC 11,274 via ultrasound. The ultrasonically aided fermentation produced 0.288 g/g in 92 hours, while the stirring produced 0.168 g/g in 120 hours. Borah *et al.* (2019) explained that the increase in solvent output in ultrasonically aided fermentation was due to increased metabolic reaction kinetics due to increased enzyme affinity for the substrate and greater resilience to substrate inhibition. Sonication has been shown in flow cytometry experiments to have high cell viability with no detrimental impacts on cell physiology [76, 77].

Nanotechnology is now essential for its broad diversity of manufacturing usages. Nanoparticles partake unique characteristics such as very strong rigorous mechanical strength along with huge explicit surface expanses. Carbon nanotubes (CNTs) are a novel nanomaterial with numerous uses [75]. Carbon nanotube (CNT) membranes have been extensively researched for a broad range of usages, together with separation of gas in addition to aquatic refinement [75]. They developed a composite membrane using CNT-PDMS for the pervaporative removal of biobutanol from the fermented mixture. CNTs serve by way of adsorption of dynamic positions with excess hydrophobicity in the membrane, enabling high-quality transport across the interior band or along smooth surfaces. The highest membrane flow rate of 10% concentration of CNT-PDMS was 244.3 g/(m^2 h) at 80°C, with a separation factor of 32.9. These disasters can be avoided by developing sustainable and alternative energy sources. Producing different biofuels from renewable vegetation could be the most feasible option to fossil fuels.

6.6 Conclusion

Biobutanol is regarded the finest option and choice for fuel and auxiliary uses due to its distinct qualities and properties. The ABF fermentation product is the most commonly used method for producing biobutanol from *Clostridium* species. There are numerous obstacles to effective commercial industrialization of biobutanol production. The major problems in the creation of techniques for ramping up biobutanol production are strain development, raw materials, progress in production, projection of bioreactor plan, separation and recovery of final bioproducts. This part goes over process enhancement and integration.

Current techniques of biobutanol fermentation and separation remained to be energy demanding. Overall, biobutanol production has a high market potential, and it will become a feasible substitute for fossil fuels. The primary issue with biobutanol is its synthesis and extraction as a result of inhibition of product and small content inside the fermented mixture. The escalation of intensified process can address these problems and trials by introducing new tools and techniques for producing and separating biobutanol. A suitable technique needs to be chosen principally on the basis of its technological in addition to its fiscal practicability.

Intensification methods using bioreactors, approaches using distillation, methods using membrane, air dragged fermented systems, adsorption procedures, liquid–liquid extraction approaches and combination systems are all addressed as alternative energy-based methods. Long fermentation periods, numerous pre-treatment steps, minimal or no expression of cellulase by microbes, contradictory conditions in a single bioreactor due to multiple steps, and inhibition of inhibitory chemicals in batch stages are some of the difficulties. These difficulties can be surmounted by using the CBP process, optimizing the fermentation process, creating new novel strains capable of making numerous enzymes, and so on. Biobutanol may become a feasible feedstock in the near future as a result of these advancements and more viable study.

References

1. Van Wyk, S. *et al.*, Pervaporative separation and intensification of downstream recovery of acetone-butanol-ethanol (ABE). *Chem. Engg Proce Proc. Inten*, 130, 148159, 2018.
2. Malode, S.J. *et al.*, Recent advances and viability in biofuel production. *Energy Convers. Manag. X*, 10, 100070, 2021.
3. Nimbalkar, P.R. *et al.*, Strategic intensification in butanol production by exoges amino acid supplementation: Fermentation kinetics and thermodynamic studies. *Bioresour. Technol.*, 288, 121521, 2019.
4. Li, S. *et al.*, Pathway dissection, regulation, engineering and application: Lessons learned from biobutanol production by solventogenic clostridia. *Biotechnol. Biofuels*, 6, 39, 1–25, 2020.

5. Gomes, A.C., Acetone–Butanol–Ethanol fermentation from sugarcane bagasse hydrolysates: Utilization of C5 and C6 sugars Electron. *J. Biotechnol.*, 42, 16–22, 2019.
6. Kujawska, A. *et al.*, ABE fermentation products recovery methods—A review. *Ren. Sust. Ener. Rev.*, 48, 648–661, 2015.
7. Lopez-Guajardo, E.A. *et al.*, Process intensification 4.0: A new approach for attaining new, sustainable and circular processes enabled by machine learning. *Chem. Eng. Process*, 180, 108671, 2022.
8. Klemes, J.J. *et al.*, New directions in the implementation of pinch methodology (PM). *Renew. Sustain. Energy Rev.*, 98, 439–468, 2018.
9. *Global biofuel market report 2022: Global initiatives are being launched to promote biofuel as an environmentally friendly fuel*, 2022, https://www.globenewswire.com/en/news-release/2022/07/18/2480809/28124/en/Global-Biofuel-Market-Report-2022-Global-Initiatives-are-being-Launched-to-Promote-Biofuel-as-an-Environmentally-Friendly-Fuel.html).
10. Burhani, D. *et al.*, *Second generation biobutanol: An update*, 19, 101–110, Reaktor, 2019.
11. Ibrahim, M.F. *et al.*, Cellulosic biobutanol by *Clostridia*: Challenges and improvements. *Renew. Sustain. Energy Rev.*, 79, 1241–1254, 2017.
12. ASTM, D., 6751-15a Standard specification for biodiesel fuel blend stock (B100) for distillate fuels. *Am. Soc Test Mater.*, 05.03, 10, 2015.
13. Kolesinska, B. *et al.*, Butanol synthesis routes for biofuel production: Trends and perspectives. *Mater*, 12, 350, 2019.
14. Darmayanti, R.F. *et al.*, Novel biobutanol fermentation at a large extractant volume ratio using immobilized Clostridium saccharoperbutylacetonicum N1-4. *J. Biosci. Bioeng.*, 126, 750–757, 2018.
15. Iyyappan, J. *et al.*, Overview of current developments in biobutanol production methods and future perspectives. *Rev. Methods Mol. Biol.*, 2290, 3–21, 2021.
16. Patrascu, I. *et al.*, Eco-efficient downstream processing of biobutanol by enhanced process intensification and integration. *ACS Sustain. Chem. Engg*, 6, 5452–5461, 2018.
17. Ndaba, B. *et al.*, n-Butanol derived from biochemical and chemical routes: A review. *Biotechnol. Rep.*, 5, 1–9, 2015.
18. Rosales-Calderon, O. and Arantes, V., A review on commercial-scale high-value products that can be produced alongside cellulosic ethanol. *Biotechnol. Biofuels*, 12, 240, 2019.
19. Patakova, P. *et al.*, Acidogenesis, solventogenesis, metabolic stress response and life cycle changes in *Clostridium beijerinckii* NRRL B-598 at the transcriptomic level. *Sci. Rep.*, 9, 1371, 2019.
20. Xin, F. *et al.*, Exploitation of novel wild type solventogenic strains for butanol production. *Biotechnol. Biofuels*, 18, 252, 2018.
21. Nawab, S. *et al.*, Genetic engineering of non-native hosts for 1-butanol production and its challenges: A review. *Microb. Cell Fact.*, 19, 79, 2020.

22. Dong, H. *et al.*, A systematically chromosomally engineered Escherichia coli efficiently produces butanol. *Metab. Eng.*, 44, 284–294, 2017.
23. Liu, X. *et al.*, Modular engineering for efficient photosynthetic biosynthesis of 1butanol from CO_2 in cyanobacteria. *Energy Environ. Sci.*, 12, 2765–2777, 2019.
24. Wang, M. *et al.*, Enhanced 1butanol production in engineered *Klebsiella pneumoniae* by NADH regeneration. *Energy Fuels*, 29, 1–19, 2015.
25. Lian, J. *et al.*, Construction of plasmids with tunable copy numbers in *Saccharomyces cerevisiae and* their applications in pathway optimization and multiplex genome integration. *Biotechnol. Bioeng*, 113, 2462–2473, 2016.
26. Wasewar, K.L., Process intensification in biobutanol production, in: *Biofuels and Bioenergy*, B. Gurunathan and R. Sahadevan (Eds.), Elsevier, USA, 2022.
27. Guo, Y. *et al.*, Production of butanol from lignocellulosic biomass: Recent advances, challenges, and prospects. *RSC Adv.*, 12, 18848–18863, 2022.
28. Valles, A. *et al.*, Fed-batch simultaneous saccharification and fermentation including *in-situ* recovery for enhanced butanol production from rice straw. *Bioresour. Technol.*, 342, 126020, 2021.
29. Keil, F., *Modelling of process intensification – An introduction and overview, in: Modeling of Process Intensification*, Keil, F. (Ed), pp. 1–7, Wiley, USA, 2017.
30. Sitter, S. *et al.*, An overview of process intensification methods. *Curr. Opin. Chem. Eng.*, 25, 87–94, 2019.
31. Ranjan, A., Biobutanol: Science, engineering, and economics. *Int. J. Energy Res.*, 36, 277–323, 2012.
32. Outram, V. *et al.*, Applied *in situ* product recovery in ABE fermentation. *Biotech. Progr.*, 33, 563579, 2017.
33. Rom, A. *et al.*, Energy saving potential of hybrid membrane and distillation process in butanol purification: Experiments, modelling and simulation. *Chem. Engg. Proce. Proc. Inten.*, 104, 201211, 2016.
34. Yang, O. *et al.*, Economic analysis of batch and continuous biopharmaceutical antibody production: A Review. *J. Pharm. Innov.*, 14, 1–19, 2019.
35. Gurunathan, B. and Sahadevan, R., *Biofuels and bioenergy: A technoeconomic approach*, Elsevier, USA, 2022.
36. Bharathiraja, B. *et al.*, Biobutanol – An impending biofuel for future: A review on upstream and downstream processing techniques. *Renew. Sustain. Energy Rev.*, 68, 788–807, 2017.
37. Azimi, H., *Pervaporation separation of butanol using PDMS mixed matrix membranes. Thesis For the Doctorate in Philosophy*, Chemical Engineering degree Department of Chemical and Biological Engineering Faculty of Engineering, University of Ottawa, Canada, 2017.
38. Sahrash, R. *et al.*, PVDF based ionogels: Applications towards electrochemical devices and membrane separation processes. *Heliyon*, 4, e00847, 2018.

39. Azimi, H. et al., Separation of organic compounds from abe model solutions via pervaporation using activated carbon/pdms mixed matrix membranes. *Membranes*, 8, 40, 2018.
40. Koter, S. et al., Modeling of transport and separation in a thermopervaporation process. *J. Membr. Sci.*, 480, 129138, 2015.
41. Borisov, I. et al., Intensification of acetone-butanolethanol fermentation via products recovery: Thermopervaporation assisted by phase separation. *Chem. Eng. Trans.*, 64, 4348, 2018.
42. Olea, F. et al., Separation of vanillin by perstraction using hydrophobic ionic liquids as extractant phase: Analysis of mass transfer and screening of ILs via COSMO-RS. *Sep. Purif. Technol.*, 274, 119008, 2021.
43. Liu, H. and Fang, Z. et al., The promising fuel-biobutanol, in: *Liquid, Gaseous and Solid Biofuels- Conversion Techniques*, pp. 175–198, Intechopen, UK, 2013.
44. Venturelli, M.G. et al., Separation of fermentation products from ABE mixtures by perstraction using hydrophobic ionic liquids as extractants. *J. Membr. Sci.*, 537, 333–343, 2017.
45. Knozowska, K. et al., Membrane assisted processing of acetone, butanol, and ethanol (ABE) aqueous streams. *Chem. Eng. Process*, 166, 108462, 2021.
46. Chen, H. et al., Novel distillation process for effective and stable separation of high-concentration acetone–Butanol–Ethanol mixture from fermentation–pervaporation integration process. *Biotechnol. Biofuels.*, 11, 286, 2018.
47. Raftery, J.P. and Karim, M.N., Process intensification of large-scale continuous biobutanol production via a multi-feed bioreactor with *in situ* gas stripping, in: *AIChE Annual Meeting*, AIChE, 2017.
48. Zhang, Q.W. et al., Techniques for extraction and isolation of natural products: A comprehensive review. *Chin. Med.*, 17, 20, 2018.
49. Jimenez-Bonilla, P. and Wang, Y., *In situ* biobutanol recovery from clostridial fermentations: A critical review. *Crit. Rev. Biotechnol.*, 38, 469–482, 2017.
50. Stoffers, M. and Andrzej, G., Continuous multi-stage extraction of n-butanol from aqueous solutions with 1-hexyl-3-methylimidazolium tetracyanoborate. *Sep. Purif. Technol.*, 120, 415–422, 2013.
51. Kurkijarvi, A.J. et al., Comparison of reactive distillation and dual extraction processes for the separation of acetone, butanol, and ethanol from fermentation broth. *Ind. Eng. Chem. Res.*, 55, 1952–1964, 2016.
52. Chen, W.-H. et al., Methylimidazolium-based ionic liquids influence the biobutanol production by solvent-producing Clostridium. *Int. Biodeterior. Biodegrad.*, 129, 163–169, 2018.
53. Motghare, K.A. et al., Phosphonium based ionic liquids: Potential green solvents for separation of butanol from aqueous media. *Korean J. Chem. Eng.*, 39, 39, 2736–2742, 2022.
54. Sanchez-Segado, S. et al., Evaluation of ionic liquids as *in situ* extraction agents during the alcoholic fermentation of carob pod extracts. *Fermentation*, 5, 90, 2019.

55. Faisal, A. et al., Zeolite MFI adsorbent for recovery of butanol from ABE fermentation broths produced from an inexpensive black liquor-derived hydrolyzate. *Biomass Conv. Bioref.*, 8, 679–687, 2018.
56. Abdehagh, N. et al., Adsorptive separation and recovery of biobutanol from ABE model solutions. *Adsorption*, 21, 185–194, 2015.
57. Shetty, D. et al., Calix [4] arene-based porous organic nanosheets. *ACS Appl. Mater. Interfaces*, 10, 17359–17365, 2018.
58. Lin, X. et al., Adsorption of butanol from aqueous solution onto a new type of macroporous adsorption resin: Studies of adsorption isotherms and kinetics simulation. *J. Chem. Tech. Biotech.*, 87, 924931, 2012.
59. Wiehn, M. et al., In situ butanol recovery from *Clostridium acetobutylicum* fermentations by expanded bed adsorption. *Biotech. Progr.*, 30, 6878, 2014.
60. Zhang, H. et al., *Clostridium acetobutylicum* biofilm: Advances in understanding the basis. *Front. Bioeng. Biotechnol.*, 3, 658568, 2021.
61. Babi, D.K. et al., Fundamentals of process intensification: A process systems engineering view, in: *Process Intensification in Chemical Engineering*, J. Segovia-Hernández and A. Bonilla-Petriciolet (Eds.), pp. 7–33, Springer, Cham, 2016.
62. Salleh, M.S.M. et al., Improved biobutanol production in 2-L simultaneous saccharification and fermentation with delayed yeast extract feeding and in-situ Recovery. *Sci. Rep.*, 9, 7443, 2019.
63. Al-Shorgani, N.K.N. et al., Continuous butanol fermentation of dilute acid-pretreated de-oiled rice bran by Clostridium acetobutylicum YM1. *Sci. Rep.*, 9, 4622, 2019.
64. Chacon, M.G. et al., Simultaneous saccharification and lactic acid fermentation of the cellulosic fraction of municipal solid waste using *Bacillus smithii*. *Biotechnol. Lett.*, 43, 667–675, 2021.
65. Hosseini, M., *Advanced bioprocessing for alternative fuels, bio-based chemicals, and bioproducts*, Woodhead Publisher, Elsevier, USA, 2019.
66. Guan, W. et al., Acetone-butanol-ethanol production from Kraft paper mill sludge by simultaneous saccharification and fermentation. *Bioresour. Technol.*, 200, 713–721, 2016.
67. Erdei, B. et al., SSF of steam-pretreated wheat straw with the addition of saccharified or fermented wheat meal in integrated bioethanol production. *Biofuel Bioprod. Biorefin.*, 6, 169, 2013.
68. Hasunuma, T. and Kondo, A., Consolidated bioprocessing and simultaneous saccharification and fermentation of lignocellulose to ethanol with thermotolerant yeast strains. *Process Biochem.*, 47, 1287–1294, 2012.
69. Johnravindar, D. et al., Waste-to-biofuel: Production of biobutanol from sago waste residues. *Environ. Technol.*, 38, 1725–1734, 2017.
70. Huang, J. et al., Production of n-butanol from cassava bagasse hydrolysate by engineered *Clostridium tyrobutyricum* overexpressing adhE2: Kinetics and cost analysis. *Bioresour. Technol.*, 292, 121969, 2019.

71. Li, T. *et al.*, Enhanced direct fermentation of cassava to butanol by Clostridium species strain BOH_3 in cofactor-mediated medium. *Biotechnol. Biofuels*, 12, 166, 2015.
72. Riederer, A. *et al.*, Global gene expression patterns in *Clostridium thermocellum* as determined by microarray analysis of chemostat cultures on cellulose or cellobiose. *Appl. Environ. Microbiol*, 77, 1243–1253, 2011.
73. Ibrahim, M.F. *et al.*, Advanced bioprocessing strategies for biobutanol production from biomass. *Renew. Sustain. Energy Rev.*, 91, 1192–1204, 2018.
74. Vane, L.M., Review: Membrane materials for the removal of water from industrial solvents by pervaporation and vapor permeation. *J. Chem. Technol. Biotechnol.*, 94, 343–365, 2019.
75. Xue, C. *et al.*, Butanol production in acetone-butanol-ethanol fermentation with *in situ* product recovery by adsorption. *Bioresour. Technol.*, 219, 158–168, 2016.
76. Borah, A.J. *et al.*, Mechanistic investigations in biobutanol synthesis via ultrasound-assisted ABE fermentation using mixed feedstock of invasive weeds. *Bioresour. Technol.*, 272, 389397, 2019.
77. Han, M. *et al.*, Bio-butanol sorption performance on novel porous-carbon adsorbents from corncob prepared via hydrothermal carbonization and post-pyrolysis method. *Sci. Rep.*, 7, 11753, 2017.

7
Bioprocess Development and Bioreactor Designs for Biobutanol Production

Vitor Paschoal Guanaes de Campos[1], Johnatt Oliveira[2], Eduardo Dellossso Penteado[1], Anthony Andrey Ramalho Diniz[1], Andrea Komesu[1*] and Yasmin Coelho Pio[1]

[1]Departamento de Ciências do Mar (DCMar), Federal University of São Paulo (UNIFESP), Santos, SP, Brazil
[2]Instituto de Ciências da Saúde, Faculdade de Nutrição, Federal University of Pará (UFPA), Belém, PA, Brazil

---

Abstract

Biobutanol is one of the most environmentally friendly renewable alternative fuels to reduce the energy demand for nonrenewable liquid fuels and the greenhouses gases emissions. This fuel is obtained from the fermentation of carbohydrates, called acetone, butanol, ethanol (ABE) fermentation. Despite this several advantages, the prices of feedstock and butanol toxicity/inhibition of the fermenting microorganism are the main problems associated with this bioprocess. However, there are a few advances in biotechnology and engineering that are being studying to overcame the bottlenecks in the conventional fermentation process. This work will focus on recent trends and challenges in the fermentation production of biobutanol. In addition, new trends in bioreactor design will also be discussed.

Keywords: Biobutanol, bioreactor, fermentation, bioprocess, fuel

7.1 Introduction

Fuel is one of the most important resources used in the developed world, we need it for transportation and electricity production [1]. For a long time the most used energy sources in the world are fossil fuels (approximately 80%

Corresponding author: andrea_komesu@hotmail.com

of the current global demand) [1]. But due to the increase in energy consumption of this type of fuel, in addition to the scarcity of energy resources already used, and climate change worldwide, the study of energy generation from renewable resources is a topic that is on the rise. In addition to issues, such as rising crude oil rates, fuel volatility, and government restrictions on fossil fuels that also raise concerns [2].

In the last two decades, the idea to transform wastes into different biofuels, which can be used in engine applications, have been proposed to mitigating the problems associated with fossil fuels [3]. In this sense, biobutanol becomes a sustainable competitor to surpass the need for fossil fuels [4]. Biobutanol is a promising environmentally friendly fuel that can replace fossil fuels as it has gasoline-equivalent fuel properties [3, 5, 6].

Butanol can be an alternative without engine modification to petroleum-based fuel, once its high energy content, low volatility and lower corrosivity are similar to gasoline and diesel [7, 8]. Among the butanol isomers, n-butanol can be mixed with gasoline, being the most promising, while other isomers are used as additives [4]. Butanol can be blended in greater proportions compared to ethanol, without modification to the engine system [9]. Each engine cycle using butanol releases more energy than gasoline, methanol, and alcohol [1]. Butanol has about 25% more energy than ethanol, it is also less volatile and therefore safer, as it evaporates less than ethanol and gasoline. Because it is less invasive than ethanol, butanol can be supplied with fuel pipes already used [10]. Another advantage of biobutanol is during combustion does not emit sulfur and nitrogen oxides, decreasing harmful gas emissions [11, 12].

Biological methods for butanol production are on the rise because they use residual biomass, offering a more affordable and ecologically viable production of biobutanol [4]. Several new fermentation techniques are emerging, increasingly transforming very accessible residual biomass into biobutanol [13]. In the past, corn, wheat, rice, and sugarcane biomass were used to produce biobutanol through ABE fermentation. However, in addition to the cost of raw materials, there was the conflict in using food to produce fuel, so the focus for biobutanol production was turned to agricultural residues, municipal residues and forest residues [14]. Lignocellulosics are waste sources that can be an alternative for fermentation, but they release toxins and inhibitors limiting the use of this type of biomass, affecting the production of biobutanol by clostridium species [15–17]. Another alternative was the use of algae biomass, which is very abundant in the world, having a much better production capacity than starch-based biomass [18].

7.2 Steps in Biobutanol Production

The biobutanol is obtained from acetone–butanol–ethanol (ABE) fermentation pathway using mainly Clostridium bacteria. This bioprocess occurs in two main phases: acidogenic and solventogenic phases. In the first stage, organic acids (acetic and butyric acid) are formed while in the next one they are converted in acetone, butanol and ethanol [1, 19]. The mesophilic temperature (30–40°C) is the optimal condition, and the initial pH of the fermentation is close to 6.8 to 7 [19].

Clostridia bacteria can use a variety of substrates as a carbon source from monosaccharides to polysaccharides and even glycerol [19]. Moreover, inexpensive substrates, such as agricultural, industrial, and food residues, can also be used. These residues can be hydrolyzed through extracellular enzymes that produce monomeric single chains of sugars, which, then, can be used as carbon sources for the production of butanol by different groups of clostridia [19].

Clostridium sp., which are anaerobic microorganism, assimilate sugars through the phosphotransferase (phosphoenolpyruvate-dependent) system. These microorganisms use different pathway to produce pyruvate according the type of sugar. The Embden–Meyerhof (EMP) pathway is used to metabolize the hexose sugar (glucose, fructose and galactose) producing 2 mol of pyruvate with net production of 2 mol of ATP and NADH from 1 mol of hexose. The pentose phosphate pathway is the via to convert the pentose sugars in pyruvate producing 5 mol of pyruvate, 5 mol of ATP and 5 mol of NADH from 3 mol of pentose [19].

During the acidogenic phase, there is an exponential growth of bacteria and the main products are acids. Glucose is consumed to produce pyruvate, activating the glycolytic pathway. Afterward, the pyruvate is converted into acetyl-CoA by oxidative decarboxylation using pyruvate–ferredoxin oxidoreductase, which is primarily responsible for the formation of acids and solvents [19, 20]. Acetate production takes place using acetyl-CoA as a substrate through two key enzymes: phosphate acetyl transferase and acetate kinase. Butyrate production using acetyl-CoA as a substrate occurs through phosphate butyltransferase and butyrate kinase. Other enzymes are also responsible for the formation of necessary intermediates that aid in the production of solvents. In parallel with acids production, energy molecule (ATP) is generated. The acids produced in this phase will be responsible for the solventogenic shift [19].

The switch to the solventogenic phase occurs as the cells recognize the low pH environment. In this stage, there is no microorganism growth and the previous produced acids are converted to solvents. Acetate and butyrate acids are formed from acetyl CoA and butyryl CoA, respectively, during the acidogenic phase, with the corresponding acyl phosphate as an intermediate. The aldehyde dehydrogenase catalyzes the conversion of butyryl-CoA and acetyl-CoA into acetaldehyde and butyraldehyde. After that, acetaldehyde can be converted to ethanol via acetaldehyde dehydrogenase and ethanol dehydrogenase or it can be converted to acetoacetyl CoA by acetyl CoA acetyltransferase, which is further converted to 3-hydroxybutyryl CoA via dehydrogenase enzymes [20]. The butanol dehydrogenase converts the butyraldehyde butanol [19, 21]. Coenzyme NADH and/or NADPH is necessary to be present in the medium to reach the maximum enzymatic activity, having the best yields in solvents production. At lower pH, the NADH-dependent dehydrogenase is more active than the NADPH-dependent one [19, 21]. The strain and the environmental condition affect the yield of fermentation butanol [1].

The formation of inhibitory compounds is one of the main bottlenecks of this bioprocess. Phenolics compounds, as consequence of lignin degradation; furfural, as result of pentose degradation; and butanol are substance that affect the microorganisms during the ABE fermentation, limiting the butanol yields and increasing the cost of the process [21]. Therefore, new technologies are still needed to help remove these barriers from ABE fermentation.

7.3 Feedstock Selection

The main aspect of ABE fermentation is the feedstock selection. This factor corresponds to about 60% to 70% of the final cost of the product, consequently affecting the economic feasibility of biobutanol production. The feedstock selection can be performed according to several parameters such as availability, cost and cellulose, and hemicellulose content [19, 21]. The biomass is the perfect feedstock, considering such parameters; also it is an environmentally friendly substrate. Moreover, biomass is still a renewable example, and it can generate employment in several sectors, improving rural transport infrastructure, developing the agricultural economy. Furthermore, large biomass production, if monetized enough, can serve as an attraction for farmers, ensuring a continuous supply of biomass and good returns to farmers, boosting agricultural and rural empowerment [19]. Another positive effect is that the compound obtained after the bioprocess can be used as fertilizers for crop cultivation [21]. In the past, food-derived substrates, such as corn and starch, were used to produce biofuels, including biobutanol, causing a conflict between the production of food or fuel, which can make operations unfeasible in the long run [19]. In this sense, the agriculture and municipal waste have been gained the attention, because they are renewable, unexpensive, high available and sustainable substrate for biofuel production [21].

Lignocellulosic biomass had all these characteristics, and it is the most abundant renewable resource for the production of biofuels and can be divided into [21, 22]:

- Waste biomass: Examples of waste biomass are plant residues such as sugarcane bagasse, corn stover and straw dust, as well as residues from paper mills and sawmills.
- Virgin biomass: These are plants, such as trees, shrubs, and grass.
- Energy crops: Biofuels can be produced through so-called energy crops, such as plants, such as *Panicum virgatum*, poplar trees, *Miscanthus giganteus,* and sugar cane.

Several studies used barley straw, switchgrass and wheat straw, spruce chips, corn stover, pineapple peel waste, cauliflower waste, press mud, pea pod waste as substrate in biofuel production *via* ABE fermentation pathway [19]. However, the main obstacle in using lignocellulosic biomass is the need to use pre-treatment methods (acids hydrolysis or enzymatic conversion) to convert the lignocellulose into simple sugar before using them as substrates in ABE fermentation due to its high resistance to degradation and its hydrolytic stability [21, 22].

In recently years, few research have been done using algal biomass, as feedstock for biofuel production. The algae, are photosynthetic aquatic microorganisms, that can grow easily and quickly in variety of environmental conditional (salty, coastal, urban waters) and even in areas unviable for agriculture. The alga biomass can double in 2 to 5 days, comparing to other substrate that are harvest once or twice a year [19]. In this ways, one advantage of using algae as substrate is its high yield in the generation of biomass. Others advantages are that algal biomass is more efficient during hydrolysis, as it does not contain lignin and has low levels of hemicellulose; and it has a higher yield in fermentation and consequently a lower final cost of the production [19]. However, the use of algae for biobutanol production is still in an experimental phase and further studies are needed to confirm its feasibility and efficiency on a larger scale [19].

7.4 Microbial Strain Selection

There are some factors that must be considered when selecting the microbial strains for the production of biobutanol, such as: the type of substrate used, the need for nutrients, butanol tolerance, yield and concentration and antibiotic resistance [21].

Clostridia bacteria are widely used for biobutanol production via ABE fermentation and the *Clostridium acetobutylicum* has being the most studied and manipulated strain [20]. The main reason to be vast used is that Clostridia class can use a wide variety of carbon source as substrate, since simple sugars (pentose and hexose) to complex polysaccharides (waste) [21]. Additionally, these microorganisms can degrade toxic chemicals and produce different isomers (chiral products), which are difficult to produce through chemical synthesis [13]. Regarding nutrient requirements, clostridia need complex sources of nitrogen, such as yeast extract, to ensure their growth and good solvent production [13]. High redox potential is required by Clostridia bacteria to produce butanol and providing additional reducing power results in improvement in butanol yield decreasing [13].

It is also necessary to mention that the strains of butanol-producing bacteria can be classified into wild type and genetically modified species. Of the Clostridial group, only a few species can produce a significant amount of butanol, such as *C. acetobutylicum*, *C. beijerinckii*, *C. saccharobutylicum*, and *C. saccharoperbutylacetonicuma* [13]. To improve the butanol yields and make the microorganisms more resilient to the toxicity of the solvent, genetic engineering creates genetically modified strains [1, 19]. Beside this microorganisms, other strains like *Bacillus subtilis*, *Escherichia coli*, *Pseudomonas* sp., and *Saccharomyces* sp. can be used for biobutanol production, but genetic modification for product biosynthesis is required [21].

7.5 Solvent Toxicity

The main problem of biobutanol production is the toxicity of butanol during fermentation, increasing production and recovery costs, affecting butanol yield. Production above 7.4 g/L of butanol ends up suppressing the growth of Clostridium, consequently ending the fermentation, decreasing the yield of biobutanol [23]. When toxicity or inhibition by butanol occurs, the membrane of the strains becomes fluid due to the hydrophobicity of butanol, killing bacterial cells by rupture of the membrane [1].

A solution found to this problem is the genetic modification that can increase the yield of butanol production using *Clostridia* sp. [24, 25], the insertion, knockout or overexpression of heterogeneous genes is performed, thus improving the quality of the strain [1]. Genetic engineering, in addition to improving the strain, the yield of biobutanol production and reducing its toxicity, brings an improvement by simplifying the entire biobutanol production process on an industrial scale [26] and provides a large amount of information on the cellular and genetic nature of *Clostridia* sp. [27, 28].

Due to the difficulty of using Clostridia strains, it is also possible to use *E. coli* for the production of biobutanol, being very treatable and having fast growth. *S. cerevisiae* is another alternative strain that can be used to mitigate the toxicity problem of butanol with Clostridium, as it has tolerance for high concentrations of butanol [29, 30].

7.6 Fermentation Technologies

A very important aspect, considering the demand forecast and the growth of the biofuels market, is finding the most viable bioreactor for the production of these biofuels [21]. The main method for obtaining organic solvents is anaerobic fermentation, which can be done into batch fermentation, continuous fermentation and two-stage continuous fermentation, with batch fermentation being the most used method [1].

The main factors considered when selecting the best bioreactor design are: product toxicity, which can influence the growth of microorganisms; preparation time; prolonged lag phase; and residence time of the product. From these aspects, one can choose a bioreactor that will result in a better yield or better productivity [21].

- Batch fermentation
  In batch fermentation, the biofermenter is fed with all substrate once and the products are removed only after the reaction is complete [1]. For the biobutanol production, when is performed in a batch form, various biological substrates are supplied to the fermenter. The cell population is always kept constant by being monitored at regular time intervals. Feed and withdrawal rates are varied according to the turbidity of the culture. The main disadvantage of this type of fermentation is the low yield in addition to the loss of substrate and residual medium, use of diluted sugar substrates, low

concentrations of butanol, high energy use for the butanol recovery phase and low effluent flows. Compared to other types of fermentation, it is the method that takes the longest to produce butanol [1]. Generally, batch fermentation is the most suitable method for small-scale operations and is often performed in laboratories to alleviate substrate inhibition [22, 31]. It is a very popular method because of the ease of operation and already having a well-developed technology [19].

- Continuous Fermentation
In this type of fermentation, feeding is carried out continuously and the product is also removed simultaneously, as opposed to batch fermentation. Among the types of fermentation, continuous fermentation results in better yield and productivity of butanol at pH 4.5 [32] although in some cases fed-batch fermentation obtained higher productivity, continuous fermentation showed greater stability for *C. acetobutylicum* [33]. In one study, a production of 6.5 g/L/h of total solvents was reported in continuous fermentation with a butanol–ethanol–acetone ratio of 8:6:1 [34].

In laboratory studies, continuous fermentation is very simple to perform and easy to study. In an economic comparison between types of fermentation, continuous fermentation was considered the best for large-scale butanol production because of its efficiency in relation to the time spent [5]. However, continuous fermentation also has disadvantages, such as the quantity and quality of the product, which are lower compared to the discontinuous one, as there is a dilution of the medium and problems in handling the harvest [35] but these disadvantages can be mitigated with butanol recovery techniques.

Another problem is the stress degeneration that is common in continuous processes [31] resulting in a reduction or even absence in the production of solvents. To mitigate this problem, several reactor configurations have been being studied. In this type of fermentation, cell washing can also become a problem, being solved by recycling cells or using cell immobilization [19].

There are three categories of cell immobilization techniques, namely adsorption, entrapment, and covalent bond formation resulting in significant improvements in butanol productivity, but according to laboratory experiments they demonstrate better results using the adsorption technique. Other advantages of adsorption include greater simplicity and better economic viability employing low-cost residues as support material to achieve cell immobilization [36].

The technique of immobilized bioreactors can prevent the catalytic activity of enzymes and improve cell stability, in addition to reducing shear force [21]. In these reactors, fibrous matrices can be used to improve the efficient of consumed substrates, because they have huge surface area, high permeability and mechanical strength. In addition, they are also inexpensive materials and maintain a low-pressure drop inside the reactor [21]. Stirred tank, packed bed, and fluidized bed reactors can use immobilized cell reactors to enhance the bioprocess [36].

Cell recycling occurs when cells turned back to the bioreactor using a separator unit (filter or clarified) and the clear liquid is removed [37]. In these systems, the reactor starts in batch mode and cell growth is allowed until the end of the exponential phase. Before reaching the stationary phase, the fermentation broth is circulated through the separation system [36].

Factors that affect the productivity of bioreactors include long operating times, restricted mass transfer, accumulation of unused biomass, blockages, short circuit effects. [21].

- Two-stage continuous fermentation
Two or multiple stage continuous fermentation systems can be used to overcome the problems associated with lower concentration of butanol observed in continuous single-stage reactors. In two-stage bioreactors, the inhibitory effect of butanol can be diminished by allowing successive acidogenic and solventogenic fermentations phases in separate reactors [36].

In this two-stage continuous fermentation, two reactors are used: the first one occurs the maximum cell growth rate, and the other one occurs the biobutanol synthesis. As this type of fermentation is divided into two stages, the growth phase does not occur together with the synthetic activity phase of the microorganism. During the stationary and exponential growth phases, the synthesis of metabolites is stopped [1].

7.7 Butanol Separation Techniques

One of the main challenges in biobutanol production is the toxicity of butanol to microorganism, causing inhibition of the final product. Therefore, it is necessary to use an efficient recovery method that continuously withdraw butanol during bioprocess. Finding this method is a challenge due to the high boiling point of butanol (117°C), which can lead to an azeotropic formation and affect the economic aspects of production [19]. Butanol is normally recovered through three methods: distillation, liquid–liquid extraction (LLE) and adsorption. However, distillation requires large amounts of energy and, therefore, it is not economically viable; it is widely applied for butanol recovery [19, 21]. Moreover, the performance of other several techniques, such as pervaporation and solvent extraction, have been studied because they seek to meet characteristics such as long-term stability, high selectivity, fast removal rates, simplicity of operation, and low energy requirements [19, 21].

- Liquid–liquid extraction (LLE)
  The liquid–liquid extraction (LLE) is a conventional method used to extract the target solute from the aqueous phase by contacting one or more water-insoluble organic extractors. Because of butanol has a hydrophobic nature, the LLE has a high potential to be used in the recovery of butanol. In this sense, butanol can be collected in organic phase once it is more soluble in extractor phase than in aqueous phase (fermentation medium) [19, 21].

  This technique can be used in *in situ* or *ex situ* extraction. In the first one, the separation module is coupled into the bioreactor, extracting butanol from the fermentation medium and maintaining the other components (nutrients and cells) in the bioreactor [19]. As the extractor and fermentation do not mix, after the butanol recovery, the extractor can be easily separated [37]. The extractor selection the most crucial parameter, contributing to recovery performance. For a better selection of the extractor, it must be nontoxic to microorganisms, low cost, commercially available and no emulsion formation, and have high selectivity, high distribution coefficient, high stability and low solubility in aqueous solution. [19]. In this regard, decanol and oleyl alcohol are the most common extractor used in *in situ* LLE [13]. The first has the highest distribution coefficient, although it affects the microorganisms [19]. In spite of this, the oleyl alcohol are nontoxic to inocula [19].

The other method of LLE, *ex situ* extraction, occurs at the end of the bioprocess, because the extractor is added in the fermentation broth and, following, there is the product separation [19].

- Gas stripping
  Gas stripping uses the vapor–liquid equilibrium to separate the selective compounds (solvent) from the fermentation medium and, then, the recovery of ABE solvents occurs in a condenser [1, 19]. In this method, bubbles gas is continuously supplied in a column where the fermentation medium is added in a countercurrent [1]. After passed through the condenser, the gas is recirculated to the bioreactor, keeping going the separation process until all substrate is consumed in the fermentation [13, 19]. Even though the gas stripping produce excessive foam and it has low selectivity of butanol, it is one of the separation processes with the highest yields [1, 19]. Moreover, this separation process is simple, easy, versatile and low-cost technique for *in situ* butanol recovery [1]. Gases produced during the ABE fermentation (H_2 and CO_2) can be used as separating agents for butanol recovery, reducing the operational costs [19, 37]. However, the butanol recovery depends on the type of gas used [19]. In the gas stripping project, operational parameters, such as gas flow, bubble size, agitation speed, and product concentration must to be considered for the effectiveness of separation [19].

- Pervaporation
  Pervaporation is separation technique that used the selective membrane and vacuum to remove the volatile toxic compounds during ABE fermentation [1, 13, 19]. In this separation process, the membrane is put into contact with the fermentation broth and the solvents diffused through it as a vapor, and it is subsequently removed by evaporation using a low-pressure vacuum system [1, 13, 19]. Selective sorption, diffusion across the membrane and vapor-phase desorption are the mass transportation mechanism involved in this technique [19].

The key point to the success of pervaporation is the membranes and the vacuum system [1, 19]. Although, the membranes used in pervaporation is cost effective and energy efficient, making the separation process commercially and economically competitive, the membrane fouling and the biocompatibility are one of the biggest obstacles to this technique [19]. In addition, other drawbacks are the high energy demand and the huge cost of vacuum system [1].

Several factors have been studying because they affect the effectiveness of the pervaporation process, including the membrane dilation, concentration polarization, permeate membrane interaction, molecular mass, size, shape, and flux [19].

- Perstration

Perstration use a selective membrane to separate the fermentation broth and extractant side where butanol will be extracting, combining the principles of liquid–liquid extraction (LLE) and pervaporation in unique process [19]. Because of this, problems reported in LLE (toxicity, emulsion formation, solvent loss) are minimized once there is no direct contact between immiscible phases [19, 37]. Other positive aspect of this method is the independence between the fermentation medium flowrate and the extractors, which supports the butanol diffusion through the membrane [19].

The efficient of perstration depends on the rate of butanol diffusion through the membrane [37]. Therefore, the membrane choice is a critical point to the success of this separation technique [19, 37]. In the literature, hydrophilic or hydrophobic membranes are reported using in perstration [19].

- Adsorption

Adsorption uses adsorbent where the biobutanol is adsorbed, then, the temperature is rising or a displacer is used to recovery the solvent [1, 19]. This simple process is one of the most energy efficient, although the yields are lower than other butanol recovery methods [1, 19]. A variety of adsorbent have been used for butanol separation, including silicates, resins and activated carbon which is found the best one [1, 19]. The major drawback of adsorption is the quickly adsorbent fouling, which could be mitigate using an ultrafiltration membrane before the adsorption column [19].

- Hybrid Separation Process
  None of the aforementioned techniques is 100% efficient when operated individually and also there are their advantages and disadvantages, in this sense, the hybrid technology consists of the integration of more than one separation technique. Thus, this technology seeks to focus on improving the butanol yield of each process involved [19]. The energy required of the hybrid systems is higher than others biobutanol recovery techniques, specially compared with distillation [19].

  It is worthy to mention that the recovery and purification processes are directly affected by the characteristics of the operational fermenter conditionals [1, 13, 19]. When metabolic engineered strain is used, biobutanol can be produced in a higher ratio than ethanol, simplifying the purification process. Another way to facilitate recovery processes refers to when a strain's tolerance to butanol is increased by metabolic engineering. Therefore, for the entire operation to be optimized, it is necessary to consider the development of the strain through fermentation to the downstream processes, leading to a reduction in final costs [13].

7.8 Current Status and Economics

Currently, biobutanol has been considered as one of the most potent alternatives to conventional gasoline, with sustainable production through the use of biomass as a raw material [38]. Even though, the biobutanol also proves to be a profitable biofuel that has all the properties of a future alternative fuel, better technology is still needed to make biobutanol a globally used fuel of the future [38].

The main obstacles preventing the economically viable production of butanol include inhibition of the product during fermentation, leading to a low concentration of the final product. To minimize the cost of production, alternatives are considered in the use of feedstocks, such as agricultural residues and other possible substrates [19, 21]. A better understanding of biosynthesis and the biotechnology industry can improve the results of biobutanol production by ensuring a reliable and cost-effective feedstock infrastructure and thus allowing biobutanol to be competitive with chemical synthesis based on fossil resources [13, 21]. Others factors that influence the cost of biobutanol production are: butanol recovery, use of sugar during fermentation, raw material cost, and heat recovery [19].

It is interesting to mention that the economic viability of biobutanol depends on the current market. According to market statistics, it is expected an increment of market demands after biobutanol starts to be used as an alternative liquid fuel [19].

Furthermore, several researches have been carried out to find economical alternative substrates, to develop the metabolic engineering of Clostridial strains, and efficient methods of removing products *in situ* and to develop projects for fermentation bioreactors [19]. Metabolic engineering has a large role and the target genes that will be developed to achieve these goals can be identified by genomic-scale metabolic flux studies and analysis, and for that, a genomic-scale metabolic network of clostridia needs to be built [13].

It can be said that many improvements are still needed for the biobutanol production process to be economically competitive. However, it is likely to become a competitive process through new metabolic engineering tools for clostridia allied to the development of fermentation and downstream processes [13].

7.9 Concluding Remarks

It is possible to conclude that biobutanol has excellent potential to replace gasoline and has the advantage of being a possible product from biomass. Furthermore, obtaining this agent is considered a highly reliable process that requires low investment in machinery. The biomethanol fermentation process is the second most used in the world, and the gases produced during the process can be used for the production and recovery of biobutanol itself. This process is generally characterized as a reliable and highly versatile process with low cost. From a technical point of view, biobutanol has a low vapor pressure, among other characteristics, that allow it to be used as an efficient blend for diesel. From an emissions point of view, it does not emit sulfur and nitrogen oxides, and its properties are similar to diesel but with more significant differences when compared to ethanol and methanol.

References

1. Pugazhendhi, A. *et al.*, Biobutanol as a promising liquid fuel for the future-recent updates and perspectives. *Fuel*, 253, 637–646, 2019.

2. Naik, S.N. *et al.*, Production of first and second generation biofuels: A comprehensive review. *Renewable Sustain. Energy Rev.*, 14, 2, 578–597, 2010.
3. Kothari, R. *et al.*, Waste-to-energy: A way from renewable energy sources to sustainable development. *Renewable Sustain. Energy Rev.*, 14, 9, 3164–3170, 2010.
4. Karthick, C. and Nanthagopal, K., A comprehensive review on ecological approaches of waste to wealth strategies for production of sustainable biobutanol and its suitability in automotive applications. *Energy Convers. Manag.*, 239, 114219, 2021.
5. Ranjan, A. and Moholkar, V.S., Biobutanol: Science, engineering, and economics. *Int. J. Energy Res.*, 36, 3, 277–323, 2012.
6. Thangavelu, S.K. *et al.*, Review on bioethanol as alternative fuel for spark ignition engines. *Renewable Sustain. Energy Rev.*, 56, 820–835, 2016.
7. Rakopoulos, D.C. *et al.*, Effects of butanol–diesel fuel blends on the performance and emissions of a high-speed DI diesel engine. *Energy Convers. Manag.*, 51, 10, 1989–1997, 2010.
8. Jin, C. *et al.*, Progress in the production and application of n-butanol as a biofuel. *Renewable Sustain. Energy Rev.*, 15, 8, 4080–4106, 2011.
9. Kumar, M. and Gayen, K., Developments in biobutanol production: New insights. *Appl. Energy*, 88, 6, 1999–2012, 2011.
10. Shapovalov, O.I. and Ashkinazi, L.A., Biobutanol: Biofuel of second generation. *Russ. J. Appl. Chem.*, 81, 12, 2232–2236, 2008.
11. Antoni, D. *et al.*, Biofuels from microbes. *Appl. Microbiol. Biotechnol.*, 77, 23–35, 2007.
12. Dürre, P., Fermentative butanol production: Bulk chemical and biofuel. *Ann. N. Y. Acad. Sci.*, 1125, 1, 353–362, 2008.
13. Lee, S.Y. *et al.*, Fermentative butanol production by Clostridia. *Biotechnol. Bioeng.*, 101, 2, 209–228, 2008.
14. Huzir, N.M. *et al.*, Agro-industrial waste to biobutanol production: Eco-friendly biofuels for next generation. *Renewable Sustain. Energy Rev.*, 94, 476–485, 2018.
15. Jang, Y.S. *et al.*, Butanol production from renewable biomass: Rediscovery of metabolic pathways and metabolic engineering. *Biotechnol. J.*, 7, 2, 186–198, 2012.
16. Jönsson, L.J. *et al.*, Bioconversion of lignocellulose: Inhibitors and detoxification. *Biotechnol.*, 6, 1, 1–10, 2013.
17. Lu, C. *et al.*, Fed-batch fermentation for n-butanol production from cassava bagasse hydrolysate in a fibrous bed bioreactor with continuous gas stripping. *Bioresour. Technol.*, 104, 380–387, 2012.
18. Veza, I. *et al.*, Recent advances in butanol production by acetone-butanol-ethanol (ABE) fermentation. *Biomass Bioenergy*, 144, 105919, 2021.
19. Bankar, S.B. *et al.*, Biobutanol: Research breakthrough for its commercial interest, in: *Liquid Biofuel Production*, pp. 237–283, Scrivener Publishing LLC, Beverly, MA, 2019.

20. Groeger, C. et al., 4. Introduction to bioconversion and downstream processing: Principles and process examples, in: *Biorefineries*, C. Groeger, W. Sabra, A.-P. Zeng (Eds.), pp. 81–108, De Gruyter, Berlin/Boston, 2015.
21. Bharathiraja, B. et al., Biobutanol versus bioethanol in acetone–butanol–Ethanol technology—A chemical and economical overview, in: *Second and Third Generation of Feedstocks*, A. Basile and F. Dalena (Eds.), pp. 83–99, Elsevier, Amsterdam, Netherlands 2019.
22. Bankar, S.B. et al., Biobutanol: The outlook of an academic and industrialist. *Rsc Adv.*, 3, 47, 24734–24757, 2013.
23. Chua, T.K. et al., Characterization of a butanol–acetone-producing Clostridium strain and identification of its solventogenic genes. *Bioresour. Technol.*, 135, 372–378, 2013.
24. Bankar, S.B. et al., Genetic engineering of Clostridium acetobutylicum to enhance isopropanol-butanol-ethanol production with an integrated DNA-technology approach. *Renew. Energy*, 83, 1076–1083, 2015.
25. Luo, H. et al., Recent advances and strategies in process and strain engineering for the production of butyric acid by microbial fermentation. *Bioresour. Technol.*, 253, 343–354, 2018.
26. Rubin, E.M., Genomics of cellulosic biofuels. *Nature*, 454, 7206, 841–845, 2008.
27. Minton, N.P. et al., A roadmap for gene system development in Clostridium. *Anaerobe*, 41, 104–112, 2016.
28. Nolling, J. et al., Genome sequence and comparative analysis of the solvent-producing bacterium Clostridium acetobutylicum. *J. Bacteriol.*, 183, 16, 4823–4838, 2001.
29. Dusséaux, S. et al., Metabolic engineering of Clostridium acetobutylicum ATCC 824 for the high-yield production of a biofuel composed of an isopropanol/butanol/ethanol mixture. *Metab. Eng.*, 18, 1–8, 2013.
30. Lu, C. et al., Enhanced robustness in acetone-butanol-ethanol fermentation with engineered Clostridium beijerinckii overexpressing adhE2 and ctfAB. *Bioresour. Technol.*, 243, 1000–1008, 2017.
31. Xue, C. et al., Prospective and development of butanol as an advanced biofuel. *Biotechnol. Adv.*, 31, 8, 1575–1584, 2013.
32. Li, S.Y. et al., Performance of batch, fed-batch, and continuous A-B-E fermentation with pH-control. *Bioresour. Technol.*, 102, 5, 4241–4250, 2011.
33. Lipovsky, J. et al., Butanol production by Clostridium pasteurianum NRRL B-598 in continuous culture compared to batch and fed-batch systems. *Fuel Process. Technol.*, 144, 139–144, 2016.
34. Pierrot, P. et al., Continuous acetone-butanol fermentation with high productivity by cell ultrafiltration and recycling. *Biotechnol. Lett.*, 8, 253–256, 1986.
35. Klutz, S. et al., Developing the biofacility of the future based on continuous processing and single-use technology. *J. Biotech.*, 213, 120–130, 2015.

36. Mariano, A.P. et al., Butanol production by fermentation: Efficient bioreactors, in: *Commercializing Biobased Products*, S.W. Snyder (Ed.), pp. 48–70, Royal Society of Chemistry, Cambridge, UK, 2015.
37. Ezeji, T.C. et al., Bioproduction of butanol from biomass: From genes to bioreactors. *Curr. Opin. Biotechnol.*, 18, 3, 220–227, 2007.
38. Mahapatra, M.K. and Kumar, A., A short review on biobutanol, a second generation biofuel production from lignocellulosic biomass. *J. Clean Energy Technol.*, 5, 1, 27–30, 2017.

8
Advances in Microbial Metabolic Engineering for Increased Biobutanol Production

Mansi Sharma[1], Pragati Chauhan[1], Rekha Sharma[1]* and Dinesh Kumar[2]*

[1]Department of Chemistry, Banasthali Vidyapith, Rajasthan, India
[2]School of Chemical Sciences, Central University of Gujarat, Gandhinagar, India

Abstract

It is important to promote sustainability with a low carbon footprint through a circular economy built on biobased products in this changing world. Butanol has been acquiring attention in recent decades as a liquid biofuel because of its superior characteristics compared to ethanol. Several studies have produced butanol from raw renewable biomass, like lignocellulosic materials. While it is possible to ferment butanol, the concentration, low yield of butanol production, and the high cost of raw materials limit its application. A decrease in butanol formation is also attributed to these limitations. Microbially produced biobutanol is a biofuel composed of four carbon hydrocarbons derived from biomass feedstock. It is used for making cosmetics, pharmaceuticals, textiles, fuel additives, and plastics. Metabolic engineering of microbes has been used to overcome the limitations of conventional processes, including clostridial species, cyanobacteria, and yeast, and to produce more butanol cost-effectively. This chapter discusses the upcoming perspective, critical issues, applicability, and challenges in the evolving field. This chapter also discusses how metabolic engineering can improve butanol production with microorganisms.

Keywords: Biobutanol, biofuels, clostridial, microbial engineering, microorganism

---

*Corresponding author: sharma20rekha@gmail.com
†Corresponding author: dsbchoudhary2002@gmail.com

8.1 Introduction

Metabolic engineering is the design of efficient cell factories that produce a variety of fuels, chemicals, pharmaceuticals, and food ingredients. Food and beverages brewed through microbial fermentation have been consumed since ancient times. Citric acid was the first chemical compound made by microbial fermentation around 1920, and this was the first process employed on a large scale in industrial production [1]. Louis Pasteur described how microorganisms produced butanol under anaerobic conditions in 1861, which marked the beginning of butanol's history [2]. A hydrocarbon compound, butanol, refers to an organic carbon compound with the formula C_4H_9OH. It is also known as biobutanol and butyl alcohol. Besides 1-butanol and 2-butanol, two other isomers of butanol are available, including isobutanol and tertbutyl alcohol [3].

Several chemicals, solvents, and even fuels require butanol for conversion. Biobutanol production is achieved using anaerobic and aerobic bacteria and can be done from algal biomass, inulin, cassava starch, cane molasses, cheese whey, corn stover, palm oil mill effluent, and sugarcane bagasse. Different types of processes for butanol production like chemical and biological are represented in Figure 8.1.

The sustainability of biomass feedstock for biofuel production is urgently needed because of the increase in food crop demand, which has led to higher food prices and competition for land. The extremely favourable feedstock to overcome these crises is lignocellulosic biomass, which has 30% to 55% cellulose, 25% to 55% hemicellulose, and 10% to 35% lignin. Because lignocellulosic biomass comprises complex materials, including lignin, hemicellulose, and cellulose, several steps must be taken to turn these materials into simpler, more readily fermentable sugars like pretreatment. The most effective pre-treatment techniques include enzymatic, alkali, and acid treatments. However, these treatments cause the release of undesirable inhibitory compounds that prevent acetone–butanol–ethanol (ABE) fermentation, like formic acid, acetic acid, levulinic acid, etc. ABE fermentation is fermenting ethanol, butanol, and acetone together [4–10]. Biobutanol is composed of isobutanol and 1-butanol, which are derived from microorganisms. Vibrion butyrique, presumably a mixed culture containing clostridia species, was detected by the French microbiologist to be high in butanol. ABE fermentation is a method of turning carbohydrates into solvents. It was mainly due to acetone's high demand for cordite in World War I that industrial curiosity in ABE fermentation arose during this time. A strain of plant known as *Clostridium acetobutylicum*

Microbial Metabolic Engineering for Biobutanol Production 211

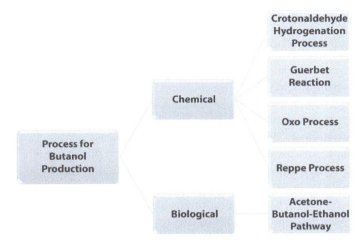

Figure 8.1 The schematic diagram represents the various processes for butanol production.

was isolated in 1922 by Chaim Weizmann, capable of producing ethanol, acetone, and butanol from sugars or starch [11–13].

Over the past decade, a significant contributor to global warming has been the emissions of greenhouse gases (GHGs) from fossil fuels. It is possible to reduce greenhouse gas, improve environmental quality and reduce emissions using biobutanol instead of fossil fuels. Ecofriendly renewable resources are crucial to solving fuel shortages, environmental problems, and climate change. Developing nations will rely on renewable energy and sustainability to meet their energy requirements and preserve the environment for future generations. Renewable energy sources, such as butanol are highly promising. Its unique properties are its high flammability, high energy content, low volatility, and explosive nature, so it differs from ethanol in these respects. There is little hygroscopic activity in butanol, as compared with other biofuels, and it is blendable with existing infrastructure and used at any concentration with gasoline. However, because of its high-level power content and insignificant particle emission, butanol can be an effective liquid transportation biofuel and a biofuel for existing vehicles [14–18].

Butanol has a low vapor pressure, so it can be stored and handled safely. Producing microbial butanol is challenging for many reasons, like fermentation microbes' sensitivity to the end target product, synthesis of by-products, low efficiency of butanol, high downstream processes to recover the product, and lignocellulose's stubborn nature [19, 20].

8.2 Metabolic Engineering

Metabolism engineering has improved the phenotypes of microorganisms in recent years. By manipulating the genetics and metabolism of microorganisms, metabolic engineering enables us to increase the synthesis of the desired metabolite by deregulating cellular metabolism. It is used effectively in several industries, including fuels, medicines, and fine chemicals. It designs native or completely new metabolic pathways for cells [21, 22].

This platform can produce biofuels by altering the metabolic pathway more efficiently. Approximately one thousand enzyme-catalyzed reactions and a selective membrane transport system enable living cells to perform their metabolic functions. Evolutionary metabolic networks in nature have not been heritably optimized for useful application. A genetic modification of cells can enhance metabolic processes. A metabolic engineer manipulates cellular regulatory, enzymatic, and transport functions to improve cellular functions through a recombinant DNA approach. The main difference between conventional genetic and metabolic engineering methods is incorporating heterologous genes and regulatory elements [23, 24].

A diverse range of chemicals can be made from simple, inexpensive starting materials through metabolic engineering. Engineering is a way to manufacture natural products or genetically intractable microbes to produce enzymes or entire metabolic pathways specific to the product in question. Biomass is an inexpensive as well as an easily accessible source of carbon that is used to produce transportation fuels via engineered metabolism. The production of alcohol has been achieved with the help of microorganisms. The production of alcohol has been reported to be possible using engineered microbes such as yeasts and bacteria.

Alkanes, esters, and alcohols can be produced through engineering microbial biosynthetic pathways. Biofuels are being produced in bioreactors with the help of advances in industrial fermentation processes. The many metabolic engineering strategies for producing advanced biofuels have been covered in this section [25–27].

8.2.1 n-Butanol

Advanced biofuel n-butanol, which can be combined with gasoline in any ratio. It contains an equal amount of energy as gasoline. The keto-acid route for the production of n-butanol represents in Figure 8.2. The keto-acid pathway produces n-butanol by reducing acetoacetyl-CoA to carbonyl-CoA and dehydrating it. By reducing crotonyl-CoA further,

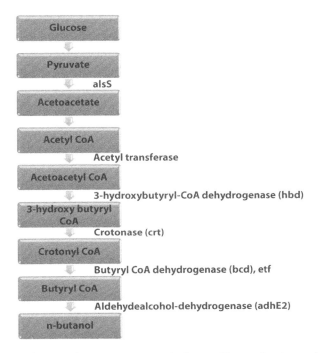

Figure 8.2 The schematic diagram represents the keto-acid route for the production of n-butanol.

butyryl-CoA is formed, and finally, alcohol dehydrogenase converts this into n-butanol [28, 29].

Acetone–butanol–ethanol (ABE) ferment via *Clostridium acetobutylicum* naturally produces high levels of n-butanol (20 g/L). While it has a slow growth rate and sporulation cycle, this is not an ideal bacterium for industrial production.

The modified cells of *Escherichia coli* have formed n-butanol through metabolic engineering attempts, such as reversing the β-oxidation pathway or inserting the n-butanol biosynthesis pathway from *Clostridium acetobutylicum*. To improve the NADH pool or increase the flow from acetyl-CoA to n-butanol, trans-enoyl-CoA reductase (ter) was used in an irreversible reaction to improve the NADH pool and produce 30 g/L of n-butanol in *Escherichia coli* after introducing the non-native n-butanol pathway. Increasing the flow through the glycerol pathway enhanced the amount of n-butanol produced in *Escherichia coli* by enhancing NADH levels. A metabolism study in *Escherichia coli* found that alcohol dehydrogenase (adhE2) can overcome CoA imbalance, increasing the production of n-butanol to 18.3 g/L [30–34].

N-butanol in trace quantity is naturally produced through *Saccharomyces cerevisiae* via the keto-acid pathway. *Streptomyces collinus'* crt and *Clostridium beijerinckii's* adhE2, crt, and hbd were heterologously formulated to improve the titer of n-butanol in *Saccharomyces cerevisiae*, resulting in 2.5 mg/L of n-butanol. Additional work raised the concentration to 6.6 mg/L by expressing trans-enoyl-CoA reductase (ter) from *Treponema denticola* in *Saccharomyces cerevisiae*.

Alcohol dehydrogenase genes (adh1-5) were removed in *Saccharomyces cerevisiae* to enhance the CoA level in the host, resulting in a titer of n-butanol of 70 mg/L. This strain was modified to create 235 mg/L n-butanol under aerobic conditions when bacterial water-forming NADH oxidase (nox) was produced. Because of *Saccharomyces cerevisiae's* large secretion of amino acids, which decreased "acid-crash" and encouraged *Clostridium acetobutylicum* to produce more butanol, coculturing the two organisms resulted in a rise in n-butanol concentrations to 16.3 g/L [35–37].

The biorefinery methods developed for ethanol are currently being used by industries like Green Biologics and Butamax to produce n-butanol commercially. Hosts still need to be further optimized for cost-effective butanol synthesis on a big scale [38].

8.2.2 Isobutanol

Isobutanol is a chemical that becomes more significant as a biofuel because it possesses features that make it less corrosive and more miscible in water than ethanol. Iso-butanol is produced using the same metabolic pathway as n-butanol. The keto-acid route for the production of isobutanol represents in Figure 8.3. Isobutanol synthesis in *Escherichia coli*, the isobutanol pathway genes alsS, kivD and adh2 (from *Bacillus subtilis*, *Lactobacillus lactis*, *Saccharomyces cerevisiae*) were overexpressed, directing the alteration of isobutanol from keto-acids, and yielding a final yield of 22 g/L isobutanol. Higher quantities of isobutanol were discovered to be harmful to the host [39, 40].

A different strain was created to withstand greater isobutanol concentrations by altering the CRP (cAMP receptor protein), producing 12 g/L isobutanol without affecting cell viability. Recently, the Entner–Doudoroff route in *Escherichia coli* was altered combined with losses in the EM route (Embden–Meyerhof route) to drive C (carbon) towards the isobutanol path giving 15 g/L of isobutanol. Increase the synthesis of isobutanol, another microbial genus, including *Enterobacter aerogenes* and *Klebsiella pneumoniae*, has also

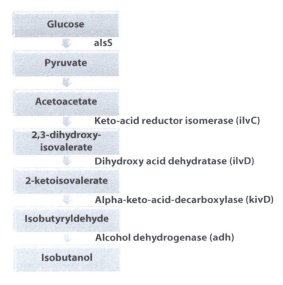

Figure 8.3 The schematic diagram represents the keto-acid route for the production of isobutanol.

been altered. Turning off the genes linked to the lactate route *Klebsiella pneumoniae*'s dormant isobutanol pathway, yielding 2.45 g/L [41, 42].

Additionally, the formate and nitrate routes were employed to produce isobutanol using the bacteria with an anoxic route like *Enterobacter aerogenes*. As a by-product of the valine and isoleucine synthesis routes in *Saccharomyces cerevisiae*, isobutanol is generated in a very small proportion (Ehrlich pathway). The competing threonine route was often deleted from *Saccharomyces cerevisiae* during engineering, and/or the valine pathway gene was overexpressed in its place. Other changes included changing the genetic factor in pyruvate shunts and ED pathways [43, 44].

Despite having high fuel qualities, isobutanol is expensive to incorporate into fuel. Further engineering and cost optimization are required to synthesize isobutanol for economically sustainable use as a substitute fuel. Industries, including Gevo, Butamax (the United States), and BASF SE (Germany), already make butanol to commercialize this chemical as a gasoline mix [45].

8.3 Microorganisms for Butanol Production

Distinct species of Clostridia naturally generate butanol, but many of these methods have poor yields and concentration, which has sparked much interest in creating novel cell factories that can be used for butanol synthesis on a biobased basis. The two main methods for manufacturing biobutanol are:

1. using a host with a natural route for butanol and enhancing its efficiency and yield.
2. reconstructing an effective butanol path in previous widely studied strains employed for commercial production of chemicals/other fuels.

The naturally occurring hosts that create butanol include yeast via the Ehrlich route. Because of its great ethanol acceptance and resistance to difficult manufacturing conditions, like strong osmotic stress and low pH, yeast is frequently used to manufacture industrial ethanol.

Different types of microbial species and substrates used for the production of butanol represent in Table 8.1. The most common species of *Clostridium* utilized to produce n-butanol are *Clostridium saccharoperbutylacetonicum, Clostridium saccharobutylicum, Clostridium acetobutylicum*, as well as *Clostridium beijerinckii*. However, because of its genetic complexity and stringent anaerobicity, this species poses issues that directly affect industrial production.

These issues include slow growth, by-products, spores' development, phage infection, or low tolerance to n-butanol. These factors led to the genetic modification of other species more frequently used in industrial settings, like *Saccharomyces cerevisiae* and *Escherichia coli*, to yield n-butanol. The primary microorganism responsible for manufacturing I-generation ethanol in North America and Brazil is the yeast *Saccharomyces cerevisiae*, employed extensively in the food sector and to produce fuel. Researchers have also looked at *Saccharomyces cerevisiae*'s ability to produce n-butanol because of its adaptation to industrial environments and the variety of hereditary tools accessible to this organism. However, the low tolerance of microbes to metabolic products impacts the expenses associated with recovering the product and prevents the manufacturing of these metabolites at an industrial scale. The microbe is necessary to survive the build-up of metabolic results in the medium once they occur, such as alcohols and acids, or else growth can slow down and eventually reach cell death. The maximum amount of n-butanol that *Saccharomyces cerevisiae* can tolerate in the medium, along with some *Clostridium* strains, is 2% (v/v). Therefore, the study of genetic engineering or evolutionary engineering to improve butanol tolerance in *Saccharomyces cerevisiae* strains has been conducted. Along with genetic advancement and the creation of strains able to produce

Table 8.1 Microbial species and substrate used for the production of butanol.

S. no.	Substrate	Species	Concentration g/L	Reference
1.	Glucose	Clostridium acetobutylicum	10.4	[46]
2.	Glucose	Clostridium aurantibutyricum	3.36	[47]
3.	Glucose	Clostridium saccharoperbutylacetonicum	16.2	[48]
4.	Glucose	Clostridium tyrobutyricum	10.0	[49]
5.	Glucose	Escherichia coli	2.0	[50]
6.	Galactose	Saccharomyces cerevisiae	0.0025	[51]
7.	Glucose	Clostridium beijerinckii	15.21	[52]
8.	Glucose	Clostridium saccharoperbutylicum	9.7	[53]
9.	Glycerol	Pseudomonas putida	0.12	[54]
10.	Crystalline cellulose	Clostridium cellulovorans	1.42	[55]
11.	Glucose	Clostridium cadaveris	0.829	[56]
12.	Rice straw	Clostridium sporogenes BE01	5.52	[57]

and tolerate greater butanol concentrations, tools and procedures have already been created.

Two methods have been used to study the generation of butanol by *Saccharomyces cerevisiae:* the amino acid absorption pathway and heterological representation of the *Clostridium* process. Because of its poor yield and productivity, particularly when compared to metabolically altered *E. coli*, yeast does not appear to be an appealing host for manufacturing biobutanol, according to several recent evaluations [58, 59].

8.3.1 The *Clostridium* Species

By using two different fermentation processes, known as acidogenesis and solventogenesis, *Clostridium* species produce butanol. The pH drops to below 5 during the acidification (acidogenic) stage because cells multiply and transform the substrate into butyric or acetic acids. Acetone, butanol, and ethanol are created in the solventogenesis process from organic acids in a 3:6:1 ratio. The unavoidable by-products of a traditional ABE fermentation are acetone and ethanol. Metabolic engineering of clostridia has been tried to increase butanol's yield and lower the cost of eliminating undesirable by-products. To do this, integrative plasmids have mostly been used to deactivate the key genes in the biosynthesis pathways of undesirable by-products [60].

The toxicity of n-butanol has also led to efforts intended to create stronger strains by mutation or metabolic engineering. The derivative *Clostridium acetobutylicum M5*, a mutation of *Clostridium acetobutylicum* ATCC 824, lacks the mega-plasmid *pSOL1*, preventing it from producing butanol. The butanol-producing capability of this bacterium was later enhanced by genetic engineering. In the presence of the *ptb* promoter, overexpression of adhE1 led to a production of 0.84 g butanol/g ABE when adhE1 was overexpressed. Compared to the wildtype, this was much higher than the amount of butanol per gram of ABE produced by wildtype clostridia. An alternative strategy to increasing production entailed disrupting the *adc* gene in *Clostridium acetobutylicum*. As a result, 0.82 grams of butanol per kilogram of ABE was selectively synthesized in a medium supplemented with methyl viologen [61].

According to metabolically altered *Clostridium acetobutylicum* PJC4BK produced by Harris *et al.*, butanol concentrations are as high as 16.7 g/L, up from 11.7 g/L for wildtype *Clostridium acetobutylicum*. An enzyme (butyrate kinase) involved in the butyrate synthesis route is encoded by the *buk* gene, which was disrupted to produce the PJC4BK. Other techniques, such as improving a microorganism's product tolerance by creating mutants, can also raise the final butanol titer. Improve the tolerance to butanol, solventogenic enzyme activity and ultimately final butanol concentration, for example, overexpressed the molecular chaperone GroESL. Recently, significant developments in clostridial 1-butanol synthesis have been examined [62, 63].

8.3.2 *Escherichia coli* Species

Compared to other recognized aerobic bacteria, solventogenic clostridia is stringent anaerobes and develops slowly. This negatively impacts the economics of producing butanol via traditional ABE fermentation. It has been challenging to manipulate *Clostridium* species. genetically, which has diverted attention to designing the clostridial fermentation path inside bacteria like *Bacillus subtilis* and *Escherichia coli* for the generation of butanol.

The *Clostridium acetobutylicum* genes thiolase (thl), 3-hydroxybutyryl-CoA dehydrogenase (hbd), crotonase (crt), butyryl-CoA dehydrogenase (bcd), acetoacetyl-CoA: acetate/butyrate: CoA transferase (ctfAB), and aldehyde/alcohol dehydrogenase (adhE2) were introduced into *Escherichia coli* to recreate a functioning butanol biosynthetic pathway. This made it possible to achieve a butanol titer of 139 mg/L under anaerobic circumstances. A metabolically modified *Escherichia coli* carrying the thiolase (thl), 3-hydroxybutyryl-CoA dehydrogenase (hbd), crotonase (crt), butyryl-CoA dehydrogenase (bcd), acetoacetyl-CoA: acetate/butyrate: CoA transferase (ctfAB), and aldehyde/alcohol dehydrogenase (adhE2) genes of *Clostridium acetobutylicum* generated 1.2 g/L of butanol from 40 g/L of glucose in another research. Despite these advances, a significant obstacle to further advancements in butanol synthesis still exists as a lack of adequate knowledge of the molecular mechanisms governing the metabolic switch from solvent to acid formation. Uncertainty exists regarding the important triggering signals, the regulators and their interactions, and the links between the regulatory networks [64–66].

8.3.3 Other Bacteria

The amino acid biosynthesis pathway's metabolic intermediates, 2-keto-acids, have also been employed to make higher alcohols by other bacteria, such as *Corynebacterium glutamicum* and *Ralstonia eutropha*. It was discovered that a *Corynebacterium glutamicum* with increased capacity for converting glucose to isobutanol had been created by overexpressing the ilvBNCD genes and deleting the pqo (pyruvate: quinone oxidoreductase), ilvE (transaminase B), and aceE (pyruvate dehydrogenase subunit E1), genes, respectively. It has been demonstrated that an engineered strain of *Ralstonia eutropha* may use an electro microbial method to transform CO_2 into higher alcohols.

An electrical current was used in a bioreactor specifically made for this operation. The modified bacteria could create 50 mg/L of 3-methyl-1-butanol and almost 19 mg/L of isobutanol using just CO_2. If there is a net energy gain, autolithotrophic microorganisms, such as *Ralstonia eutropha*, obtain their electrons from electricity rather than hydrogen. It is useful in producing biofuels [67–69].

8.3.4 Biochemistry and Physiology

The heterofermentative anaerobic bacteria known as clostridia that produce solvent have two metabolic phases: acidogenesis and solventogenesis. The clostridia exponentially proliferates throughout the acidogenesis phase, producing acetate, butyrate, hydrogen, and carbon dioxide. Pyruvate is created during the glycolysis cycle and is subsequently transformed into acetyl-CoA and CO_2 via the pyruvate: ferredoxin oxidoreductase. A group of enzymes, including 3-hydroxybutyryl-CoA dehydrogenase, thiolase, butyryl-CoA dehydrogenase, and crotonase, work together to create butyryl-CoA from two molecules of acetyl-CoA.

The metabolic route of the clostridia includes important intermediates like butyryl-CoA or acetyl-CoA, which, in the acidogenesis or solventogenesis processes, respectively, are converted into acids and solvents. The cell cycle of solvent forming clostridia represents in Figure 8.4. ATP and

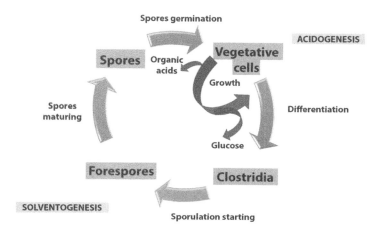

Figure 8.4 The schematic diagram represents the cell cycle of solvent forming clostridia.

NADH are produced during the glycolysis pathway's oxidation of glucose to pyruvate. However, for biosynthesis and growth, more ATP is needed. The glycolysis process should also proceed with the reoxidation of NADH to NAD1. The most helpful branch employed for ATP generation is producing acetic acid, while the NADH is oxidized using butyric acid.

In addition, clostridia have a successful method for getting rid of extra electrons/protons by utilizing the last enzyme, hydrogenase. Thus, protons can be used by the cell as terminal electron acceptors, resulting in the creation of hydrogen. The phase known as solventogenesis occurs after exponential development when the cell switches the carbon flow from routes that produce acids to those that produce solvents. Numerous factors, such as temperature, oxygen, nutritional limitation, internal and external pH, and acid end products, impact solventogenesis.

According to certain reports, undissociated butyric acid has a significant role in controlling the synthesis of solvents, and the relationship between pH and this component may be seen. Several processes, such as cell differentiation and the reutilization of acids, are connected to solventogenesis. Butyryl-CoA and Acetyl-CoA become crucial players by switching to solventogenesis, where two sets of dehydrogenases convert intermediates produced from carbon sources and butanol/ethanol from reassimilated acids. DNA microarray investigation of *Clostridium beijerinckii* NCIMB 8052 and *Clostridium acetobutylicum* ATCC 824 has looked at the changes in gene expression at the genome level that occurs when the process of solventogenesis changes from acidogenesis. The detoxifying procedure used by clostridia in reaction to the accumulating toxic substances in creating solvent by the reassimilation of acids. The by-products of solventogenesis, particularly butanol, are equally harmful to cells. Butanol enhances membrane fluidity and ruptures H-bonds between lipid tails in the plasma membrane after reaching a threshold concentration. This results in the loss of membrane ability. As a result, acetone–butanol–ethanol fermentation with the clostridia might produce a low butanol titer.

Several butanol-tolerant microorganisms were studied, and they concluded that their butanol tolerance was likely caused by increased thickness of the extracellular capsule, fatty acid compositions in membranes, and adaption through unidentified procedures [70–72].

8.4 Metabolic Engineering of Clostridia

Genetic modification of the clostridial hosts was used to improve butanol synthesis. It was difficult to produce large amounts of butanol because butanol is poisonous to bacteria. Several *in situ* product removal techniques were successfully used to solve this issue. New heterologous hosts were also investigated in place of obtaining high yield using Clostridial species.

8.4.1 Genetic Tools for Clostridial Metabolic Engineering

Clostridial metabolic engineering requires highly effective genomic tools. Considering that the gene for solvent production is overexpressed in *Clostridium acetobutylicum*, due to a DNA constraint in *Clostridium acetobutylicum*, the overexpression of solvent-generating genetic material was achieved using the shuttle vector pFNK1. It does not cleave in *Clostridium acetobutylicum* ATCC 824 because that organism cannot identify the sequence 5-GCNGC-3. The absence of effective ways for selectively knocking off genes further hinders the metabolic engineering of clostridia. There have only been four single integrations of replication-deficient plasmids used to knockout WVE genes in *Clostridium acetobutylicum* (buk, pta, adhE, solR, and spoOA). The mobile II introns of the Lactococcus lactis' ltrB gene (Ll. ltrB) has been utilized to develop a gene knockout system for Clostridium.

A significant role is played by several reporter genes from *Clostridium acetobutylicum*, including a galZ (5-galactosidase), a lucB (luciferase), and an eglA (1,4-endoglucanase) gene from *Thermoanaerobacterium thermosulfurigenes*, *Photinus pyralis*, *Clostridium acetobutylicum* P262 romoters in determining transcription levels. The promoters of certain Wc genes must be studied using a reporter gene system. With the help of these genetic tools, it is possible to engineer clostridia to produce more solvents, produce butanol with higher selectivity, and tolerate the solvent better [73, 74].

8.4.2 Optimum Selectivity Techniques for Butanol Production

The recovery of the product will be significantly simpler with selective butanol synthesis. Improved butanol synthesis selectivity was achieved using an antisense RNA (asRNA) method. On the other hand, *Clostridium acetobutylicum* showed decreased acetone and butanol titers upon downregulating ctfB. The entire aad–ctfA–ctfB gene was rapidly degraded during a sRNA treatment against ctfB. The butanol/acetone ratio was therefore raised by using a sRNA against the ctfB approach in conjunction with aad overexpression [75].

A comparison between the control and engineered strain showed that the ratio of butanol/acetone in the engineered strain was over twofold greater. The engineered strain exhibited a lower butanol concentration than the wild type, despite increased selectivity for butanol by modifying clostridia metabolically. This is most likely because CoA transferase is necessary to convert butyrate into butanol. Butyrate/acetoacetyl-CoA and butanoyl-CoA/acetoacetate undergo a reversible process catalysed by CoA transferase. CtfB downregulation inhibits not only acetoacetate synthesis but also butyryl-CoA synthesis, which is a precursor for butanol and acetone.

Acetone was reduced by inhibiting the expression of adc gene by antisense RNA. The level of acetone was not decreased by effective downregulation of the adc gene expression, though. As a result, the enzyme that limited the rate was coenzyme A-transferase (CoAT) in the C. acetobutylicum acetone production pathway instead of AADC (acetoacetate decarboxylase). The efficacy of an asRNA method for the metabolic engineering of *Clostridium acetobutylicum* was also investigated. The metabolic route in *Clostridium acetobutylicum* represents in Figure 8.5.

A polycistronic Aad–CtfA–CtfB message was chosen instead of ptb and buk. There is a connection between butyrate formation (phosphotransbutyrylase) PTB, and (butyrate kinase) BK genes, respectively. As predicted by the authors, the buk-asRNA strain produced 34% more butyrate than the control strain, but only 35% more butanol than the ptb-asRNA strain, while the control strain formed eight point four grams per litre of butanol

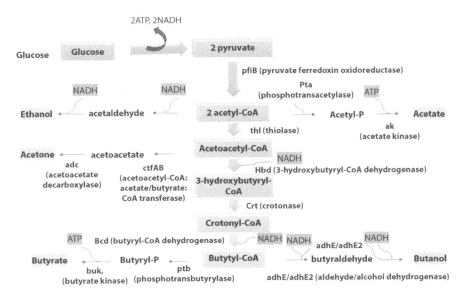

Figure 8.5 The schematic diagram represents the metabolic route in *Clostridium acetobutylicum*.

compared with 8.1 g/L of buk-asRNA. The acetate/butyrate pathways' importance for clostridia's energy metabolism may help explain these outcomes.

In clostridia, ATP is provided to microbial cells by the formation of butyrate or acetate. When the butyrate pathway is blocked, bacteria cannot obtain sufficient ATP for cell growth and biosynthesis. Clostridia likely do not prefer butanol, and its creation is done to relieve the stress brought on by low pH levels. This mutant has a deficient activity for butyraldehyde dehydrogenases, adenosine deaminase, and coenzyme A, so it cannot produce butanol or acetone. It restored butanol production and did not produce acetone by overexpressing the aad gene. Acetate and ethanol levels decreased when the thl gene was overexpressed, while butanol levels decreased.

A lot of NADH is necessary to ferment with clostridia to produce more butanol and less acetone, and the formation of solvent is directly associated with the amount of NADH present. The clostridial metabolic route is catalysed by hydrogenase that takes in H_2. The H_2 is oxidized, and by H_2-evolving hydrogenase in which H_2 is evolved, with NADH being consumed simultaneously. *Clostridium saccharoperbutylacetonicum* was downregulated by asRNAs to produce hydrogen from water.

The antisense strain had a high ratio for acetone/butanol of 0.8 g/L (5 g/L acetone, 80.6 g/L butanol), as opposed to the wildtype strain's 0.27 g/L (3 g/L acetone, 11.4 g/L butanol).

In conclusion, the downregulation of the gene for H_2 evolving hydrogenase can increase butanol production or reduce acetone production. Because of the complex structure of the clostridia metabolic regulatory system, the results show that downregulating one gene is not significantly affect the metabolic in the way that is commonly imagined. To increase the process selectivity towards butanol, an effective asRNA method should be developed based on a deeper insight into the genetics of clostridia [76–78].

8.5 Metabolic Engineering of *Escherichia coli*

The development of *Escherichia coli* metabolic engineering for butanol synthesis is gaining increasing attention despite the long history of clostridial metabolic engineering. Researchers created an artificial route in *Escherichia coli* for manufacturing particular molecules for butanol. The production of butanol in clostridia is conducted by a group of genes, which were expressed as well as cloned in *Escherichia coli*. These genes include thiolase (thl), 3-hydroxybutyryl-CoA dehydrogenase (hbd), crotonase (crt), butyryl-CoA dehydrogenase (bcd), acetoacetyl-CoA: acetate/butyrate: CoA transferase (ctfAB), and aldehyde/alcohol dehydrogenase (adhE2). The entire process of producing 1-butanol in modified *Escherichia coli* is outlined. This synthesis route in the anaerobic condition generates 139 mg/L butanol. The butanol titer (373 mg/L) was further increased by deleting a few host genes that were in competition. Furthermore, cells generated five times as much 550 mg/L butanol in TB medium supplemented with glycerol as they did 130 mg/g butanol in M9 medium. Nevertheless, after the bcd gene was replicated into *Escherichia coli*, there was no evidence of BCD activity. This outcome might be explained by the produced protein's incorrect folding and *Escherichia coli*'s inability to use it.

Genes from *Clostridium acetobutylicum* ATCC 824 that produce butanol were also cloned and expressed in *Escherichia coli*. Unlike, they substituted adhe1 or adhe for adhe2 when butanol dehydrogenase was overexpressed. Four times as much butanol was generated by the *Escherichia coli* strain with adhe1 as it was by the strain with adhe2, at 1.2 g/L. The poor efficiency of multigene expression may be responsible for the low butanol titer. Because of this, co-overexpression of 6 genes in the identical host continues to be difficult. As a result, absent from the development of an effective multigene expression approach, it is challenging to increase the butanol concentration via inducing the clostridial butanol production path in *Escherichia coli*.

The synchronization of various gene expressions and improving culture conditions to obtain larger butanol titers might be the main areas of future research. James C. Liao's study team recently produced alcohol utilizing the extremely active amino acid biosynthetic pathway of *Escherichia coli* rather than the naturally developed mechanism for butanol production in clostridia. When 2-keto acids and alcohol dehydrogenases (ADHs) are overexpressed in *Escherichia coli*, 2-keto acids are intermediates in the pathways that produce amino acids and can be transformed into alcohols. But 1ol's precursor, 2-ketovalerate, a 2-keto acid, is not a frequently occurring metabolite in *Escherichia coli*. The IlvA-LeuABCD pathway must thus be encoded via an operon that must be cloned into *Escherichia coli*.

L-threonine is converted to 2-ketobutyrate via the enzyme threonine dehydratase, which is produced by the ilvA gene. The protein from the leuABCD gene facilitates the transformation of 2-ketobutyrate to 2-ketovalerate. Compared to the strain without overexpression of this system, the strain with overexpression of the ilvA-leuABCD route generated about 30 mg/L (five times as much butanol). The primary pathway for 2-ketovalerate biosynthesis was used by overexpressing ilvA and leuABCD, which increased the metabolic conversion of Xux to 2-ketobutyrate. A feedback-resistant

ThrA (ThrAfbr) operon thrAfbrBC was cloned and overexpressed as a consequence of the discovery that threonine biosynthesis was the bottleneck in alcohol accumulation, leading to a 3- to 4-fold greater concentration of 1-propanol (175 mg/L and 1-butanol (100 mg/L).

By further inactivating the host genes metA and tdh, which represent threonine dehydrogenase and homoserine O-succinyltransferase, respectively, the correlated synthesis of 1-butanol and 1-propanol increased to 1.2 g/L. 1-propanol contributed the most to this rise. Butanol levels increased almost threefold (to about 0.8 g/L) when the isoleucine, leucine, and valine biogenesis route was disrupted, while propanol was not significantly affected. The accessibility of acetyl-CoA and 2-ketobutyrate was enhanced, LeuABCD was liberated from their natural substrates, LeuA or IlvA were relieved by inhibiting the biogenesis of isoleucine as well as leucine. These factors together likely contributed to the improvement in selectivity.

The method with the best chance of success is undoubtedly the development of a butanol production route that is nonclostridial. As an example, *Escherichia coli* is well known for its quick growth. It also shows remarkable benefits in the selective manufacture of butanol and has a considerably better power generation efficiency, and this efficiently encourages *Escherichia coli* to overexpress unknown genes. Since we have a good understanding of the genetics of *Escherichia coli*, it is workable to create strains of bacteria that are better designed to produce butanol. In this situation, technical obstacles remain for metabolic engineering, like raising the concentration of butanol [79–82].

8.6 Microbial Strain

The acetone–butanol–ethanol fermentation method was created via Chaim Weizmann, who is regarded as the pioneer of industrial fermentation, using *Clostridium acetobutylicum*, one of the best-known biobutanol-producing bacteria. Over several years, several developments in the ABE fermentation process have improved yields and productivity. Using gene silencing to manage the metabolic regulation of cell cycles has also become more popular. The infrequent loss of genes with host generation cycles and incorporation of target genes presents a barrier to the widespread application of plasmid transfer techniques.

ABE fermentation is a conventional method used on an industrial scale since the 20th century to produce butanol from biomass. It is yeast-based ethanol fermentation in terms of formation volume. The isolation of novel bacterial strains and metabolic engineering from the discovered strains to boost their biobutanol production capacity have been made possible by developments in microbial engineering and a rise in interest in biobutanol production.

Some clostridial species can produce butanol, such as *Clostridium acetobutylicum, Clostridium thermupapyrulyticum,* and others. Cellulosomes are cellulolytic enzyme complexes that cannot use cellulose as a substrate. A complicated substrate molecule is broken down by clostridia that has been metabolically modified to produce an effective and functional cellulosome complex. When a scaffolding protein called miniCipA is produced in the bacterium *Clostridium acetobutylicum*, it demonstrates the *in vivo* development of a miniature cellulosome. Soucaille or Sabathe' achieved the first *in vivo* construction of a recombinant micro cellulosome, although the resulting strain could not break down cellulose. Heterological minicellulosomes with dual distinct cellulases coupled to a mini scaffolding were produced in *Clostridium acetobutylicum* to improve the consumption of crystalline cellulose.

A different work produced heterologous minicellulosomes made from *Clostridium thermocellum* rather than *Clostridium acellulolyticum*. Full-length mannanase was released because of the coexpression of the mannanase-encoding man5K gene from *Clostridium cellulolyticum* with the scaffolding-encoding cipC1 gene in *Clostridium acetobutylicum*, which helped to improve the stability of the enzyme. Through the use of the diverse array of substrates present, such as sucrose, glycerol, syngas, and alternate carbon sources for ABE fermentation. The excretion and the development of extracellular heterological minicellulosomes may be observed to be well facilitated by *Clostridium acetobutylicum* [83, 84].

8.7 Butanol Tolerance Improvement Through Genetic Engineering

Increasing butanol concentrations inhibit microbial growth because they disrupt the cytoplasmic membrane's structure and partition into it, interfering with a cell's ability to operate normally. Butanol-producing strains need to be made to be less sensitive to butanol toxicity to improve the viability of the manufacturing process. In response to butanol stress, *Clostridium acetobutylicum* overexpressing the Spo0A gene shows an increased butanol tolerance. As another example, overexpression of the groESL gene in *Clostridium acetobutylicum* improves butanol tolerance by preventing aggregation and helping to fold proteins. Genes in *Escherichia coli* that are tolerant to butanol were screened using a genetic database enrichment method.

In experiments using transgenic mice, overexpression of feoA (encoding iron transport-related protein) and entC (encoding isochorismate synthase) led to an increase in tolerance to butanol of 32.8% and 49.8%, respectively. Compared to this, a 48.7% rise in butanol resistance was reported for the astE gene (encoding a succinylglutamate desuccinylase protein). A 3% butanol concentration has been found to be suitable for allowing *Lactobacillus* strains to grow.

There has been reported that modified *Pseudomonas putida* strains can survive up to 6% butanol, the maximum concentration of butanol that a microorganism can tolerate. The butanol-tolerant genetically engineered strains established the capability of butanol production; however, the yields and concentrations of butanol obtained were low [85–87].

8.8 Economic Viability

During World War I, the first two major industrial butanol production facilities were built in Canada and the United States. When Weizmann's butanol patent expired in the 1930s, new manufacturing facilities were established in Japan, South Africa, and further areas of the United States. One difficulty that still has to be solved is the twofold increase in feedstock usage in ABE compared to ethanol production.

The cost-effectiveness of the technique is achieved by improvements in the fermentation process and molecular level alterations. Butanol's good formation was curtailed in several regions of Europe and the US in the ninety fifties because it was more expensive to produce than its synthetic substitutes. Until the 1980s, certain production facilities using the fermentation technique were still in operation in China, South Africa, and Russia.

While commercial ethanol production typically ranges between 5% and 9% and can be raised to 16%, the highest overall butanol concentration during ABE fermentation is 2% to 4%. *Clostridium acetobutylicum* produces around 12 g/L of butanol on an industrial scale, but this strain can mutate to generate up to 20 g/L of butanol. Using integrated strategies from other sectors might be helpful in the future to accomplish this [88–90].

8.9 Problems and Limitations of ABE Fermentation

It is challenging to produce biobutanol economically using ABE fermentation because of high feedstock costs, butanol recovery problems, low butanol concentrations, low butanol yields, and low butanol productivity caused by butanol inhibition.

Various approaches have been used to solve these problems, including innovative fermentation methods, and hereditarily engineered metabolic butanol synthesis from renewable biomass as an addition to electron carriers (Ecs).

Fermentations are sluggish, butanol titers are low, the accuracy of the butanol in the fermenting broth is poor, and phage infections or prevention of microbial growth that produces butanol contributes to low product recovery. Research must be conducted in this field to make biofuel production workable. It would be possible to make biological production processes more economically competitive by improving process recovery and creating bioprocess approaches for genetic engineering and applied microbiology.

ABE fermentation is being investigated for alternative substrates, removing inhibitors in biomass hydrolysate, recovering products, improving analytic techniques for procedure observing, and the metabolic engineering design of *Clostridium* species [91].

8.10 Future Outlook

It has been reported that ABE fermentation can be made more cost-effective by simultaneously producing commercial products like riboflavin, isopropanol, and others. The following provides a future outlook:

- ❖ When creating sustainable alcoholic biofuels, biobutanol is an excellent substitute for fossil fuels for transportation.
- ❖ Biobutanol is the best contender among alcoholic biofuels for future use as a gasoline substitute. Biobutanol has similar features to gasoline, so it can be used without modification in existing engines and pipelines.
- ❖ There are several difficulties in producing butanol using biological methods, such as fermentation of acetone–butanol–ethanol in *Clostridium* species. Generating several by-products (acids and gases), which reduces the biobutanol yield, is one difficulty seen when biological processes produce butanol. Conversely, the biobutanol production and concentration may be increased through genetic modification and altering the fermentation conditions.
- ❖ It is anticipated that market demand for butanol manufacturing will rise in the future years.

8.11 Conclusion

Butanol production is more interesting than challenging from an economic perspective. Many efforts have been made to boost butanol formation from Clostridia yet minimizing production costs is also essential. Further development in this field could employ modern technologies such as gene editing, molecular engineering, etc. It has been possible to reduce acetone and convert it to extra effective by-results, such as isopropanol, through the development of metabolic flux regulation. ABE fermentation, combined with the production of by-products, would be a promising concept.

Despite this, there is still scope for improving butanol titer because of the intricate metabolic regulatory mechanism of the relevant enzymes and proteins. Although butyl alcohol can be produced in batch fermentation mode at 19 g/L, it is still less competitive than ethanol at 100 g/L. The development would reduce production costs in downstream *in situ* catalyst recovery and biological separation.

Biobutanol production can be scaled up ecofriendly using economical substrate selection. A novel and potentially economical way to convert biomass into liquid transportation fuels can be achieved by deploying these strategies. Because of its physicochemical characteristics and the ongoing dispute over food fuels, biofuels of I-generation, such as ethanol, are viewed as unsustainable finally. Because they have qualities like more significant intensity/energy amounts and are more resilient, fuels that include alcohols like butyl alcohol and propyl alcohol are recommended.

However, the limited production of these hydrocarbons from biological sources is a significant barrier to commercialization. Thanks to significant advancements in microbiology and biotechnology, an admirable basis for insight into the metabolic rate and cellular activities has been established. Recent advancements in metabolic engineering have made it possible to improve the cellular phenotypes of microbes. To improve the generation of biofuels, metabolic engineering offers a platform for altering metabolic pathways.

In reality, around 1000 enzyme-catalyzed processes and specific membrane transport systems work together to actualize the metabolic activities of live cells in a tightly controlled, connected network. The fundamental building blocks of metabolism is enzymes. The sequential action of many enzymes creates a route. Metabolic engineering's main pillars are the engineering of enzymes and pathways. Our ability to thoroughly examine, create, and improve an organism's metabolism will determine the effectiveness and success of metabolic engineering. Biomass is a readily available, inexpensive renewable source of carbon that may be used in the designed metabolism to produce transportation fuels.

Modern biofuels like alcohols, esters, and alkanes may be produced by manipulating microbial metabolic pathways.

New possibilities for manipulating microorganisms to create biofuels have emerged because of recent advances in synthetic biology and metabolic engineering. Hence, the ability of microorganisms to generate biofuels at an affordable, beneficial, and ecological level while also producing them more quickly and at a reduced cost is crucial for the effective and long-term usage of microbes in the biofuel industry. This chapter highlights the necessity of merging systems biology and synthetic biology to improve host species and circumvent the biosynthetic route. Metabolic engineering is discussed concerning the generation of biofuels.

Acknowledgment

Mansi Sharma, Pragati Chauhan, and Rekha Sharma are thankful to the Banasthali Vidyapith University, Rajasthan, and Dinesh Kumar is thankful to the Central University of Gujarat, Gandhinagar for providing infrastructure facility.

References

1. Hong, K.K. and Nielsen, J., Metabolic engineering of Saccharomyces cerevisiae: A key cell factory platform for future biorefineries. *Cell. Mol. Life Sci.*, 69, 16, 2671–2690, 2012.
2. Zhang, J., Wang, S., Wang, Y., Biobutanol production from renewable resources: Recent advances, in: *Advances in Bioenergy*, vol. 1, pp. 1–68, 2016.
3. Gao, K., Li, Y., Tian, S., Yang, X., Screening and characteristics of a butanol-tolerant strain and butanol production from enzymatic hydrolysate of NaOH-pretreated corn stover. *World J. Microbiol. Biotechnol.*, 28, 10, 2963–2971, 2012.
4. Lu, C., Yu, L., Varghese, S., Yu, M., Yang, S.T., Enhanced robustness in acetone-butanol-ethanol fermentation with engineered Clostridium beijerinckii overexpressing adhE2 and ctfAB. *Bioresour. Technol.*, 243, 1000–1008, 2017.
5. Ellis, J.T., Hengge, N.N., Sims, R.C., Miller, C.D., Acetone, butanol, and ethanol production from wastewater algae. *Bioresour. Technol.*, 111, 491–495, 2012.

6. Sarchami, T. and Rehmann, L., Optimizing enzymatic hydrolysis of inulin from Jerusalem artichoke tubers for fermentative butanol production. *Biomass Bioenergy*, 69, 175–182, 2014.
7. Becerra, M., Cerdán, M.E., González-Siso, M.I., Biobutanol from cheese whey. *Microb. Cell Fact.*, 14, 1, 1–15, 2015.
8. Li, S., Guo, Y., Lu, F., Huang, J., Pang, Z., High-level butanol production from cassava starch by a newly isolated Clostridium acetobutylicum. *Appl. Biochem. Biotechnol.*, 177, 4, 831–841, 2015.
9. Li, H.G., Luo, W., Gu, Q.Y., Wang, Q., Hu, W.J., Yu, X.B., Acetone, butanol, and ethanol production from cane molasses using Clostridium beijerinckii mutant obtained by combined low-energy ion beam implantation and N-methyl-N-nitro-N-nitrosoguanidine induction. *Bioresour. Technol.*, 137, 254–260, 2013.
10. Gao, M., Tashiro, Y., Yoshida, T., Zheng, J., Wang, Q., Sakai, K., Sonomoto, K., Metabolic analysis of butanol production from acetate in Clostridium saccharoperbutylacetonicum N1-4 using 13 C tracer experiments. *Rsc Adv.*, 5, 11, 8486–8495, 2015.
11. Dürre, P., Fermentative butanol production: Bulk chemical and biofuel. *Ann. N.Y. Acad. Sci.*, 1125, 1, 353–362, 2008.
12. Buehler, E.A. and Mesbah, A., Kinetic study of acetone-butanol-ethanol fermentation in continuous culture. *PLoS One*, 11, 8, e0158243, 2016.
13. Ndaba, N., Chiyanzu, I., Marx, S., Butanol derived from biochemical and chemical routes: A review. *Biotechnol*, 8, 1–9, 2015.
14. Hussain, A., Shahbaz, U., Khan, S., Basharat, S., Ahmad, K., Khan, F., Xia, X., Advances in microbial metabolic engineering for the production of butanol isomers (isobutanol and 1-butanol) from a various biomass. *Bioenergy Res.*, 15, 4, 1–18, 2022.
15. Lonti, N. and Popescu, F., Clean energy production using biofuels, in: *Ecology of Urban Areas*, vol. 493, 2014.
16. Nayyar, A., Sharma, D., Soni, S.L., Mathur, A., Experimental investigation of performance and emissions of a VCR diesel engine fuelled with n-butanol diesel blends under varying engine parameters. *Environ. Sci. Pollut. Res.*, 24, 25, 20315–20329, 2017.
17. Lamani, V.T., Yadav, A.K., Gottekere, K.N., Performance, emission, and combustion characteristics of twin-cylinder common rail diesel engine fuelled with butanol-diesel blends. *Environ. Sci. Pollut. Res.*, 24, 29, 23351–23362, 2017.
18. Kattela, S.P., Vysyaraju, R.K.R., Surapaneni, S.R., Ganji, P.R., Effect of n-butanol/diesel blends and piston bowl geometry on combustion and emission characteristics of CI engine. *Environ. Sci. Pollut. Res.*, 26, 2, 1661–1674, 2019.
19. Baral, N.R. and Shah, A., Techno-economic analysis of cellulosic butanol production from corn stover through acetone–butanol–ethanol fermentation. *Energy Fuels*, 30, 7, 5779–5790, 2016.

20. Lee, S.Y., Park, J.H., Jang, S.H., Nielsen, L.K., Kim, J., Jung, K.S., Fermentative butanol production by Clostridia. *Biotechnol. Bioeng.*, *101*, 2, 209–228, 2008.
21. Lee, S.Y., Park, J.H., Jang, S.H., Nielsen, L.K., Kim, J., Jung, K.S., Fermentative butanol production by Clostridia. *Biotechnol. Bioeng.*, *101*, 2, 209–228, 2008.
22. Lee, J.W., Na, D., Park, J.M., Lee, J., Choi, S., Lee, S.Y., Systems metabolic engineering of microorganisms for natural and non-natural chemicals. *Nat. Chem. Biol.*, *8*, 6, 536–46, 2012.
23. Nielsen, J., Metabolic engineering. *Appl. Microbiol. Biotechnol.*, *55*, 3, 263–283, 2001.
24. Adegboye, M.F., Ojuederie, O.B., Talia, P.M., Babalola, O.O., Bioprospecting of microbial strains for biofuel production: Metabolic engineering, applications, and challenges. *Biotechnol. Biofuels*, *14*, 1, 1–21, 2021.
25. Liu, Y. and Nielsen, J., Recent trends in metabolic engineering of microbial chemical factories. *Curr. Opin. Biotechnol.*, *60*, 188–197, 2019.
26. Jawed, K., Yazdani, S.S., Koffas, M.A., Advances in the development and application of microbial consortia for metabolic engineering. *Metab. Eng. Commun.*, *9*, e00095, 2019.
27. Singh, A., Jasso, R.M.R., Gonzalez-Gloria, K.D., Rosales, M., Cerda, R.B., Aguilar, C.N., Ruiz, H.A., The enzyme biorefinery platform for advanced biofuels production. *Bioresour. Technol. Rep.*, *7*, 100257, 2019.
28. Steen, E.J., Chan, R., Prasad, N., Myers, S., Petzold, C.J., Redding, A., Keasling, J.D., Metabolic engineering of Saccharomyces cerevisiae for the production of n-butanol. *Microb. Cell Fact.*, *7*, 1, 1–8, 2008.
29. Atsumi, S. and Liao, J.C., Metabolic engineering for advanced biofuels production from Escherichia coli. *Curr. Opin. Biotechnol.*, *19*, 5, 414–419, 2008.
30. Ezeji, T.C., Qureshi, N., Blaschek, H.P., Butanol fermentation research: Upstream and downstream manipulations. *Chem. Rec.*, *4*, 5, 305–314, 2004.
31. Inui, M., Suda, M., Kimura, S., Yasuda, K., Suzuki, H., Toda, H., Yukawa, H., Expression of Clostridium acetobutylicum butanol synthetic genes in Escherichia coli. *Appl. Microbiol. Biotechnol.*, *77*, 6, 1305–1316, 2008.
32. Dellomonaco, C., Clomburg, J.M., Miller, E.N., Gonzalez, R., Engineered reversal of the β-oxidation cycle for the synthesis of fuels and chemicals. *Nature*, *476*, 7360, 355–359, 2011.
33. Saini, M., Wang, Z.W., Chiang, C.J., Chao, Y.P., Metabolic engineering of Escherichia coli for production of n-butanol from crude glycerol. *Biotechnol. Biofuels*, *10*, 1, 1–8, 2017.
34. Shen, C.R., Lan, E.I., Dekishima, Y., Baez, A., Cho, K.M., Liao, J.C., Driving forces enable high-titer anaerobic 1-butanol synthesis in Escherichia coli. *Appl. Environ. Microbiol.*, *77*, 9, 2905–2915, 2011.
35. Saini, M., Li, S.Y., Wang, Z.W., Chiang, C.J., Chao, Y.P., Systematic engineering of the central metabolism in Escherichia coli for effective production of n-butanol. *Biotechnol. Biofuels*, *9*, 1, 1–10, 2016.

36. Schadeweg, V. and Boles, E., n-Butanol production in Saccharomyces cerevisiae is limited by the availability of coenzyme A and cytosolic acetyl-CoA. *Biotechnol. Biofuels*, 9, 1, 1–12, 2016.
37. Schadeweg, V. and Boles, E., Increasing n-butanol production with Saccharomyces cerevisiae by optimizing acetyl-CoA synthesis, NADH levels and trans-2-enoyl-CoA reductase expression. *Biotechnol. Biofuels*, 9, 1, 1–11, 2016.
38. Luo, H., Zeng, Q., Han, S., Wang, Z., Dong, Q., Bi, Y., Zhao, Y., High-efficient n-butanol production by co-culturing Clostridium acetobutylicum and Saccharomyces cerevisiae integrated with butyrate fermentative supernatant addition. *World J. Microbiol. Biotechnol.*, 33, 4, 1–10, 2017.
39. Das, M., Patra, P., Ghosh, A., Metabolic engineering for enhancing microbial biosynthesis of advanced biofuels. *Renewable Sustain. Energy Rev.*, 119, 109562, 2020.
40. Fenkl, M., Pechout, M., Vojtisek, M., N-butanol and isobutanol as alternatives to gasoline: Comparison of port fuel injector characteristics, in: *EPJ Web of Conferences*, vol. 114, EDP Sciences, p. 02021, 2016.
41. Generoso, W.C., Brinek, M., Dietz, H., Oreb, M., Boles, E., Secretion of 2, 3-dihydroxyisovalerate as a limiting factor for isobutanol production in Saccharomyces cerevisiae. *FEMS Yeast Res.*, 17, 3, fox029, 2017.
42. Noda, S., Mori, Y., Oyama, S., Kondo, A., Araki, M., Shirai, T., Reconstruction of metabolic pathway for isobutanol production in Escherichia coli. *Microb. Cell Fact.*, 18, 1, 1–10, 2019.
43. Gu, J., Zhou, J., Zhang, Z., Kim, C.H., Jiang, B., Shi, J., Hao, J., Isobutanol and 2-ketoisovalerate production by Klebsiella pneumoniae via a native pathway. *Metab. Eng.*, 43, 71–84, 2017.
44. Jung, H.M., Kim, Y.H., Oh, M.K., Formate and nitrate utilization in Enterobacter aerogenes for semi-anaerobic production of isobutanol. *Biotechnol. J.*, 12, 11, 1700121, 2017.
45. Feng, R., Li, J., Zhang, A., Improving isobutanol titers in Saccharomyces cerevisiae with over-expressing NADPH-specific glucose-6-phosphate dehydrogenase (Zwf1). *Ann. Microbiol.*, 67, 12, 785–791, 2017.
46. Lee, S.H., Yun, E.J., Kim, J., Lee, S.J., Um, Y., Kim, K.H., Biomass, strain engineering, and fermentation processes for butanol production by solventogenic clostridia. *Appl. Microbiol. Biotechnol.*, 100, 19, 8255–8271, 2016.
47. George, H.A., Johnson, J.L., Moore, W.E.C., Holdeman, L.V., Chen, J., Acetone, isopropanol, and butanol production by Clostridium beijerinckii (syn. Clostridium butylicum) and Clostridium aurantibutyricum. *Appl. Environ. Microbiol.*, 45, 3, 1160–1163, 1983.
48. Thang, V.H., Kanda, K., Kobayashi, G., Production of acetone–butanol–ethanol (ABE) in direct fermentation of cassava by Clostridium saccharoperbutylacetonicum N1-4. *Appl. Biochem. Biotechnol.*, 161, 1, 157–170, 2010.

49. Zhang, J., Yu, L., Xu, M., Yang, S.T., Yan, Q., Lin, M., Tang, I.C., Metabolic engineering of Clostridium tyrobutyricum for n-butanol production from sugarcane juice. *Appl. Microbiol. Biotechnol.*, *101*, 10, 4327–4337, 2017.
50. Shen, C.R. and Liao, J.C., Metabolic engineering of Escherichia coli for 1-butanol and 1-propanol production via the keto-acid pathways. *Metab. Eng.*, *10*, 6, 312–320, 2008.
51. Steen, E.J., Chan, R., Prasad, N., Myers, S., Petzold, C.J., Redding, A., Keasling, J.D., Metabolic engineering of Saccharomyces cerevisiae for the production of n-butanol. *Microb. Cell Fact.*, *7*, 1, 1–8, 2008.
52. Qureshi, N., Saha, B.C., Hector, R.E., Dien, B., Hughes, S., Liu, S., Cotta, M.A., Production of butanol (a biofuel) from agricultural residues: Part II– Use of corn stover and switchgrass hydrolysates. *Biomass Bioenergy*, *34*, 4, 566–571, 2010.
53. Berezina, O.V., Brandt, A., Yarotsky, S., Schwarz, W.H., Zverlov, V.V., Isolation of a new butanol-producing Clostridium strain: High level of hemicellulosic activity and structure of solventogenesis genes of a new Clostridium saccharobutylicum isolate. *Syst. Appl. Microbiol.*, *32*, 7, 449–459, 2009.
54. Liu, S., Bischoff, K.M., Qureshi, N., Hughes, S.R., Rich, J.O., Functional expression of the thiolase gene thl from Clostridium beijerinckii P260 in *Lactococcus lactis* and *Lactobacillus buchneri*. *New Biotechnol.*, *27*, 4, 283–288, 2010.
55. Yang, X., Xu, M., Yang, S.T., Metabolic and process engineering of Clostridium cellulovorans for biofuel production from cellulose. *Metab. Eng.*, *32*, 39–48, 2015.
56. George, H.A., Johnson, J.L., Moore, W.E.C., Holdeman, L.V., Chen, J., Acetone, isopropanol, and butanol production by Clostridium beijerinckii (syn. Clostridium butylicum) and Clostridium aurantibutyricum. *Appl. Environ. Microbiol.*, *45*, 3, 1160–1163, 1983.
57. Gottumukkala, L.D., Parameswaran, B., Valappil, S.K., Mathiyazhakan, K., Pandey, A., Sukumaran, R.K., Biobutanol production from rice straw by a non acetone producing Clostridium sporogenes BE01. *Bioresour. Technol.*, *145*, 182–187, 2013.
58. Algayyim, S.J.M., Wandel, A.P., Yusaf, T., Al-Lwayzy, S., Butanol–acetone mixture blended with cottonseed biodiesel: Spray characteristics evolution, combustion characteristics, engine performance and emission. *Proc. Combust. Inst.*, *37*, 4, 4729–4739, 2019.
59. Azambuja, S.P. and Goldbeck, R., Butanol production by Saccharomyces cerevisiae: Perspectives, strategies and challenges. *World J. Microbiol. Biotechnol.*, *36*, 3, 1–9, 2020.
60. Bao, T., Feng, J., Jiang, W., Fu, H., Wang, J., Yang, S.T., Recent advances in n-butanol and butyrate production using engineered Clostridium tyrobutyricum. *World J. Microbiol. Biotechnol.*, *36*, 9, 1–14, 2020.

61. Kolek, J., Patáková, P., Melzoch, K., Sigler, K., Řezanka, T., Changes in membrane plasmalogens of Clostridium pasteurianum during butanol fermentation as determined by lipidomic analysis. *PLoS One*, *10*, 3, e0122058, 2015.
62. Xue, C., Zhao, J., Chen, L., Yang, S.T., Bai, F., Recent advances and state-of-the-art strategies in strain and process engineering for biobutanol production by Clostridium acetobutylicum. *Biotechnol. Adv.*, *35*, 2, 310–322, 2017.
63. Willson, B.J., Kovács, K., Wilding-Steele, T., Markus, R., Winzer, K., Minton, N.P., Production of a functional cell wall-anchored minicellulosome by recombinant Clostridium acetobutylicum ATCC 824. *Biotechnol. Biofuels*, *9*, 1, 1–22, 2016.
64. Tomas, C.A., Welker, N.E., Papoutsakis, E.T., Overexpression of groESL in Clostridium acetobutylicum results in increased solvent production and tolerance, prolonged metabolism, and changes in the cell's transcriptional program. *Appl. Environ. Microbiol.*, *69*, 8, 4951–4965, 2003.
65. Jang, Y.S., Lee, J.Y., Lee, J., Park, J.H., Im, J.A., Eom, M.H., Lee, S.Y., Enhanced butanol production obtained by reinforcing the direct butanol-forming route in Clostridium acetobutylicum. *MBio*, *3*, 5, e00314–12, 2012.
66. Zheng, J., Tashiro, Y., Wang, Q., Sonomoto, K., Recent advances to improve fermentative butanol production: Genetic engineering and fermentation technology. *J. Biosci. Bioeng.*, *119*, 1, 1–9, 2015.
67. Jang, Y.S., Lee, J., Malaviya, A., Seung, D.Y., Cho, J.H., Lee, S.Y., Butanol production from renewable biomass: Rediscovery of metabolic pathways and metabolic engineering. *Biotechnol. J.*, *7*, 2, 186–198, 2012.
68. Liang, Y.F., Long, Z.X., Zhang, Y.J., Luo, C.Y., Yan, L.T., Gao, W.Y., Li, H., The chemical mechanisms of the enzymes in the branched-chain amino acids biosynthetic pathway and their applications. *Biochimie*, *184*, 72–87, 2021.
69. Mund, N.K., Liu, Y., Chen, S., Advances in metabolic engineering of cyanobacteria for production of biofuels. *Fuel*, *322*, 124117, 2022.
70. Su, H., Lin, J., Wang, G., Metabolic engineering of Corynebacterium crenatium for enhancing production of higher alcohols. *Sci. Rep.*, *6*, 1, 1–20, 2016.
71. Huffer, S., Clark, M.E., Ning, J.C., Blanch, H.W., Clark, D.S., Role of alcohols in growth, lipid composition, and membrane fluidity of yeasts, bacteria, and archaea. *Appl. Environ. Microbiol.*, *77*, 18, 6400–6408, 2011.
72. Kanno, M., Katayama, T., Tamaki, H., Mitani, Y., Meng, X.Y., Hori, T., Kamagata, Y., Isolation of butanol-and isobutanol-tolerant bacteria and physiological characterization of their butanol tolerance. *Appl. Environ. Microbiol.*, *79*, 22, 6998–7005, 2013.
73. Chandgude, V., Välisalmi, T., Linnekoski, J., Granström, T., Pratto, B., Eerikäinen, T., Bankar, S., Reducing agents assisted fed-batch fermentation to enhance ABE yields. *Energy Convers. Manag.*, *227*, 113627, 2021.
74. Milestone, N.B. and Bibby, D.M., Concentration of alcohols by adsorption on silicalite. *J. Chem. Technol. Biotechnol.*, *31*, 1, 732–736, 1981.
75. Pereira, J.P., Overbeek, W., Gudiño-Reyes, N., Andrés-García, E., Kapteijn, F., Van Der Wielen, L.A., Straathof, A.J., Integrated vacuum stripping and

adsorption for the efficient recovery of (biobased) 2-butanol. *Ind. Eng. Chem. Res.*, 58, 1, 296–305, 2018.
76. Tummala, S.B., Junne, S.G., Papoutsakis, E.T., Antisense RNA downregulation of coenzyme A transferase combined with alcohol-aldehyde dehydrogenase overexpression leads to predominantly alcohologenic Clostridium acetobutylicum fermentations. *J. Bacteriol.*, 185, 12, 3644–3653, 2003.
77. Tummala, S.B., Welker, N.E., Papoutsakis, E.T., Design of antisense RNA constructs for downregulation of the acetone formation pathway of Clostridium acetobutylicum. *J. Bacteriol.*, 185, 6, 1923–1934, 2003.
78. Sillers, R., Chow, A., Tracy, B., Papoutsakis, E.T., Metabolic engineering of the non-sporulating, non-solventogenic Clostridium acetobutylicum strain M5 to produce butanol without acetone demonstrate the robustness of the acid-formation pathways and the importance of the electron balance. *Metab. Eng.*, 10, 6, 321–332, 2008.
79. Nakayama, S.I., Kosaka, T., Hirakawa, H., Matsuura, K., Yoshino, S., Furukawa, K., Metabolic engineering for solvent productivity by downregulation of the hydrogenase gene cluster hupCBA in Clostridium saccharoperbutylacetonicum strain N1-4. *Appl. Microbiol. Biotechnol.*, 78, 3, 483–493, 2008.
80. Atsumi, S., Cann, A.F., Connor, M.R., Shen, C.R., Smith, K.M., Brynildsen, M.P., Liao, J.C., Metabolic engineering of Escherichia coli for 1-butanol production. *Metab. Eng.*, 10, 6, 305–311, 2008.
81. Shen, C.R. and Liao, J.C., Metabolic engineering of Escherichia coli for 1-butanol and 1-propanol production via the keto-acid pathways. *Metab. Eng.*, 10, 6, 312–320, 2008.
82. Zheng, Y.N., Li, L.Z., Xian, M., Ma, Y.J., Yang, J.M., Xu, X., He, D.Z., Problems with the microbial production of butanol. *J. Ind. Microbiol. Biotechnol.*, 36, 9, 1127–1138, 2009.
83. Vivek, N., Nair, L.M., Mohan, B., Nair, S.C., Sindhu, R., Pandey, A., Binod, P., Bio-butanol production from rice straw–Recent trends, possibilities, and challenges. *Bioresour. Technol. Rep.*, 7, 100224, 2019.
84. Jiang, Y., Lv, Y., Wu, R., Lu, J., Dong, W., Zhou, J., Jiang, M., Consolidated bioprocessing performance of a two-species microbial consortium for butanol production from lignocellulosic biomass. *Biotechnol. Bioeng.*, 117, 10, 2985–2995, 2020.
85. Wang, S., Sun, X., Yuan, Q., Strategies for enhancing microbial tolerance to inhibitors for biofuel production: A review. *Bioresour. Technol.*, 258, 302–309, 2018.
86. Capilla, M., San-Valero, P., Izquierdo, M., Penya-roja, J.M., Gabaldón, C., The combined effect on initial glucose concentration and pH control strategies for acetone-butanol-ethanol (ABE) fermentation by Clostridium acetobutylicum DSM 792. *Biochem. Eng. J.*, 167, 107910, 2021.
87. Zheng, J., Tashiro, Y., Wang, Q., Sonomoto, K., Recent advances to improve fermentative butanol production: Genetic engineering and fermentation technology. *J. Biosci. Bioeng.*, 119, 1, 1–9, 2015.

88. Green, E.M., Fermentative production of butanol—The industrial perspective. *Curr. Opin. Biotechnol.*, 22, 3, 337–343, 2011.
89. Waqas, M., Rehan, M., Khan, M.D., Nizami, A.S., Conversion of food waste to fermentation products, in: *Encyclopedia of Food Security and Sustainability*, vol. 1), p. 501, 2018.
90. Rosales-Calderon, O. and Arantes, V., A review on commercial-scale high-value products that can be produced alongside cellulosic ethanol. *Biotechnol. Biofuels*, 12, 1, 1–58, 2019.
91. Sarangi, P.K. and Nanda, S., Recent developments and challenges of acetone-butanol-ethanol fermentation, in: *Recent Advancements in Biofuels and Bioenergy Utilization*, pp. 111–123, 2018.

9

Advanced CRISPR/Cas-Based Genome Editing Tools for Biobutanol Production

Narendra Kumar Sharma[1]*, Mansi Srivastava[1] and Yogesh Srivastava[2]

[1]*Department of Bioscience and Biotechnology, Banasthali Vidyapith, Tonk, Rajasthan, India*
[2]*Department of Genetics, University of Texas MD Anderson Cancer Center, Houston TX, USA*

Abstract

In the evolutionary world heading toward the brink of the technological pinnacle, the use of resources needs to be as efficient as possible. Taking charge of the situation, scientists have been focusing on nature-based solutions for the fulfilment of these growing needs. Biofuels, like ethanol, biodiesel, and butanol, are being utilized as safer and more ecofriendly options in the industry and for automobiles. Butanol especially has gained the spotlight recently due to its physiochemical properties and ease of production. Cost-effective methods for this butanol production rely on studies in various fields, such as bioengineering, microbiology, bioprocess, and chemistry. In the course of this chapter, we take a closer look at the role of microbial biotechnology in making the phenomenon of biobutanol production efficient using genome editing tools, such as CRISPR–Cas9. CRISPR systems have revolutionized the idea of genetic editing and are widely used in plant biotechnology, fermentation technology, and other gene therapy studies for the biomedical industry. The issues faced during the traditional ABE fermentation process in bacteria, such as *Clostridia,* can be resolved through genome editing technologies and help reduce off-target consequences. Modification of the host, as well as the biomass, has been explored previously, which is condensed into a chapter here. Using technological advancements, the aim is to increase butanol production and minimize losses, in order to save the world from pollution.

Keywords: Butanol, CRISPR–Cas, genetic engineering, biofuels

Corresponding author: snarendrakumar@banasthali.ac.in; drnarendraks@gmail.com

9.1 Introduction

The utilization of hydrocarbon fuels still holds importance and will continue to do so for many decades to come. Up to 30% of the emissions of greenhouse gases are made by the transportation sector [1]. The rising demand of fuels Biofuels are viable alternates of conventional fossil fuels used in the industry and for automobiles. They are made from renewable resources and aim at providing ecofriendly options in order to sustain the ecological balance.

The production of biofuel started with the use of sugars, starch and oil as substrates, but has evolved over the years to now use lignocellulosic biomass for bioethanol, biobutanol and biodiesel production. The evolution of biofuel production can be attributed to the simple nature of the process of fermentation occurring naturally in organisms. The use of microbes for biofuel production utilizing metabolic pathways of carbohydrate catabolism, fatty acid metabolism and isoprenoid pathway has paved a way for improvement and progress in the industrial process.

The technological marvels of genetic engineering have contributed immensely to modifying the strains of microorganisms for their efficient use in the production of biofuels. Bio alcohols have been used the most in biofuels due to their high-octane rating, enthalpy of vaporization, and various other physiochemical properties [2]. Our biofuel of interest, biobutanol is generally produced by the bacterial genus *Clostridia*. The most commonly used strain for genome editing-based technology applications for biobutanol production include *Clostridium acetobutylicum* and *Clostridium beijerinckii*.

Genetic engineering techniques have been recognized in Plant Biology, and successfully contributed to efficient yield, disease resistance and greater tolerance in plants for better agricultural progress and crop development. Similarly, these techniques have created a platform for growth and advancements in biofuel production. Techniques like Zinc finger nucleases (ZFN), transcription activator-like effector nucleases (TALEN), RNA interference (RNAi) and mutagenesis, have been contributors for the creation of recombinant or transgenic organisms; however, the revolution in genome editing did not occur until the clustered regularly interspaced short palindromic repeats CRISPR-associated proteins (CRISPR–Cas) system was elucidated. Current research paradigm focuses on the advancements in the genetic regulation and editing through the CRISPR–Cas system in bacteria. However, certain challenges are specific to this system, but it is one of the most efficient techniques for minimizing off-targets and improving efficiency [3].

In order to produce butanol, there are several approaches that have been used till date. Manipulation of the ABE fermentation cycle of the bacterium and fungus, editing the metabolic proteins in the host organisms, tweaking the biomass used as substrate, etc. The one problem that has been persistent in natural butanol production using microorganisms in the intolerance of most species to high levels of butanol and inability to host to process the substrate. Majority of these issues become the target of the study using CRISPR systems for genome editing and increasing yield [4]. In this chapter, we will focus on the mechanisms of the biobutanol production pathway of the microbes, and how CRISPR/Cas can help in the challenges faced in the production of biobutanol.

9.2 Microorganisms as the Primary Producer of Biobutanol

The production of biobutanol and other bioalcohols is predominantly carried out by microorganisms in nature. There are several microbes used as producers of the fuel and have had a great impact on their efficiency. Both bacteria and fungi contribute to this bioproduction. A list of species that contribute to biobutanol production has been enlisted in Table 9.1.

Clostridium acetobutylicum is regarded as the most indispensable species for the biogenesis of butanol. It is heterofermentative, anaerobic, rod shaped and nonpathogenic gram positive bacteria [5]. It is used in the industry to produce products from starch at a large scale. These microbes

Table 9.1 List of microorganisms that produce biobutanol in nature.

S. no.	Organisms	References
1.	*Clostridium acetobutylicum*	[12]
2.	*Clostridium beijerincki*	[7]
3.	*Escherichia coli*	[13]
4.	*Cyanobacteria*	[14]
5.	*Clostridium tyrobutyricum*	[11]
6.	*Klebsiella pneumoniae*	[9]
7.	*Saccharomyces cerevisiae*	[15]
8.	*Pseudomonas putida*	[16]

involve a stationary phase where complete conversion of organic acids to solvents takes place. *Clostridia acetobutylicum* concerns two stages of acidogenic and solventogenic. The exponential growth during the acidogenic stage leads to activation of acid forming pathways, which decreases the pH to 5. In the solventogenic stage, acetone, butanol, and ethanol are produced using organic acid [6].

C. beijerinckii utilizes sugars and carbohydrates like hexose and pentose as substrate [7]. It is a rod-shaped bacterium, which is flagellated. It is mesophilic and saccharolytic due to which it can ferment starch. Its growth cycle differs slightly, due to which the second stage of growth, production of solvents and formation of endospore both get limited by the end product.

Apart from the natural production of butanol in microorganisms various genetically modified organisms have been successful in building mechanisms for biobutanol production and tolerance. This process of generation of mutants has been carried out for *Escherichia coli*. *E. coli* is a gram-negative bacterium, which is an anaerobe and has lateral flagella or pili for motility. Various studies have confirmed the high production of butanol using *E. coli*. It has been effective in carrying out both aerobic and anaerobic fermentation of butanol, and its doubling time is also around 20 minutes.

Another bacterium used for the process is Cyanobacterium which is a photosynthetic organism also known as Cyanophyta. It has found its application in the production of isobutyraldehyde, isobutanol, and ethylene. It does not require expensive substrates and nutrients. They grow fast and show compliance. It utilizes CO_2 for the production of biofuel. Efficiency in butanol production using *Cyanobacteria* can be promoted by the help of acetyl Co-A–dependent pathways. Recent studies involving the genomic sequences of the cyanobacterium have extended its application in metabolic engineering [8].

Klebsiella pneumoniae is another bacterium involved in biobutanol production. It is also rod-shaped and is gram-negative. It consumes carbon as a source and is grown on a glycerol substrate. Butanol production was recorded up to 1,030 mg/L [9]. It is a unicellular organism and has various applications in the food industry. After genetic manipulation, it can withstand high concentrations of n-butanol. Butanol production was recorded up to 2.1 mg/L [10].

Clostridium tyrobutyricum is another rod-shaped bacterium that lacks the acetone-forming genes leading to higher butanol production. The yield of biobutanol from this bacterium is around 4.645 mg/L [11]. Pseudomonas is a gram-negative bacterium has a terminal oxidation pathway that contributes to butanol. It is also extensively used in bioremediation.

Another host, microalgae can be used for biofuel production. It requires a high growth rate and consumes waste water contributing to efficient biofuel production. Simple nutrients are required for growth and produces lipid in greater amounts. It is an antioxidant and is used in the pharmaceuticals and nutraceutical industry due to the presence of phytochemicals chlorophyll and carotenoids. All the abovementioned organisms are used for the bioproduction of butanol, and genome editing technology has immensely helped in the process.

9.3 Acetone–Butanol–Ethanol Producing *Clostridia* and Its Limitations

Clostridia undergoes fermentation that produces solvents such as acetone, butanol and ethanol, which is termed as ABE fermentation. It derives solvents from the use of carbon and electron for metabolic processes. The Emden–Meyerhof–Parnas (EMP) pathway converts glucose to pyruvate. 2 mol of ATP is produced during glycolysis. Pyruvate is followed by production of carboxylic acid which leads to lowering of the pH of the fermentation medium while also provides energy to the strains, resulting in exponential cell growth. Phase of exponential growth is thus termed as acidogenic phase. The second phase known as solventogenic phase in which ABE is produced. The pH in this phase goes down due to the uptake of external acids. The abrupt change in gene expression patterns causes the transition between the two phases [17].

Various strategies have been employed in these bacteria causing disruption in the butyrate, acetone, lactate and acetate pathways for increasing overall butanol production. Insertion/deletion/overexpression of *SpoA* and *groESL* genes has been utilized [8]. Butanol production experiences most of the issues during ABE fermentation process due to high feedstock and butanol recovery costs. Butanol concentration recorded in the final stage is also found to be low (around 20 g/L) along with the yield, which is approximately 0.33 g/g, which decreases the productivity to a mere 0.5 g/L/h. This is attributed to butanol inhibition. Biobutanol consequently loses its economic benefits in comparison with petrochemicals and other fuel production [17]. In order to breakdown the process and make it economically viable and useful in the long run, using current technology research should focus on creating an efficient protocol for the production. Genomic editing of the host organisms and the biomass will unravel better opportunities for mass production of butanol contributing to increase in its usage.

9.4 CRISPR–Cas System for Genome Editing

CRISPR Cas9 is a genome editing technique used by the mechanism of the immune system found in bacteria naturally for cleaving foreign DNA. Using the Cas9 protein and sgRNA guiding systems, it is used to customize DNA segments for recombinant technology. Currently, this genome editing tool is based on the known genes found in the CRISPR–Cas systems which are found in many prokaryotic and archaea strains. To carry out the process till date 1,302 bacterial and archaeal strains have been classified as having putative CRISPR arrays in the CRISPRdb online database, demonstrating the pervasive nature of the system [18].

These strains have been classified into Class I or Class II systems depending on their configuration with the nuclease built. Cas proteins are endonucleases that perform the catalytic cleavage, e.g. Cas9, Cpf1, and Cas12a. Single guide RNA or sgRNA is a fused form of intrinsic bacterial CRISPR RNA (crRNA) and transactivating RNA which is also called as transcRNA that offers selectivity. Further subdivisions of the CRISPR systems on the number of nuclease complexes are in six broad categories. Mechanisms of crRNA processing and targeting are also another criteria for classification [19].

In the CRISPR Cas system, protospacer adjacent motifs (PAM) are present right after the target gene to make sure the binding to genomic DNA is successful. The Cas complex and single guide RNA contribute to Cas9 localization by binding with the target sequence. Cas9 after binding cleaves the DNA (both strands) within the range of 3 to 4 nt from the PAM sequence. This double-stranded break structure formed can be repaired using two strategies—Nonhomologous end joining DNA repair (NHEJ) or Homology directed repair pathway (HDR) [20]. NHEJ has shown evidence to produce insertion or deletion mutations. HDR pathway requires the presence of a DNA template as well as sgRNA. With such a variety of CRISPR–Cas machinery available, modifications in the system can help generate biofuel using a variety of conventional and nonconventional bacterial hosts (Figure 9.1).

A CRISPR experiment is considered an effortless process which is utilized by Synthego and Genescript, which are leading genetic engineering companies that offer CRISPR customizations for expression systems. Another accessible route is to use the Addgene vector database where Scientists across the world deposit their plasmids and share them for a low cost. It is a very helpful platform for developing nations (www.addgene. com). It contributes to the affordability and has been viable mainly for

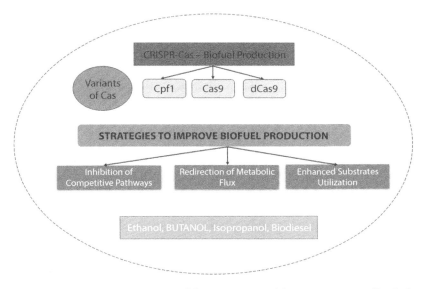

Figure 9.1 Schematic representation of the strategies used for improvement of biofuel production using genome editing technology.

microbial stains for biofuel production. In the following sections, we bring to light the various ways in which CRISPR has altered the production of biofuels, such as alcohols, biodiesel, and lipids from various microorganisms. Table 9.2 enlists some existing studies on the use of CRISPR–Cas system for butanol production.

9.4.1 CRISPR–Cas Mediated Strategies for Genome Editing for Biobutanol Production in Microorganisms

9.4.1.1 Inhibition of Contentious Pathways

Recent study conducted by Xue *et al.* consolidated the genetic modifications in *Clostridia* species [21]. In previous research genome editing in Clostridia yielded lower transformation efficiency leading to low or no transformants. Lowered recombination efficiency, Cas9 early expression-related toxicity, and issues with vector integration all posed as challenges for CRISPR–Cas system-based modifications. The solution to this problem was the use of plasmid-based editing, which replaced DNA template with a linear template. Issues pertaining to the toxicity of Cas9 expression at a premature state were dealt with the use of inducible promoters.

Table 9.2 CRISPR Cas9 systems for butanol production.

Microbial strain	Target genes	CRISPR Cas9 machinery			Butanol produced	References
		gRNA promoter	Cas variant	Cas promoter		
C. saccharoperbutyla cetonicum N1-4	Δpta, Δbuk	P_{j23119}	Cas9	Lac	19.0 g/L	[23]
E. coli	gltA	-	Cas9	-	1.08 g/L	[24]
C. tyrobutyricum	Cat1	small RNA promoter	Cas	P_{lac}	26.2 g/L	[25]
E. coli BW25113	pta, frdA, ldhA, adhE	P_{j23119}	dCas9	P_{rhaBAD}	1.06 g/L	[26]
C. cellulovorans DSM743B	Δhyd, ΔClocel-	P_{j23119}	dCas9	P_{thl}	11.5 g/L	[27]

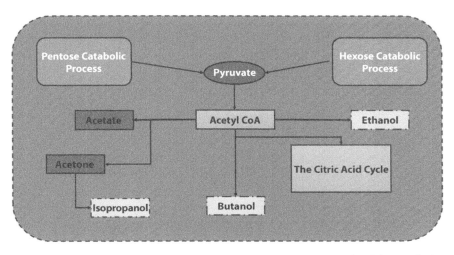

Figure 9.2 Outline of the pathways found in bacteria that are being utilized for metabolic engineering of the genomes for improved biofuel production using CRISPR–Cas.

Hyperbutanol producing *Clostridia saccharoperbutylacetonicum* N1–4 was one of the hosts that was edited using the inhibition technique to produce butanol and select the butanol production pathway [9]. Here, a previously customized system of CRISPR was used from *C. beijerinckii*. In the host system, the targeted phosphotransacetylase *(pta)* and butyrate kinase *(buk)* genes were edited as single or double mutants. The double deletion was found to be more efficient with P_{J23119} promoter from *E. coli* and up to 19 g/L biobutanol production was observed with an increased selectivity for ethanol (20.8%) [22]. Figure 9.2 is an outline of pathways found in *Clostridium* and *E coli* around the butanol production.

9.4.1.2 Redirection of the Flux of Metabolic Pathways for Better Solvent Production

Restoring and redirecting carbon flux have been found to be a good strategy for efficient biobutanol production in microbial system. EMJ50 *E. coli* strain that produces butanol using biological pathways with glucose as substrate utilises an overexpressing machinery. Endogenously expressing acetoacetyl-CoA thiolase *(thl)*, alcohol dehydrogenase *(adhE2)*, and formate dehydrogenase *(fdh1)* contributes to its machinery [28]. This helps in the redirection of the butyryl pathway to produce butanol via oxygen-tolerant pathway using *PduP* from *S. enteric*. Alcohol dehydrogenase adhA from *L. lactis* was fused to fdh1 from *C. boidinii* for the remodeling to produce butanol.

Another mutant with *gltA* knockdown was found to produce butanol at the yield of 0.120 g/g glucose, which revealed that the 5′-UTR citrate synthase when undergoes modification due to Cas9, carbon flux is redirected to acetoacetyl-CoA from TCA [22].

9.4.1.3 Enhancement of Substrate Uptake

It is thought to be a significant step toward lowering production costs for industrial *Clostridia* strains to be adapted to the use of inexpensive feedstock to produce higher alcohol [29]. The presence of glucose in the feedstock inhibits utilization of other sugars via repression of other metabolites containing carbon, which can be overcome using the modification in genes involved in carbohydrate utilization, thus further helping the utilization of other sugars [30]. By suppressing the kinase/phosphorylase (hprK) gene, Bruder, Pyne, Moo-Young, Chung, and Chou repressed the carbon catabolite pathway in *C. acetobutylicum* DSM792 and *C. pasteurianum* ATCC6013, resulting in coutilization of glucose and xylose from lignocellulosic feedstock. The production of biobutanol with the help of glycerol which is a significant by-product of biofuel production, had also been highlighted by this study on the suppression of carbon catabolite [31].

Typically, CRISPR–Cas9 methods for increasing biofuel production explained above used a single-plasmid system with antibiotic-resistant genes, and replication origin. Although due to the limitations pertaining to the size of a large plasmid as well as their reduced transformation efficiency, *Clostridium* or *E. coli* which have been used in genome engineering conventionally result in a smaller number of transformants which threatens the overall success of the engineering process.

9.4.2 Improvement of the Biofuel Production

9.4.2.1 Off Targets in CRISPR–Cas System

CRISPR–Cas systems have had various significant off target consequences, which lead to depletion of its efficiency during gene therapy. Although in microbial systems, the off-target effects are not as severe or drastic as in gene therapy, it cannot be overlooked. Cas9 protein has no known function in a eukaryotic system, as opposed to other genomic editing tools like ZFN and TALEN, thus the off-target effects cannot easily be identified. This process on the contrary is simpler for prokaryotes due to their small genomic size and variability, the possibility of off targets is also reduced. Thus, it is more preferable to use prokaryotic systems for biobutanol production than eukaryotes [32].

9.4.2.2 Using sgRNA Design to Reduce Off Target Effects

The success rate of targeted genomic edits (TGE) is similar for sgRNA as well; however, it is highly dependent on the selection of the right target site, which will provide zero or few genetically similar segments. CHOPCHOP, E-CRISP, and CRISP design are some of the algorithm-based tools developed to navigate the range of factors, including sequence similarity, number, and location of mismatches. The ratio of sgRNA to Cas9 has been studied in various studies and it has been proven that gRNAs that are truncated (17–18 nts) lead to lower number of off-target effects maintaining the efficiency, which reveals that smaller genome size and complexity as discussed in the previous section is indeed beneficial for reducing off-targets, and here leads to efficient base pairing [33, 34].

9.4.2.3 Cas9 Modifications to Reduce Off-Target Effects

The expression control of Cas9 protein at the temporal, locus specific, and spatial level is what defines the success of the CRISPR procedure. Cas9 protein expression is not required continuously, it is only useful when Cas9 and gRNA are coexpressed. When the expression of Cas9 protein is not modulated in the host, it may lead to DNA damage and increases the chances for off-target effects. To reduce the toxicity of Cas9 overexpression and avoid the consequences, inducible promoters are used (Figure 9.3a). The "codon-optimized" processes are the composition of nucleotides of appropriate species of the microorganisms in order to adjust the Cas9 expression further (Figure 9.3b) [35].

Another strategy for the minimization of off-targets is the use of ZFN and TALEN elements, FokI nuclease. When inactive Cas9 was bound with FokI domain, the TGE efficiency and specificity was found increased which led to enhanced binding of the target through the requirement for dimerization of FokI as compared to Cas9 (Figure 9.3c). A modified Cas9 SpCas9–HF1 is synthesized by substitution of quadruple alanine placing it at the site of hydrogen bond formation for binding of the genomic DNA.

A point mismatch at 5′end of the gRNA could become a cause of unwanted cleavage by the Cas9 enzyme, to remedy the situation, a single inactive catalytic domain called "nickases" is added. Cas9 nickase cleaves one strand creating a single strand nick or break with its active domain. In dCas9, the Cas9 nickase is capable of binding the DNA based on the gRNA

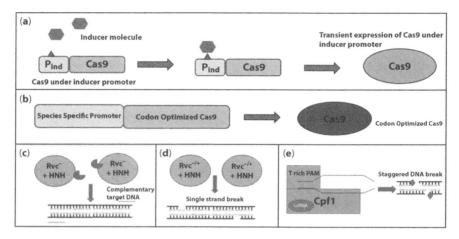

Figure 9.3 Strategies for reducing off target effects by modifying Cas9. (a) Inducible promoters for maintaining the cytotoxicity of Cas9 transient expression. (b) Use of codon optimization for specificity of Cas9. (c) Stringent dimerization of dCas9 (inactive) for with the help of FokI for catalysis. (d) Dual nickase activity by causing breaks at proximal and opposite ends. (e) Cpf1 that causes nuclease activity upon binding with crRNA and produces sticky ends. Abbreviations P_{ind} inducible promoter; HNH and Rvc-nuclease factors in Cas9 protein and FokI nuclease domain of ZFN and TALEN.

specifics. Most vectors for CRSIPR systems are obtained from *S. pyogenes*. Inactivation of *RuvC* by D10A mutation and Inactivation of *HNH* by H480A mutation is carried out. On the other hand, two proximal, opposite strand nicks introduced by a Cas9 nickase are treated as a DSB and are often referred to as a "double nick" or "dual nickase" CRISPR system (Figure 9.3d) [36]. Moreover, Slaymaker *et al.* created a further noteworthy Cas9 variant known as enhanced specificity Cas9 (eSpCas9), which suggested that off-target cleavage was caused by Cas9's proclivity to untangle DNA at untargeted sites [37].

9.4.3 Efficient and Modified Biomass "Designed" for Biobutanol Production

According to Zheng *et al.*, 2015, "designed biomass" is a term used for biomass which after hydrolysis contain saccharides and other inexpensive products as cosubstrates [45]. Thus, biomass is used for efficient production of butanol and reduces the cost of fermentation. The metabolic flux increases by using glucose and acetone as a cosubstrate, which further improves the production of solvent in start of the growth phase [46].

Table 9.3 List of different substrates for butanol production using genetically modified organisms.

Substrates	Strain	Butanol conc. (g/L)	References
Sugarcane molasses	C. beijerinckii TISTR 1461	14.13	[38]
Lettuce	C. acetobutylicum DSM 792	1.10	[39]
Palm kernel cake	C. saccharoperbutylacetonicum N1-4	4.15	[40]
Bamboo	C. beijerinckii ATCC 55025-E604	6.45	[41]
SCB	C. acetobutylicum XY16	9.30	[42]
Microalga	C. acetobutylicum ATCC824	13.10	[9]
Palm kernel cake	C. saccharoperbutylacetonicum N1-4	3.27	[43]
Potato peel	C. saccharoperbutylacetonicum DSM 2152	8.11	[44]

Addition of acetate significantly increases butanol and ABE production in C. saccharobutylacetonicum N1-4 by including lactic acid, which acts as a designed biomass. A few other designed biomasses are listed in Table 9.3, with details about the studies conducted and how cosubstrates can increase the butanol produced.

9.5 Conclusion

Strategies for reducing carbon footprint are being valued now more than ever. With the advantages of Butanol over other biofuels there are certain limitations to its production process. This has contributed to lowered consumption. However, with the help of cutting edge CRISPR–Cas systems, various strategies and tool kits are being designed to support the growth of butanol in the biofuel industry. Using microorganisms such as *Clostridia* and using recombinant biology techniques many variants and mutants have been designed for the betterment of the process of production of butanol. Improving the metabolic pathways by inhibiting the pathways of competitive nature, redirecting carbon flux and enhancing the substrate uptake are the approaches that have been studied so far. Due to single plasmid systems used thus far, transformation efficiency and off target effects have been prominent, but researchers are using varied approaches such as using sgRNA to guide the nuclease and modifying Cas9. Designed biomass has contributed to the story as well by giving a newer outlook to solving the issues related to the substrate uptake. The above mentioned, and many more forthcoming initiatives by the scientific community will bring forward approaches for better butanol production which will furthermore reduce the environmental carbon burden.

References

1. Awad, O. I. *et al.*, Overview of the oxygenated fuels in spark ignition engine: Environmental and performance. *Renew. Sustain. Energy Rev.*, 91, 394–408, 2018.
2. Erdiwansyah, *et al.*, An overview of higher alcohol and biodiesel as alternative fuels in engines. *Energy Rep.*, 5, 467–479, 2019.
3. Ran, F.A. *et al.*, Genome engineering using the CRISPR-Cas9 system. *Nat. Protoc.*, 8, 2281–2308, 2013.
4. Shanmugam, S. *et al.*, Advanced CRISPR/Cas-based genome editing tools for microbial biofuels production: A review. *Renew. Energy*, 149, 1107–1119, 2020.
5. Majidian, P. *et al.*, Metabolic engineering of microorganisms for biofuel production. *Renew. Sustain. Energy Rev.*, 82, 3863–3885, 2018.
6. Patakova, P. *et al.*, Acidogenesis, solventogenesis, metabolic stress response and life cycle changes in Clostridium beijerinckii NRRL B-598 at the transcriptomic level. *Sci. Rep.*, 9, e1371, 2019.
7. Qureshi, N. *et al.*, Butanol production from wheat straw hydrolysate using Clostridium beijerinckii. *Bioprocess Biosyst. Eng.*, 30, 419–427, 2007.
8. Rao, A. *et al.*, Genetic engineering in bio-butanol production and tolerance. *Braz. Arch. Biol. Technol.*, 59, e16150612, 2016.
9. Wang, M. *et al.*, CRISPRi based system for enhancing 1-butanol production in engineered Klebsiella pneumoniae. *Process Biochem.*, 56, 139–146, 2017.
10. Krivoruchko, A. *et al.*, Improving biobutanol production in engineered Saccharomyces cerevisiae by manipulation of acetyl-CoA metabolism. *J. Ind. Microbiol. Biotechnol.*, 40, 1051–1056, 2013.
11. Yu, M. *et al.*, Metabolic engineering of Clostridium tyrobutyricum for n-butanol production. *Metab. Eng.*, 13, 373–382, 2011.
12. Kudahettige-Nilsson, R.L. *et al.*, Biobutanol production by Clostridium acetobutylicum using xylose recovered from birch Kraft black liquor. *Bioresour. Technol.*, 176, 71–79, 2015.

13. Atsumi, S. et al., Metabolic engineering of Escherichia coli for 1-butanol production. *Eng. Metab. Pathw. Biofuels Prod.*, 10, 305–311, 2008.
14. Lan, E.I. et al., Metabolic engineering of cyanobacteria for 1-butanol production from carbon dioxide. *Metab. Eng.*, 13, 353–363, 2011.
15. Steen, E.J. et al., Metabolic engineering of Saccharomyces cerevisiae for the production of n-butanol. *Microb. Cell Factories*, 7, 36, 2008.
16. Rühl, J. et al., Selected Pseudomonas putida strains able to grow in the presence of high butanol concentrations. *Appl. Environ. Microbiol.*, 75, 4653–4656, 2009.
17. Abo, B.O. et al., Production of butanol from biomass: Recent advances and future prospects. *Environ. Sci. Pollut. Res.*, 26, 20164–20182, 2019.
18. Grissa, I. et al., The CRISPRdb database and tools to display CRISPRs and to generate dictionaries of spacers and repeats. *BMC Bioinf.*, 8, 172, 2007.
19. Mohanraju, P. et al., Diverse evolutionary roots and mechanistic variations of the CRISPR-Cas systems. *Science*, 353, (6299):aad5147, 2016.
20. Jiang, W. et al., RNA-guided editing of bacterial genomes using CRISPR-Cas systems. *Nat. Biotechnol.*, 31, 233–239, 2013.
21. Xue, C. et al., Recent advances and state-of-the-art strategies in strain and process engineering for biobutanol production by Clostridium acetobutylicum. *Biotechnol. Adv.*, 35, 310–322, 2017.
22. Jang, Y.-S. et al., Enhanced butanol production obtained by reinforcing the direct butanol-forming route in Clostridium acetobutylicum. *mBio*, 3, e00314–12, 2012.
23. Shaohua, W. et al., Genome Editing in Clostridium saccharoperbutylacetonicum N1-4 with the CRISPR-Cas9 system. *Appl. Environ. Microbiol.*, 83, e00233–17, 2017.
24. Heo, M.-J. et al., Controlling citrate synthase expression by CRISPR/Cas9 genome editing for n-butanol production in Escherichia coli. *ACS Synth. Biol.*, 6, 182–189, 2017.
25. Zhang, J. et al., Exploiting endogenous CRISPR-Cas system for multiplex genome editing in Clostridium tyrobutyricum and engineer the strain for high-level butanol production. *Metab. Eng.*, 47, 49–59, 2018.
26. Kim, S.K. et al., CRISPR interference-guided balancing of a biosynthetic mevalonate pathway increases terpenoid production. *Metab. Eng.*, 38, 228–240, 2016.
27. Wen, Z. et al., Enhanced solvent production by metabolic engineering of a twin-clostridial consortium. *Metab. Eng.*, 39, 38–48, 2017.
28. Saini, M. et al., Systematic engineering of the central metabolism in Escherichia coli for effective production of n-butanol. *Biotechnol. Biofuels*, 9, 69, 2016.
29. Al-Shorgani, N.K.N. et al., Isolation of a Clostridium acetobutylicum strain and characterization of its fermentation performance on agricultural wastes. *Renew. Energy*, 86, 459–465, 2016.

30. Gu, Y. et al., Utilization of economical substrate-derived carbohydrates by solventogenic clostridia: Pathway dissection, regulation and engineering. *Cell Pathw. Eng.*, 29, 124–131, 2014.
31. Bruder Mark, R. et al., Extending CRISPR-Cas9 Technology from genome editing to transcriptional engineering in the genus Clostridium. *Appl. Environ. Microbiol.*, 82, 6109–6119, 2016.
32. Kim, J.-S., Genome editing comes of age. *Nat. Protoc.*, 11, 1573–1578, 2016.
33. Fu, Y. et al., Improving CRISPR-Cas nuclease specificity using truncated guide RNAs. *Nat. Biotechnol.*, 32, 279–284, 2014.
34. Ran, F.A. et al., Double nicking by RNA-guided CRISPR Cas9 for enhanced genome editing specificity. *Cell*, 154, 1380–1389, 2013.
35. Nagaraju, S. et al., Genome editing of Clostridium autoethanogenum using CRISPR/Cas9. *Biotechnol. Biofuels*, 9, 219, 2016.
36. Mali, P. et al., CAS9 transcriptional activators for target specificity screening and paired nickases for cooperative genome engineering. *Nat. Biotechnol.*, 31, 833–838, 2013.
37. Slaymaker, I.M. et al., Rationally engineered Cas9 nucleases with improved specificity. *Science*, 351, 84–88, 2016.
38. Wechgama, K. et al., Enhancement of batch butanol production from sugarcane molasses using nitrogen supplementation integrated with gas stripping for product recovery. *Ind. Crops Prod.*, 95, 216–226, 2017.
39. Procentese, A. et al., Pre-treatment and enzymatic hydrolysis of lettuce residues as feedstock for bio-butanol production. *Biomass Bioenergy*, 96, 172–179, 2017.
40. Shukor, H. et al., Saccharification of polysaccharide content of palm kernel cake using enzymatic catalysis for production of biobutanol in acetone–butanol–ethanol fermentation. *Bioresour. Technol.*, 202, 206–213, 2016.
41. Kumar, S. et al., Simultaneous pre-treatment and saccharification of bamboo for biobutanol production. *Ind. Crops Prod.*, 101, 21–28, 2017.
42. Kong, X. et al., Biobutanol production from sugarcane bagasse hydrolysate generated with the assistance of gamma-valerolactone. *Process Biochem.*, 51, 1538–1543, 2016.
43. Shukor, H. et al., Enhanced mannan-derived fermentable sugars of palm kernel cake by mannanase-catalyzed hydrolysis for production of biobutanol. *Bioresour. Technol.*, 218, 257–264, 2016.
44. Hijosa-Valsero, M. et al., Industrial potato peel as a feedstock for biobutanol production. *New Biotechnol.*, 46, 54–60, 2018.
45. Zheng, J. et al., Recent advances to improve fermentative butanol production: Genetic engineering and fermentation technology. *J. Biosci. Bioeng.*, 119, 1–9, 2015.
46. Gao, M. et al., Metabolic analysis of butanol production from acetate in Clostridium saccharoperbutylacetonicum N1-4 using 13C tracer experiments. *RSC Adv.*, 5, 8486–8495, 2015.

10

Role of Nanotechnology in Biomass-Based Biobutanol Production

Pragati Chauhan[1], Mansi Sharma[1], Rekha Sharma[1]* and Dinesh Kumar[2]†

[1]*Department of Chemistry, Banasthali Vidyapith, Rajasthan, India*
[2]*School of Chemical Sciences, Central University of Gujarat, Gandhinagar, India*

Abstract

The depletion of conventional resources and the prohibitive cost of alternative plant-based materials for the manufacture of biofuels are major barriers to mechanizing mobility. To address these concerns, biofuel has developed as a viable option for reducing pollutants produced by greenhouse warming emissions. Biobutanol is growing in popularity as a viable, sustainable, and cost-effective renewable fuel by many renewables. However, because of the rising worry regarding food supply throughout the globe, the use of traditional farming commodities as fuel is delicate and contentious. When certain biofuels are manufactured using a nanotechnology-based technique, their long-term supply can be increased. Several nanomaterials and nanocomposites have been employed to boost bioenergy generation efficiencies. Various kinds of nanomaterials and other factors impact biofuel generation, like biobutanol. This chapter discusses current developments in the usage of nanomaterials in the research and manufacture of biobutanol. Furthermore, biobutanol synthesis requires effective and better biotechnology.

Keywords: Biofuel, biobutanol, nanocomposites, nanomaterials, biotechnology

10.1 Introduction

The commercial entry of technologies and the continuing improvement of first-generation methods are all included in the estimates for biofuel production. Among the advances for upcoming biofuels is nanotechnology

---

*Corresponding author: sharma20rekha@gmail.com
†Corresponding author: dsbchoudhary2002@gmail.com

along either side. Nanostructures are a broad field with countless applications. Creating single-molecule devices and gadgets that are a few nanometres wide much less than a cell can be summed up as nanomaterials. The effect of nanoparticles (NPs) on the creation and manufacturing of biofuels is currently estimated using a variety of nanomaterials, nanotubes, nanofibers and nanometals [1]. A fantastic accomplishment is the practical utilization of nanotechnology in the bioenergy business using a cost-effective and efficient procedure. The phases of microbiology methods are those where nanomaterials have been effectively employed to enhance energy recovery. Nanoparticles are a viable option for boosting the biofuel sector as opposed to the fossil-fuel industry since it has the flexibility to employ a variety of nanomaterials. Because of its incredibly various benefits and tiny particle size, nanostructures have emerged as the most emerging tool used in numerous sectors. High surface to volume ratio leads to develop much more reactive groups, which are essential for conducting numerous procedures. The nanomaterials may take on many geometries and are widely used in various specialized disciplines, drug delivery applications, environmental clean-up, including bioimaging, water treatment method, manufacture of decent value of constituents, etc. [2]. Compared to deformable or macrosized molecules, substances in nanotechnology interact with some other substances more quickly. The nanocomposite also exhibits several crucial qualities, strong catalysts, including increased adsorption capacity, surface quality treatments, better crystalline phase levels, and outstanding thermal durability. Nanoparticles are mainly generated by top-down and bottom-up approaches.

Moreover, nanoparticles fall into several categories, such as semiconducting materials, hydrogels, polymeric nanocomposites, and metal nanoparticles. Nanomaterials are used as a catalytic and play a crucial role in boosting the functioning of anaerobic consortia, reducing antagonistic composites, and transmitting electrons because of their many critical unique qualities. Despite a few studies, the application of nanomaterials in manufacturing biodiesel is still in its infancy. As a result, further study is needed to increase the manufacturing of biofuels by using various nanoadditives and nanographene powder, like nanotubes, nanofibers, and nanobuds. Using several nanoparticle varieties and their uses to improve the yields, quality, and pace of biomass transformation to fluid biodiesel are covered in the latest study [3].

10.2 Nanoparticles for Producing of Biofuel

10.2.1 Magnetic Nanoparticles

In general, magnetic nanoparticles (MNPs) are categorised as a subclass of nanoparticles that the electromagnetic field may manipulate. This form of nanoparticles often comprises two parts: a biochemical constituent with utility and a ferromagnetic element (typically Fe, Ni, or Co). MNPs have several uses in various disciplines, biotech, including the production of biofuels, agriculture, environmental science, material sciences, etc. For the following factors, scientists are currently interested in using MNPs as a catalyst [4]:

- A high surface-to-volume ratio
- One advantage of MNPs over similar basic nanoparticles would be that they may be used as highly effective catalysts that can immobilize readily detachable materials with the aid of necessary magnetism, often having no toxic effects
- Due to their small size, MNPs can transport other commodities, such as pharmaceuticals in the delivery system.

Electromagnetic nanosized elements of different varieties are currently being researched and used as effective catalysts. Typically, such materials are composed of platinum alloys, Co, Fe, Ni, and other metal oxides [5]. MNPs are amazingly effective for biodiesel production. Bioethanol can be made from the lignocellulosic mass by immobilizing enzymes like cellulases and hemicelluloses. Likewise, spent immobilized proteins may be reused and magnetized retrieved for future use. However, in the absence of a magnetic field, MNPs scatter similarly to specific other simple nanoparticles. MNPs are not just used as a catalyst for the immobilization of enzymes. They are also coated with other catalytic active nanomaterials. Functionalized MNPs may function as nanocatalysts for various significant bioprocesses to produce biodiesel. With an elevated magnetism, nanoparticles have several interesting applications, including hydrogenation, photooxidation, and induced heating.

MNPs permit simple manipulation inside the magnetism and function as enzymatic transport. Complexes have been observed to have better thermally than regular mobile catalysts [6]. The immobilized catalysts have improved thermally compared to the conventional loose enzyme and thus are durable even when heated at 70°C. Also, making them suitable for the lipase's immobilization enzyme as during the manufacture of biodiesel from palm oil. Besides immobilization, using a magnetism may also remove the use of biocatalyst exhibits excellent stability, catalytic activity, and up to 80%. Co, Fe, and Ni instantly produce robust sulfate precipitation in the gasification furnace while also displaying novel, better molecular differentiation characteristics [7]. Because of its interaction with dissolved organic material of microbes, the Co nanomaterials are permeable. The inorganic materials sulfide minerals had an exceptionally low water solubility. As a result, Fe and Ni species produced a variety of patterns in which communication governed and regulated the binding affinity with sulphide groups and different soluble organic linkers. In the process of anaerobic digestion to create biogas, numerous types of nanomaterials perform in various ways. With consecutive improvements in twitch reflexes, the employment of nanoparticles boosts the quantity of biogas generated.

The most effective MNPs that contribute to a rise in the production of biogas and the proportion of CH_4 are Ni, Co, Fe_3O_4, and zero-valent Fe. While including Co and Fe_3O_4 improves the amount of methane volume by dual and introducing zero-valent metal increases the amount of methane by approximately 1.67 times. Then, introducing Ni as MNPs increases the capacity of biogas production by around 2.17 times. Since CH_4 is the sole source of energy in bioenergy that can burn. The more significant the proportion of CH_4, the higher the quality of the biogas fuel, which is why biogas fuel with a more significant percentage of CH_4 is utilized as a fuel in motor engines. Since CH_4 has no smell and produces no hazardous gases during burning, it has a detrimental effect on the ecosystem. Therefore, Ni > Fe_3O_4 > Co > Fe is the sequence in which MNPs are most effective in producing biofuels. MNPs exhibit a variety of unique properties while producing gases from agricultural residues. Heavy metals significantly influenced the glycolysis of human tissue, such as Cu, Zn, Fe, Mo, Ni, and Co. The inorganic compounds are added to the biogas digesters as micronutrients, which improves the efficiency of the biofuel made from agricultural waste [8]. MNPs of Co, Fe, and Ni during enzyme synthesis significantly impact the development and reproduction of methanogenic bacteria. The biological process of producing CH_4 involves the MNPs of Co and Ni.

10.2.2 Carbon Nanotubes

Carbon nanotubes (CNTs) are nanoparticles that may be created using arc discharge, chemical vapor deposition, and laser ablation. These nanoparticles often comprises many cylindrically shaped graphene sheets that have been packed tightly. It has a great biocompatibility and nanosized dimension, making it a well-known performing proper for immobilizing enzymes. In electrocatalytically systems and fusion reactors, CNTs are employed as catalysts. When compared to other nanoparticles, CNTs have remarkable structural, mechanical, biocompatible, and thermal qualities. They are widely employed in biotechnological applications, particularly in developing synthesizing bioenergy and biosensors via enzymatic crosslinking [9]. The large surface area of CNTs contributes to the great enzymatic unloading capacity and minimal dispersion barrier. According to current reports, the durability and activity of CNTs are increased by enzymatic coupling.

Similarly, surface modification can improve the performance of CNTs. The formation of proteins can alter their catalytic performance when they are immobilized. Furthermore, through the transmission of electrons, CNTs have the potential to affix to the firmly established catalyst surface of proteins. The primary reason CNTs are used in developing and manufacturing biofuels is because of their 3D electrical area, which increases the density of the enzymatic and other redox chemicals on their interface. Nevertheless, they may be effectively employed to immobilize proteins because of several exceptional qualities of CNTs, like permeability and electrical properties. Multiwalled carbon nanotubes (MWCNTs) nanomaterial improves both the combustion the folding achievement and vehicle's ability of biofuel [10]. To maximize the efficiency of the biofuel, 10–50 mg/l of MWCNTs particulate having tube lengths of 1–10 microns and sizes of 10–15 nm were introduced to the mix of biofuel and gasoline using an ultrasonic cleaner. All engine metrics perform better when MWCNT naps are added at any dose during gasoline bleed. However, the greatest output features are seen at a dosage reaction of 30 mg/l. CNTs were created when fatty-acid methyl esters, or biodiesel, were burned. When alcohol and biofuel are combined, a fuel injector that provides heat and emissions is created. Alcohol is a fuel that is made from the cellulose of crops. Ferrocene is a well-known catalyst to form CNTs, while Mo/S is a booster in blended fuels [11]. By trans esterifying triglycerides, hard acidic catalyst on nanotechnology plays a crucial part in the research and production of rising biofuels. The breakdown of lipids methyl ester to produce biofuel uses sulphonated MWCNTs (S-MWCNTs) as a reliable catalyst for solid acids. High acidification of S-MWCNTs with stopping their polycyclic textural matrix power accounts for higher efficiency. To satisfy our current energy requirements in the coming years, we seek a viable alternative to ecological energy owing to the ongoing depletion of hydrocarbon fossil energy and the rising ecological deterioration. A viable answer to this critical problem is biofuel, created from a feedstock of livestock or vegetable oils, as well as through renewable materials or solid wastes [12]. Eventually, it turned out that the two major common pathways are better suited for biodiesel synthesis.

- Volatile fatty compound fatty acid esters from renewable feedstock materials
- Lipid transesterification from bioenergy materials

Transesterification procedure–When the hydroxyl groups interact with mono, di, or triglycerides, it does so with the help of an enzyme, base, or acid as catalysis.

Supersonic treatments, ultrasonic treatments, and microwave-assisted extraction help the esters' reaction processes. However, a recent study discovered that using CNTs with just an alkaline motivator significantly improves the manufacturing quality and efficiency of the biofuel generated [13].

However, in a recent study CNT nanomaterial have been used with just an alkaline motivator significantly improves biofuel's manufacturing quality and efficiency. While highly porous titanium oxides and certain other iron oxides, such as silica, zirconia, alumina, etc., help produce biofuel, using nanomaterials of CNTs improves both production rates and biofuel composition.

- Carboxylic compounds may be effectively joined with heterocyclic texture matrices using CNTs, and the composition turns highly acidic, increasing the solubility and stability. Because of the hydrophobicity of the catalyst's surface, CNTs may attack organic reagents and impede undesirable chemical changes when hydrogen ions are present [14]. Clavulanic acid was changed into an ester (ethyl levulinate) using sulfated nanotechnology as a catalyst. It was shown that clavulanic acid might consume the reactive groups of the sulfated CNTs.

As a result, S-MWCNTs are used as a catalyst to transesterification in ethanol between 130 and 170°C.

10.2.3 Graphene and Graphene-Derived Nanomaterial for Biofuel

The transformation of bioenergy materials serves a significant and fascinating role for the production of fuels and compounds with the additional value from sustainable organic materials without using petrochemicals. The method of collection and the location of the biomass distribution influence the characteristics of the bio-oil and gas produced. The biofuel made by lignocellulose, which contains 20%–30% lignin, 40%–50% cellulose, and 20%–40% hemicellulose, is more efficient. Bio-oil is produced because of the hydrolyzed or pyrolysis of biofuels. Furfurals and quinones could be removed from generated crude bio-oil to purify it further because they have easily polymerized alternatively [15, 16]. Now that energy and gasoline consumption from physical reserve supplies is being considered, a responsible extracting method must be chosen. An appropriate catalyst must be designed for the sustained extracting method; it must be effective, reliable, selective, and stable.

Microalgae have been thoroughly studied and might be an excellent feed for biofuel synthesis. The biocatalyst is used to create biofuel, which comprises immobilized protease on polymeric saturation magnetization graphene oxide (MGO) sheets through a transesterification process. Fe_3O_4 MGO and a few-layered graphene oxide (GO) are used to create a composite polymeric polymer, and MGO was functionalized with 3-aminopropyl triethoxysilane (MGO-AP) before being combined with AP and glutaraldehyde (MGO–AP–GA). With both the aid of MGO-AP and MGO via ionic attraction and crosslinking utilizing MGO-AP-GA, Rhizopus oryzae lipase (ROL) was immobilized [17]. This nanobiocatalyst produces biofuel with incredibly high efficiency, as measured by kinetic models, comparatively higher activation, superior storage time, and maximum load capacity. The catalyst often employed in transforming cellulose to biodiesel comprises carbon-based catalysts, zeolites, solid acids, metal oxides with or without metal nanoparticles, etc. However, a graphene-based nanocatalyst has just been created, and because of its superior efficiency, it has replaced this catalysis for the transformation of bioenergy. It has the advantage of being a double substance since carbon materials products may also be made from cellulose and compounds connected to it. Thus, it was determined that reduced graphene oxide (RGO), graphene, and graphene oxide are all useful for the transformation. While graphite and products built on it are effective catalysts for biodiesel creation, chemists have a hard time consistently producing high-quality graphite and its by products [18].

Synthesis techniques are often divided into two groups:

- Top-down method
- Bottom-up method

In the top-down approach, the mixture is removed into the graphene layers and, through oxidation or other treatments, defeats the van der Waals forces of attraction within the layers. Whenever the biofuel is put

and through slow pyrolysis in isometric view and the decomposition byproduct is turned into the derivatives of graphene in bottom-up, these methods are always used concurrently [19].

10.2.4 Other Nanoparticles Applied in Heterogeneous Catalysis for Biofuel Production

It has been noticed that activated carbon, rather than a homogeneous direct cause, is more successful in producing biofuel. The activated carbon has many advantages, including that the biodiesel generated is not contaminated, is simpler to isolate, is noncorrosive, environmentally friendly, has a longer lifespan, and has high selectivity. The employment of based catalysts in the manufacture of biofuel indicates an outcome for transforming lignocellulosic materials into the creation of biodiesel or biofuel. Analyzing cellulose or creating FAMEs, or fatty acids methyl esters requires the insertion of nanomaterials into polymeric media (basic nanoscale catalysts). Because of their significant characteristics, such as tiny pore size, specific surface area, acidic quantity, and acidic strength, inorganic nanoparticles are occasionally used as catalysts in various industries [20]. Gasoline may be made using a variety of various nanomaterials, including H-form zeolites, cation-exchange resins, transition-metal oxides, solid acids, heteropoly compounds, and solid carbonaceous acids. It has been shown that the utilization of mixed nanostructure and ionic liquid (IL) forms in heterogeneous catalysts methods yields positive outcomes. Table 10.1 shows the comparison of fuel properties of gasoline and alcohol-based fuel.

Table 10.1 Comparison of fuel properties for gasoline and alcohol-based fuel.

Fuel	Energy density (MJ/L)	Average octane (AKI rating/RON)
Methanol	~16	98.65/108.7
Propanol	~24	108/118
Gasoline	~33	85–96/90–105
Ethanol	~20	99.5/108.6
Butanol	~30	97/103

10.3 Factors Affecting the Performance of Nanoparticles in Biofuel's Manufacturing

Various variables affect the effectiveness and purity of the nanomaterials used in creating biofuel. These variables include synthesis method, pH, pressure, temperature, and synthesis media. The research has a variety of synthesis techniques, such as the coprecipitation method, microemulsion, thermal decomposition, synthesis using living things (fungi and algae), hydrothermal synthesis, and synthesis using plant materials, among others. Even if each synthesis technique has its own advantages and disadvantages, biological approaches are widely favoured and advised. Since they employ sustainable, nonhazardous ingredients and exhibit minimal inhibitory effects on biocatalysts throughout the manufacturing steps of biodiesel. The nanomaterials made from vegetation and bacterial biomass are more suited since they require less money and energy. The following are several variables influencing the creation of nanoparticles and nanocatalysts [21].

10.3.1 Synthesis Temperature

The heating rate significantly influences the production of nanoparticles. Based on the manufacturing procedure, the metallic nanomaterials' carbonization ranges from 100°C–700°C. The biology procedures need a frequency of less than 100°C or even weather, but the physical and chemical processes often need a temperature of over 300°C. The architecture of the nanomaterials, like pore size, shape, size, and stability, is also influenced by heat [22].

10.3.2 Synthesis Pressure

Force influences the creation of nanomaterials. The force used during production controlled the diversity of nanomaterials generation. Applying force to the interaction media causes nanomaterials to aggregate and take on a certain size and form. The dimension of the naps has increased because of greater pressure [23].

10.3.3 Synthesis pH

The pH variation considerably impacts the production of Au, Cu, Ag, Zn, and Pd nanoparticles. The nanoparticles aggregate and increase the lifetime of the nanoparticles at pH levels lower than 7. Thus, a shift in the pH even during production of the nanoparticles may readily influence and manage the shape and dimensions of the particles [24].

10.3.4 Size of Nanoparticles

Producing biodiesel better successfully uses nanomaterials with diameters ranging from 5–100 nm. The dimension of the nanoparticles and the ideal combination of operating parameters are just two variables that affect the yield of the product of biodiesel. The manufacturing process is influenced by the size and concentration of the nanoparticles [25].

10.4 Role of Nanomaterials in the Synthesis of Biofuels

The current process for making biofuels has several limitations, which is why the industrialization of biodiesel is constrained. Research must eliminate these restrictions to achieve the practical and economical deployment of diverse biodiesel as a sustainable energy resource. In addition to nanoparticles, microorganisms may produce more hydrolysate to a detectable level. The protein used is exceptionally thermally stable, a range of 80°C will not affect it in half-lifetime. It has once again been observed that employing the composite Fe_3O_4/alginate can enhance the production of enzyme production and its heat resistance. The inclusion of nanoparticles of nickel cobaltite improves the thermal durability of enzyme production [26]. According to a recent investigation, adding ZnO nanoparticles can boost the pH durability of raw lignocellulose. Once more, it was discovered that including nanomaterials, which have a half-life at 70°C, can increase the heat transfer of the b-glucosidase protein. Thus, it was determined that nanoparticles significantly alter the entire phase of biomass conversion.

10.5 Utilization of Nanomaterials in Biofuel Production

10.5.1 Production of Biodiesel Using Nanocatalysts

Biodiesel is often a fuel source made from macroalgal or microalgal oil, vegetable oils, or mono ethyl or methyl esters with extended chained fatty acids. Since triglycerides often have a large molecular mass and are a less flammable liquid, applying them directly to diesel engines is difficult. The ignition assembly's generation of minute combustion products and other unwanted coatings is to blame for this. The transesterification process, catalyzed by the bases or acid's brief alcohols (often CH_3OH), which can produce fatty acids with methyl esters (FAME) can resolve this issue. These are favored because they are less viscous and environmentally benign, especially the potassium and sodium alkoxides and hydroxides [27]. However, the main drawback is that the supplies of triglyceride include significant amounts of free fatty acids (FFAs) that neutralize the base groups in it and make a large quantity of base and CH_3OH necessary as a catalyst to produce high biofuels. Utilizing an acid catalyst to preesterify the substrate is always necessary; otherwise, these two conversions can be carried out using a specific type of acidic solution.

The catalyst is active throughout the preesterification process but less so during the transformation. With several benefits, including a straightforward and environmentally friendly manner of treatment, the heterogeneous type of catalytic in a crystalline structure offers a viable replacement for both basic and acid catalysts. While various catalysts have been described, including hydroxides including, alkoxides of alkali metals, and hydrocarbonates, potential industrial uses are still in the early stages of development. The application of nanomaterials in this area opens more possibilities. For example, CaO's regular form is much less active than its nanocrystalline form, which is active in biodiesel production with a 99% triglyceride conversion [28].

A further benefit is that this nanoparticle is easy to recycle and may even be used up to five times. Because of the availability of organic contaminants and the synthesis of enolate from the procedure of deprotonation at the carbon of carboxyl group existing in the FAMEs or triglyceride, the manufacturing cycle was partially deactivated. The nanoform of Al_2O_3 demonstrates remarkable activity upon insemination with the 15% weight percent KF, yielding up to 97.7% of methyl ester. This results from the interaction between the elevated nucleophiles of the catalyst's surface and the increased surface-area-to-volume ratio of Al_2O_3 nanoparticles. Filling KF with oxides of Ca–Al–Bi metal at the interface of the highly porous form of Mg–Fe also results in a greater efficiency percentage [29]. Even in relatively moderate conditions, the potent protein protease in the format of nanoparticles is utilized effectively as a catalytic for biodiesel production. The lipase is grafted with a Fe_3O_4 magnetic nanomaterials using the bridging compound 1-ethyl-3-(3-dimethylaminopropyl) carbodiimide, the outputs of the resultant altered lipase nanomaterials are increased by up to 94%. One of the best processes for producing biofuel in the United States and Brazil is enzyme coupled with extreme paramagnetic nanomaterials in soybean oil. This is a further significant catalyst. Soybean oil is vulnerable to breakdowns because it includes significant amounts of polyunsaturated fats. As a result, a particular method must be used to increase the antioxidant capacity of biodiesel made from soybeans. Among the cleanest, most effective, and precise ways to trigger partial protonation of the oleic acid alkyl compounds made from soy protein is the *in situ* synthesis and stabilization of palladium (Pd) nanomaterials in the cationic liquid state [30].

Hydrogenated lipid acids may be converted into alkenes by combining the Pd metals with oxygen (a diesel-like substance). An encouraging outcome is that nanomaterials may be produced and dispersed using bioenergy as a medium. For example, dispersed Fe_3O_4 nanomaterials in vegetable oil, it is possible to create nanomaterials with superior performance. While employing an organism as a catalyst to produce biodiesel is an intriguing process. However, it is not cost-effective to produce it on an industrialized basis. Because of the preceding, using catalytic systems in this situation is more suitable:

- The activated carbons can be reused by separating the goods and decomposition products produced from them.
- The separating process makes it simple to decrease the trash produced. The strong ground and hydrochloric motivators are more appropriate in this circumstance because of their special qualities, which include greater surface area, higher stability, nontoxicity, renewability, and ease of purification [31]. Although the solid foundation form of motivators outperforms the strong acid shape, the latter cannot esterify copious amounts of FFAs in the nonperishable materials' fuel sources.

Solid-acid catalysts also have the advantage of being more tolerant to higher concentrations of FFAs and the ability for concurrently esterification and transesterification of lipids to produce biodiesel. Currently, a variety of different catalysts (natural, synthetic polymer materials) in the nanosized form are being established for inexpensive fuel sources. However, these components continue to be conceived as catalytic systems because of issues like high reaction temperatures, prolonged reaction times, low rates of reactions, decreased stability, and small pore sizes. By improving the yield and lowering the response time, introducing customized nanocatalysts enhances the efficiency of manufacturing [32].

There are a variety of challenges involved in producing biodiesel, including:

- Because oil and alcohol mix insoluble, there is less mass transfer.
- Manufacturing method uses more catalyst, takes longer to react, has a greater methanol-to-oil molar ratio, stirs at a faster pace, and operates at a higher temperature.

Recently, a sophisticated technique for manufacturing biodiesel quickly and affordably, employing heterogeneous catalysts and an ultrasonic aided transesterification process, was devised. Using a catalyst, reaction duration, methanol-to-oil ratio, and reaction temperature were all postponed by the emulsification of the reactants by ultrasonic energy. Because of some of its key characteristics, like its accessibility, low gas emissions, nonhazardous nature, renewability, and biodegradability, they commonly regarded biodiesel as an acceptable source of biofuel [33]. However, today, approximately 90% of biodiesel is produced by transesterifying low molecular weight triglycerides and alcohols with a homogeneous base or acid as a catalyst. However, the following significant issues are confronted by the biodiesel-producing industry:

- The cost of the biodiesel feedstock, which accounts for around 75%–85% of the entire production cost, is expensive.
- The high price of biofuel is influenced by factors such as the purification, separation, neutralization, and processing of industrial wastes and by products.

These issues are caused using inexpensive feedstocks, yet the oil's catalysts are tolerant of the FFAs and humidity. This is because the existence of FFAs and humidity in inexpensive materials has a detrimental impact on the efficiency of the process. It works well to employ SO_4^{2-}–TiO_2–SiO_2 as nanoparticles as a catalyst for solid acids for the concurrent transesterification and the esterification of inexpensive feedstocks with high FFA levels [34]. The main benefit of the SO_4^{2-}–TiO_2–SiO_2 catalysts is that they may be recycled a maximum of four times without losing its effectiveness. However, a fresh and recent development in the manufacturing of biodiesel is the use of heterogeneous catalysts in using green technology. Typically, the uneven catalytic process is less efficient, takes longer, and is resistant to mass transport. Using catalysts with a significant covering area at the nanotechnology produces an overwhelming rise in catalytic activity and produces biodiesel economically and effectively. Even yet, choosing the right fuel to produce biodiesel for commercial use remains a difficulty for us.

10.5.2 Application of Nanomaterials for the Pre-Treatment of Lignocellulosic Biomass

Beyond lignin, hemicellulose and cellulose make up two-thirds of the lignocelluloses, which, with the right processing, may be effectively converted into biofuels. Using nanomaterials as the catalyst, the cost of the reagents used in the biomass pre-treatment may be reduced. An effective technique may produce sugar with a greater yield percentage, and it is also being used to produce biofuels. Therefore, to increase the effectiveness of the biomass conversion of the lignocellulosic materials, researchers are concentrating on appropriate pre-treatment technology. It was discovered that using maize stover as the substrate and adding Fe_3O_4 MNPs as a pre-treatment approach enhanced the sugar yield. It was discovered that controlling and regulating using nanoparticles instead of managing and regulating without them can create 13–19% more xylose and glucose. Therefore, the experimental finding shows that using Fe_3O_4 nanocomposites in the pre-treatment procedure may produce more sugar at 100°C than regular metallic iron. Using the composites, Fe_3O_4–RGO–SO_3H has also been reported to improve the synthesis of sugar by enzymatic hydrolysis. As a result, this work suggests a new possibility for producing biofuels using more affordable methods by including appropriately effective nanomaterials. As a result, it is simpler to put biofuel production into practice as an alternative energy source [34, 35].

10.5.3 Application of Nanomaterials in Synthesis of Cellulase and Stability

Cellulase's optimal, efficient system can function in extreme conditions to hydrolyze waste biomass more effectively. The substance ions play a crucial function in boosting cellulase efficiency and putting its manufacturing on an industrial scale. Using nanocrystals and nanomaterials now generates a brand new, innovative method for producing biofuels that encourage the durability of enzymes. The addition of hydroxyapatite nanoparticles to bacteria can increase the percent yield of enzyme production. The cellulase used in this approach is a very thermostable enzyme with a half-life of 80°C and incorporating Fe_3O_4/alginate nanoparticles can boost the thermal properties and efficiency of cellulase. Figure 10.1 shows the different nanocatalysts for biomass conversion.

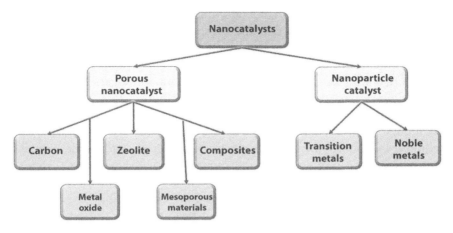

Figure 10.1 Different nanocatalysts for biomass conversion.

10.5.4 Application of Nanomaterials in the Hydrolysis of Lignocellulosic Biomass

Enzyme degradation at the typical temperature of 45°C–50°C has several problems, including partial hydrolysis, a sluggish reaction, a simulative susceptibility to microbial contamination, etc. With increasing enzyme loading rate, the lower proportion of sugar fermentation that is often experimented with may be raised, as can the production percentage. It was discovered that using thermophilic or thermostable enzymes produces better results for removing restrictions with an improvement in cellulose stability. Thus, cellulase's thermal stability may be increased by adding calcium hydroxyapatite nanoparticles at a range of 80°C and then increasing the synthesis of sugar [36].

10.5.5 Use of Nanotechnology in Bioethanol Production

Today, nanotechnology is a crucial and rapidly growing area of science and technology that is replacing older technologies and finding success in various applications. It is currently the most well-known field of research in contemporary science and is effectively used in bioethanol's cellular and molecular synthesis. However, it was demonstrated that the tertiary structure of enzymes changes randomly during the immobilization process. It is important for the performance or function of the biocatalyst. It has been discovered that various nanomaterials, such as magnetic nanotubes, metal oxide nanomaterials, nanobuds, and other nanomaterials, are better suited for environmentally friendly bioenergy generation. Typically, carbon sources, including wheat, sugarcane juice, rice husks, and other biomass products, are used to produce ethanol or bioethanol. These types of alcohol may now be extracted using the fermentable sugars derived from lignocellulosic sources. Hemicellulose and cellulose, two carbohydrates with a structure like that of polymeric materials, lignin, make up most of the biomass (fermentable sugars). For the proportion of hemicellulose and cellulose in sugars, which can break the bonds in the stoppable biomass and dissolve the proportion of the monomers of the polymer composites present in the bioenergy product. Several pre-treatment techniques are required for it. Therefore, after undergoing some initial processing, a portion of the lignocellulosic materials is degraded by a particular set of enzymes. Compared to alternative nonfermentable or chemical processes, the employment of enzymes is less expensive. Hydrolysis of lignocellulosic biomass products by utilizing the enzyme cellulases accounts for 18% of the overall expenditures associated with producing biodiesel.

Therefore, if we can employ some of the planned and improved techniques, we can reuse and collect the used enzymes, lowering the manufacturing cost. However, it was demonstrated that the tertiary structure of enzymes changes randomly during the immobilization process and is important for the performance or function of the biocatalyst. An excellent technique to make information readily available is using MNPs to immobilize enzymes. Immobilizing enzymes using MNPs is beneficial since it enables simple enzyme recovery using the magnetic field, which is accountable for enzyme recovery and many enzyme reuses [37]. As a result, MNPs can immobilize enzymes, which makes it easier to produce bioethanol. Covalent bonding or physical adsorption typically accomplishes the immobilization of an enzyme using nanoparticles. However, developing covalent links between nanoparticles and enzymes makes the covalent bonding process more acceptable since it results in less protein disruption. The materials must be changed by coating with appropriate polymeric materials that establish a link between the protein and chemical bonding in the polymer composites to increase the stability of the immobilization of the enzyme using nanomaterials.

Lignocellulose, which has a dry weight percentage of 40–60% cellulose, 20–40% hemicellulose, and 10–25% lignin, is the main component of the bioenergy of plant cells. A long-chain biopolymer with the monomer-glucose, cellulose is interconnected to form microfibril bundles. Hydrogen bonds hold xylenes or xyloglucans to microfibrils together to form hemicelluloses. Lignins, often phenolic aromatic compounds, are created when three monomers, including p-coumaryl, synapyl alcohols, and coniferyl, copolymerize. Because cellulose and hemicellulose's microbial degradation into simple and tiny fermentable sugars, the hydrolysis of lignocellulosic biomass is regarded as the second most important stage in producing bioethanol. However, it is only a pre-treatment. Other techniques, including grinding and milling, extrusion, hydrodynamic cavitation, microwave, freeze pre-treatment, etc., carry out the pre-treatment process. Ionic liquid therapy, ozonolysis, and other chemical pre-treatment techniques include acid and alkaline treatment. Hot water is used to treat a variety of physicochemical pre-treatment procedures, including recycling precipitation, wet oxidation, CO_2 explosion, ammonia fiber explosion, etc. In the biological pre-treatment process, many microorganisms and enzymes are used. However, the effectiveness of the process was measured by the lignocellulosic biomass during hydrolysis, and it was more easily accomplished owing to the degradation processes by employing an enzyme as a biocatalyst [38].

Various acidic catalysts, including H_2SO_4 and HCl, are used for the chemical hydrolysis of many materials. However, this catalyst performs less well when converting glucose to hexose because hydrolysis produces several unwanted byproducts. Although employing biocatalysts helps green technology, the manufacturing cost is high without chemicals, and the stability and reusability of the enzymes are reduced, which is a significant disadvantage when using them in hydrolysis lignocellulosic biomass. However, the immobilization of the enzymes encourages covalent bonding, physical adsorption, and crosslinking and affects the stability of the enzymes against the assault of environmental and chemical variables, which boosts the catalyst's effectiveness. Recent advances in the field of ground-breaking nanotechnology have made it possible to immobilize enzymes using a variety of nanomaterials. It has been shown and widely accepted that immobilizing enzymes using nanoscale materials is a viable and cutting-edge technique that boosts the enzymes catalytic activity. The term "nanobiocatalyst" refers to all immobilized enzymes by nanoparticles. Nanobiocatalysts are used as a crosslinking agent to conjugate the enzyme with various nanomaterials, supporting the necessary substrate specificity and chemical kinetics. The crosslinking among solid supports and enzymes during the immobilization of enzymes results in increased space or gap, which increases the suppleness and durability of the immobilized enzyme. However, it was demonstrated that the tertiary structure of the enzymes shifts randomly throughout the immobilization process and is crucial to the performance or function of the biocatalyst. The method of immobilizing an enzyme employing a biocatalyst is still unclear. However, the procedure entirely depends on the type of biocatalyst utilized, the characteristics of the solid support used, and the immobilization technique used. We need to add a suitable indication that may generate complex forms of chemicals when combined with enzymes during the immobilization process and easily dissociate again once immobilization is complete to prevent the blocking of the catalyst surface of the additional biocatalyst. A larger surface area of nanomaterials makes it easier to load more enzymes and reduces the resistance of the substrate molecules to mass transfer [35]. This is regarded as a crucial requirement for the development of nanobiocatalysts. The MNPs in the magnetic nanobiocatalysts make it simple to immobilize enzymes, and they also make it simple to separate them and utilize them repeatedly on a large scale. Therefore, lowering the cost considerably extends the life span and efficiency of the biocatalyst. Once more, it was shown that if cellulase was entirely physiosorbed on the nanoparticles as opposed to the free enzyme, the ethanol production rate was substantially doubled. The enzymatic hydrolysis process primarily uses glucose, which creates bioethanol.

Nanotechnology has recently shown promising applications in advancing ethanol production through continuous fermentation. MNPs can be used to collect the alginate matrix acquired by Saccharomyces cerevisiae cells. Another significant biofuel is biobutanol, which is now receiving attention since it has a lower vapor pressure, a larger blending capacity, a greater energy density, and is more hydrophobic than bioethanol. Therefore, the industrial manufacture and commercial usage of 1-butanol by utilizing nanobiocatalysts devote significant attention to biobutanol fermentation. Organophilic polymers like as polydimethylsiloxane (PDMS), poly (1-trimethylsilyl-1-propyne) (PTMSP), and poly (ether-block-amide) are utilized to make the various membranes that are used to recover butanol (PEBA). Inorganic fillers, such as clay materials, are injected into the constructed matrix film to improve the polymeric membranes' evaporation performance. Nanotechnology has improved the recovery of 1-butanol by impregnating membranes with organ silane-functionalized SiO_2 nanoparticles, which considerably boosts 1-butanol's sorption coefficient. Nanotechnology is also used more successfully and effectively to produce biofuel cell electrodes and directly transforms alcohol into electrical energy with better efficiency and less pollution emission [36].

10.5.6 Upgradation of Biofuel by Using Nanotechnology

The biofuel sector is currently constrained by its lack of consistency, barrier properties, and corrosive character; thus, it should be mixed with other crude oil to increase its performance. Determining the method and technique for upgrading bio-oils, which comprises hydrodeoxygenation, hydrogenation, steam reforming, esterification, and emulsification of the bio-oil with diesel, is thus the subject of active study on a global scale. Most of these technologies' processes are pricy, challenging, and include vulgar catalysts as processing aids. However, it was subsequently discovered that solid nanohybrid material particles had an intriguing outcome, as they could both maintain the emulsion between oil and water and speed up the bilayer hydrodeoxygenation at the interaction between the two liquids. By coupling Pd metal with CNTs and nanoscale oxides as catalysts, the emulsification process is prolonged, and this hybrid system reaction can occur even at low temperatures. Compared to a monophasic system, the biphasic system displays increased activity because of the correct particle dispersion at the liquid–liquid interface, which raises the amount of hydrogen just at interaction since hydrogen is much more mobile in the organic layer than in the aqueous phase. The hydrogenation of phenols is a process that benefits from the usage of nanoparticles. The complicated catalyst, which comprises an acidic ionic liquid, Ru, Pt, and Pd nanoparticles, can catalyze dehydration and hydrogenation processes, which are necessary for turning phenols into cyclohexane. Another way to improve bio-oil is by reforming. This approach currently uses CNTs strengthened with Ni catalysts, is made using the homogeneous precipitation–deposition method. It is successful at reducing the reforming of the organic components in bio-oil. The ideal Ni-loading capacity is close to 15% of the weight. At 550°C, the production of H_2 over a catalyst of 15 weight percent Ni–CNTs is about 92.5% [36, 37].

10.5.7 Nanoparticle Use in Biorefineries

A notable number of nanoparticles in previous generations of biorefineries was often associated with aiding the breakdown of biomass products. The dispersion of the additional nanomaterials throughout the biomass product is the main downside, and several components are acting as a catalytic activity in this process at a quick rate. There has to be an appropriate method of creating clean energy since the need for energy sources is always rising because of the world's growing middle class and the constant expansion of the global population. The extraction and utilization of these waste materials, which are limited in supply, have a significant role in determining the usage of elements in the treatment and purification of biomass. Metals, including Si, Al, Fe, and Ni, are much more efficient for ensuring a uniform mixture within the bioenergy product of plant leftovers during controlled catalyzed pyrolysis. Approximately 10% of the dry matter of the plant is deposited with silica, which is often found in the plant residues that make up the cell walls. This silica deposition affects the plants' structural stiffness and cellulose content. Unfortunately, it was discovered that plant residues have substantially less inorganic mass concentration and cannot withstand interaction with large levels of these elements. Some plant biomasses, referred to as hyperaccumulators, can tolerate high metal concentrations. Compared to other plant species living in the same environment that do not accumulate metals, this type of species of plant can acquire substantially greater metal concentrations. Streptanthus polygaloides, Berkheya coddii, and Alyssum bertolonii, for instance, can absorb Ni concentrations up to 1% of the final (dry weight) plant biomass product. According to estimates, 40% of the world's arable land has acidic soil with a higher-than-average concentration of aluminum than normal soil, which has a detrimental effect on plant development. As a result, this terrain cannot be used for agriculture. The capacity of some plant species to tolerate higher metal concentrations is helpful for the plant to flourish in a metal-contaminated soil, and these plant species may be utilized to produce chemicals and fuel but not food. As a result, as time goes on and crops are continuously harvested, the amount of heavy poisonous metal concentration steadily drops along with the soil's toxicity. According to a recent investigation, plants like Brassica juncea can easily absorb metals like Cu, Au, and Ag and thrive in mining wastes when metal nanoparticles are present [38, 39].

10.6 Nanotechnology in Bioethanol/Biobutanol Production

"Nanotechnology" has permeated every field of technology and sciences over the past several years, with numerous applications in everyday life. These nanoparticles are widely used in various industries, including the manufacturing of biofuels, solid rocket propellants, thin-film solar cells, textiles, coatings, biofuel cells, medication delivery, water treatment, wastewater treatment, and thin-film solar cells. Researchers are also becoming interested in other new domains like nanotoxicology and nanodiagnostics, which are soon expected to dominate the use of nanotechnology. These have been discovered to be beneficial for making target-specific alterations in biocatalysts to help the biotransformation of lignocellulosic materials into biodiesel healthier by researchers in the biological and chemical sciences. Nanofibers, nanorods, nanowires, metal and metal oxide nanoclusters, and other structures are examples of nanoparticles. Because of their size, which offers a very high area-to-volume ratio and the resulting enormous utilization of existing for chemical reactions to occur, these nanocatalysts vary from traditional bulk catalysts. Nanomaterials can speed up the reaction by offering reagents in solid, liquid, or gaseous phase active spots [40]. The creation of various nanomaterials, including nanospheres, nanowires, and nanotubes, which may be used in various sectors are necessary for fully exploiting nanotechnology. The application of nanomaterials in the bioenergy industry for long-term environmental preservation and sustainable energy supply has garnered interest internationally. Using nanoparticles in the manufacturing of alcohol enhances its overall efficacy. By raising the efficacy of preprocessing, and enzymatic hydrolysis, as well as increasing the reaction rate during the fermentation stage, the use of nanomaterials even during the alcohol production process improves the process's overall performance. Particle size and shape, surface area, the makeup of the nanoparticles, and the biomass used are the main contributors to the production of final products and enable efficient regulation of the reaction rate. Low reaction rates, high processing costs for biomass, and relatively low yield are the primary limitations of traditional techniques for producing bioethanol/biobutanol. Nanomaterials have been successfully employed to produce bioethanol to solve these problems and are now being introduced to increase productivity. However, there has not been anything documented up to this point on biobutanol. Using Nanomaterials to Prepare Raw Biomass Particularly for increasing the effectiveness of biological processes, nanoparticles are of interest. Their use in the handling of second- and third-generation biomass and the production of various liquid biofuels (biodiesel, bioethanol/butanol), as well as gaseous biofuels, is a developing field with the potential to reduce processing and manufacturing costs while raising the quality and output of the final product [41].

Researchers are becoming more interested in using nanocatalysts for manufacturing bioalcohols because of their reusability regarding using nanomaterials in producing bioalcohols, which is still a lot to be learned.

- Preprocessing biomass is a necessary but expensive phase that must be significantly improved to produce bioethanol and biobutanol at a cost-effective rate. The method is made more robust by using nanomaterials for this reason in combination with other alternative ways of pretreating raw biomass. Besides eradicating the pollution brought on by chemical pre-treatment, nanoparticles may effectively be used during the pre-treatment to improve the chemistry at the molecular level and to enable the precise and targeted alteration of biocatalysts. Because of their tiny size and facile

interaction with biomolecules, metal nanoparticles are effective at piercing the cell walls of raw biomass and releasing carbohydrates that may be utilized to make bioethanol or biobutanol [42].
- Preprocessing biomass is a necessary but expensive phase that must be significantly improved to produce bioethanol and biobutanol at a cost-effective rate. The method is made more robust by using nanomaterials for this reason in combination with other alternative ways of pretreating raw biomass. Besides eradicating the pollution brought on by chemical pre-treatment, nanoparticles may effectively be utilized during the pre-treatment to improve the chemistry at the molecular level and to enable the precise and targeted alteration of biocatalysts. Because of their tiny size and facile interaction with biomolecules, metal nanoparticles are effective at piercing the cell walls of raw biomass and releasing carbohydrates that may be utilized to make bioethanol or biobutanol. Within 40 minutes of processing at 100 rpm, over 15.26% of the increased carbohydrate output from Chlorella biomass was generated by biological method) was produced. Higher nanoparticle concentrations resulted in a shorter incubation period before a significant amount of the cell wall region ruptured, releasing intracellular components (carbohydrate/lipid). Because cellulose and proteins, which make up the cell wall, interact strongly with nanoparticles and provide a broad surface area on which to operate on cells, efficient cell wall degradation may follow. Wheat straw is pretreated using MNPs that have been alkaline (perfluoroalkylsulfonic (PFS) and alkylsulfonic (AS)) at 80°C for 24 hours and 160°C for 2 hours. Nanoparticles' magnetic cores were made using cobalt spinel ferrite.

For lignocellulosic biomass, this phase is required; however, it is not necessary for the generation born of fuel sources (algal biomass). Cellulases, hemicellulases, b-glucosidases, and other enzymes must be used effectively to convert cellulose and hemicellulose into monomeric sugars. These enzymes must be purchased for each cycle of hydrolysis, which is incredibly expensive and impractical on an industrial scale. Cellulase is the primary enzyme needed for the enzymatic breakdown of the material during the

fermentation process to produce alcohol. Enzyme stability and activity are the main issues during processing, and the potential use of nanoparticles for these purposes has generated a lot of attention [43, 44].
- Problems with the manufacturing of commercial biobutanol
Even while advertising biobutanol production is expanding quickly, several obstacles still need to be addressed before production can be profitable. Several issues must be addressed, including the creation of butanol-tolerant strains, overall cost competitiveness, reduced yields, slow fermentation, unprofitable product recovery, and microbe separation. Production of biobutanol based on algae is still in its infancy. Despite all its benefits, several obstacles prevent the commercial application of this biomass. A genetic modification algal bacterium with such property of significant carbohydrate buildup, lesser biomass yield of different microalgae and cyanobacteria that can be used commercially. Unavailability is hampered by the accessibility of suitable green algae with large carbohydrate contents and limited understanding of algal genomes [45].

10.7 Future Perspective

Making ABE fermentation workable on a wide scale still requires much study. It is unethical to use grains for cereal. Despite being a renewable and environmentally friendly source, lignocellulosic biomass has limitations in such a vast area for growing and producing toxins, which has caused attention to turn to third-generation microalgae biomass. Although stream and downstream processes are continually improving at low volume, more effective R&D and scale-up techniques are still needed to make cellulosic and algal biomass an attractive feedstock. To lower the cost, efforts are needed to use microorganisms that promote the *in situ* production of component-specific enzymes coupled with nanotechnology. The catalytic preparation, in combination with other approaches, leads to higher sugar release. The two fundamental issues lower butanol content and product toxicity can be overcome using metabolic engineering, cocultivating microorganisms for optimal raw-biomass utilization, and integrated composting systems to lower end-product toxicity. In this direction, an in-depth study is required. Using various nanomaterials for biomass processing has been investigated and offers enormous potential. Research is still needed to investigate its applications during fermentation to increase production and lessen problems like the ones listed below. The creation of more adaptable nanocatalysts for several types of biomasses. The creation of nanomaterials that can operate as biocatalysts in the biobutanol production process from extracted sugars. The marketing of these nanoparticles boosts industrial scale biobutanol production efficiency and biomass processing effectiveness.

10.8 Conclusion

In today's world, the manufacturing of several types of biofuels is aided by nanotechnology. Because of its enormous surface area compared to other forms of immobilized enzyme used for biosorption, the nanomaterials form of the enzyme demonstrates good catalytic activity throughout the bioremediation process. An expanding quantity of liquid fuels will be required to meet the rising energy demand. In this context, biofuel is essential, but to continue a sustainable endeavor, it is necessary to use technology that allows for large-scale biofuel extraction from nonfood feedstocks. Nanotechnology is the main technology that will play a significant part in future green energy generation. Future commercialization of various liquid fuels, including bioethanol, biodiesel, sustainable hydrocarbons, bio-oil, and biobutanol, as well as certain fatty esters and biogas, will be made possible by nanotechnology. Nanomaterials play a critical role in supplanting petroleum-based liquid fuel by employing biofuels with improved nanodroplet combustion. This chapter focuses more on using nanotechnology to sustainably and economically increase biodiesel generation. Cultivating microalgae with nanomaterials is a superior supplement that significantly improves the biomass's ability to produce and store CO_2. In a culture system using MNPs, biomass production is labor intensive, highly efficient, and economical. The usage of nanotechnology has led to an enzyme system that is more resilient to the fluid's adverse effects during biodiesel production. Nanoparticles have the unique ability to weaken and shatter the cell membranes of algae. This approach offers an excellent prospect for reducing the cost of biodiesel production in the future by combining nanotechnology with biodiesel technologies. As a result, in the industry that produces biodiesel, solid-state acid catalysts are frequently used as a replacement for other homogeneous catalysts to induce esterification and epoxidation from inexpensive feedstocks or sources that are not food in a typical environmental setting. Therefore, additional development is desired to prevent the deactivation of the catalyst's reactive sites and to create effective, affordable, stable, and long-lasting catalysts with solid acids to solve the current issues with biodiesel synthesis.

Acknowledgment

Pragati Chauhan, Mansi Sharma, and Rekha Sharma are thankful to the Banasthali Vidyapith University, Rajasthan and Dinesh Kumar are thankful to the Central University of Gujarat, Gandhinagar for providing infrastructure facility.

References

1. Mandotra, S.K., Kumar, R., Upadhyay, S.K., Ramteke, P.W., Nanotechnology: A new tool for biofuel production, in: *Green Nanotechnology for Biofuel Production*, pp. 17–28, Springer, Cham, 2018.
2. Kumar, G., Mathimani, T., Rene, E.R., Pugazhendhi, A., Application of nanotechnology in dark fermentation for enhanced biohydrogen production using inorganic nanoparticles. *Int. J. Hydrogen Energy, 44*, 26, 13106–13113, 2019.
3. Biswal, T. and Shadangi, K.P., Application of Nanotechnology in the Production of Biofuel, in: *Liquid Biofuels: Fundamentals, Characterization, and Applications*, pp. 487–515, 2021.
4. Santillan-Jimenez, E., Morgan, T., Loe, R., Crocker, M., Continuous catalytic deoxygenation of model and algal lipids to fuel-like hydrocarbons over Ni–Al layered double hydroxide. *Catal. Today, 258*, 284–293, 2015.
5. Gardy, J., Rehan, M., Hassanpour, A., Lai, X., Nizami, A.S., Advances in nano-catalysts based biodiesel production from non-food feedstocks. *J. Environ. Manag., 249*, 109316, 2019.
6. Photong, N. and Wongthanate, J., Biofuel production from bio-waste by biological and physical conversion processes. *Waste Manag. Res., 38*, 1, 69–77, 2020.
7. Khoo, K.S., Chia, W.Y., Tang, D.Y.Y., Show, P.L., Chew, K.W., Chen, W.H., Nanomaterials utilization in biomass for biofuel and bioenergy production. *Energies, 13*, 4, 892, 2020.
8. Jeerapan, I. and Ma, N., Challenges and opportunities of carbon nanomaterials for biofuel cells and supercapacitors: Personalized energy for futuristic self-sustainable devices. *C, 5*, 4, 62, 2019.
9. Hossain, N., Mahlia, T.M.I., Saidur, R., Latest development in microalgae-biofuel production with nano-additives. *Biotechnol. Biofuels, 12*, 1, 1–16, 2019.
10. Suzuki, S. and Mori, S., Synthesis of carbon nanotubes from biofuel as a carbon source through a diesel engine. *Diamond Relat. Mater., 82*, 79–86, 2018.

11. Galadima, A. and Muraza, O., From synthesis gas production to methanol synthesis and potential upgrade to gasoline range hydrocarbons: A review. *J. Nat. Gas Sci. Eng.*, 25, 303–316, 2015.
12. Jeerapan, I. and Ma, N., Challenges and opportunities of carbon nanomaterials for biofuel cells and supercapacitors: Personalized energy for futuristic self-sustainable devices. *C*, 5, 4, 62, 2019.
13. Ingle, A.P., Chandel, A.K., Philippini, R., Martiniano, S.E., da Silva, S.S., Advances in nanocatalysts mediated biodiesel production: A critical appraisal. *Symmetry*, 12, 2, 256, 2020.
14. Sharma, S., Saxena, V., Baranwal, A., Chandra, P., Pandey, L.M., Engineered nanoporous materials mediated heterogeneous catalysts and their implications in biodiesel production. *Mater. Sci. Energy Technol.*, 1, 1, 11–21, 2018.
15. Adzmi, M.A., Abdullah, A., Naqiuddin, A., Combustion characteristics of biodiesel blended with Al2O3 and SiO2 nanoparticles, in: *AIP Conference Proceedings*, vol. 2059, AIP Publishing LLC, p. 020052, 2019, January.
16. Gardy, J., Rehan, M., Hassanpour, A., Lai, X., Nizami, A.S., Advances in nanocatalysts based biodiesel production from non-food feedstocks. *J. Environ. Manag.*, 249, 109316, 2019.
17. Galadima, A. and Muraza, O., From synthesis gas production to methanol synthesis and potential upgrade to gasoline range hydrocarbons: A review. *J. Nat. Gas Sci. Eng.*, 25, 303–316, 2015.
18. Zuliani, A., Ivars, F., Luque, R., Advances in nanocatalyst design for biofuel production. *ChemCatChem*, 10, 9, 1968–1981, 2018.
19. Rai, M., Ingle, A.P., Pandit, R., Paralikar, P., Biswas, J.K., da Silva, S.S., Emerging role of nanobiocatalysts in hydrolysis of lignocellulosic biomass leading to sustainable bioethanol production. *Catal. Rev.*, 61, 1, 1–26, 2019.
20. Khoshnevisan, K., Poorakbar, E., Baharifar, H., Barkhi, M., Recent advances of cellulase immobilization onto magnetic nanoparticles: An update review. *Magnetochemistry*, 5, 2, 36, 2019.
21. Alsaba, M.T., Al Dushaishi, M.F., Abbas, A.K., A comprehensive review of nanoparticles applications in the oil and gas industry. *J. Pet. Explor. Prod. Technol.*, 10, 4, 1389–1399, 2020.
22. Shak, K.P.Y., Pang, Y.L., Mah, S.K., Nanocellulose: Recent advances and its prospects in environmental remediation. *Beilstein J. Nanotechnol.*, 9, 1, 2479–2498, 2018.
23. Kushwaha, D., Upadhyay, S.N., Mishra, P.K., Nanotechnology in bioethanol/biobutanol production, in: *Green Nanotechnology for Biofuel Production*, pp. 115–127, Springer, Cham, 2018.
24. Deuss, P.J., Barta, K., de Vries, J.G., Homogeneous catalysis for the conversion of biomass and biomass-derived platform chemicals. *Catal. Sci. Technol.*, 4, 5, 1174–1196, 2014.
25. Abada, E., Al-Fifi, Z., Osman, M., Bioethanol production with carboxymethylcellulase of Pseudomonas poae using castor bean (Ricinus communis L.) cake. *Saudi J. Biol. Sci.*, 26, 4, 866–871, 2019.

26. Arora, A., Nandal, P., Singh, J., Verma, M.L., Nanobiotechnological advancements in lignocellulosic biomass pre-treatment. *Mater. Sci. Energy Technol.*, 3, 308–318, 2020.
27. Rosales-Calderon, O. and Arantes, V., A review on commercial-scale high-value products that can be produced alongside cellulosic ethanol. *Biotechnol. Biofuels*, 12, 1, 1–58, 2019.
28. Nguyen, M.K., Moon, J.Y., Bui, V.K.H., Oh, Y.K., Lee, Y.C., Recent advanced applications of nanomaterials in microalgae biorefinery. *Algal Res.*, 41, 101522, 2019.
29. Huang, C., Dong, H., Su, Y., Wu, Y., Narron, R., Yong, Q., Synthesis of carbon quantum dot nanoparticles derived from byproducts in bio-refinery process for cell imaging and *In Vivo* bioimaging. *Nanomaterials*, 9, 387–398, 2019.
30. Shanmugam, S., Ngo, H.H., Wu, Y.R., Advanced CRISPR/Cas-based genome editing tools for microbial biofuels production: A review. *Renewable Energy*, 149, 1107–1119, 2020.
31. Onay, M., The effects of indole-3-acetic acid and hydrogen peroxide on Chlorella zofingiensis CCALA 944 for bio-butanol production. *Fuel*, 273, 117795, 2020.
32. Ashraful, A.M., Masjuki, H.H., Kalam, M.A., Fattah, I.R., Imtenan, S., Shahir, S.A., Mobarak, H.M., Production and comparison of fuel properties, engine performance, and emission characteristics of biodiesel from various non-edible vegetable oils: A review. *Energy Convers. Manag.*, 80, 202–228, 2014.
33. Lee, S.K., Chou, H., Ham, T.S., Lee, T.S., Keasling, J.D., Metabolic engineering of microorganisms for biofuels production: From bugs to synthetic biology to fuels. *Curr. Opin. Biotechnol.*, 19, 6, 556–563, 2008.
34. Sun, C., Zhang, S., Xin, F., Shanmugam, S., Wu, Y.R., Genomic comparison of Clostridium species with the potential of utilizing red algal biomass for biobutanol production. *Biotechnol. Biofuels*, 11, 1, 1–15, 2018.
35. Shanmugam, S., Hari, A., Kumar, D., Rajendran, K., Mathimani, T., Atabani, A.E., Pugazhendhi, A., Recent developments and strategies in genome engineering and integrated fermentation approaches for biobutanol production from microalgae. *Fuel*, 285, 119052, 2021.
36. Biswal, T. and Shadangi, K.P., Application of nanotechnology in the production of biofuel, in: *Liquid Biofuels: Fundamentals, Characterization, and Applications*, pp. 487–515, 2021.
37. Lee, H.V. and Juan, J.C., Nanocatalysis for the conversion of nonedible biomass to biogasoline via deoxygenation reaction, in: *Nanotechnology for Bioenergy and Biofuel Production*, pp. 301–323, Springer, Cham, 2017.
38. Kushwaha, D., Srivastava, N., Mishra, I., Upadhyay, S.N., Mishra, P.K., Recent trends in biobutanol production. *Reviews. Chem. Eng.*, 35, 4, 475–504, 2019.
39. Shuttleworth, P.S., Parker, H.L., Hunt, A.J., Budarin, V.L., Matharu, A.S., Clark, J.H., Applications of nanoparticles in biomass conversion to chemicals and fuels. *Green Chem.*, 16, 2, 573–584, 2014.

40. Arun, N., Sharma, R.V., Dalai, A.K., Green diesel synthesis by hydrodeoxygenation of bio-based feedstocks: Strategies for catalyst design and development. *Renewable Sustain. Energy Rev.*, 48, 240–255, 2015.
41. Kushwaha, D., Upadhyay, S.N., Mishra, P.K., Nanotechnology in bioethanol/biobutanol production, in: *Green Nanotechnology for Biofuel Production*, pp. 115–127, Springer, Cham, 2018.
42. Kushwaha, D., Upadhyay, S.N., Mishra, P.K., Nanotechnology in bioethanol/biobutanol production, in: *Green Nanotechnology for Biofuel Production*, pp. 115–127, Springer, Cham, 2018.
43. Huang, C., Dong, H., Su, Y., Wu, Y., Narron, R., Yong, Q., Synthesis of carbon quantum dot nanoparticles derived from byproducts in bio-refinery process for cell imaging and *in vivo* bioimaging. *Nanomaterials*, 9, 3, 387, 2019.
44. Rosales-Calderon, O. and Arantes, V., A review on commercial-scale high-value products that can be produced alongside cellulosic ethanol. *Biotechnol. Biofuels*, 12, 1, 1–58, 2019.
45. Singh, S.B. and Tandon, P.K., Catalysis: A brief review on nanocatalyst. *J. Energy Chem. Eng.*, 2, 3, 106–115, 2014.

11

Commercial Status and Future Scope of Biobutanol Production from Biomass

Arunima Biswas

Department of Microbiology, Raidighi College, West Bengal, India

Abstract

As world population continues to increase, so does the demand for energy consumption. Constant, relentless depletion of nonrenewable energy sources, like the traditional fossil fuels, is leading to their alarmingly increasing scarcity. Overuse and over exploitation of petroleum products have taken global pollution to a disturbing level. Escalating cost of raw materials is another serious matter of concern for common people. Hence, the search for renewable, affordable, ecofriendly, alternate energy sources. Biofuel derived from biomass seems to be the most prospective candidate due to its sustainability, commercial feasibility, and environment-friendly nature. Biobutanol is highly preferred among different groups of scientists, industrialists, and environmentalists as a sustainable additive or complete biofuel. Compared to bioethanol, it has higher energy density, lower flammability, hydrophobicity, minimum corrosive nature, and is well miscible with gasoline. Biobutanol can be efficiently produced through microbial fermentation from various types of industrial, agricultural, and domestic waste materials, which would otherwise add to the pollution level. According to a recent report, biobutanol market size is estimated to reach US$1.8 billion by 2027 after growing at a compound annual growth rate of around 7.3% from 2022 to 2027. However, technological challenges in production are currently dampening its demand. Also, the growing demand for electric vehicles may hinder the growth of the biobutanol market. This chapter presents a strength, weakness, opportunity, challenge (SWOC) analysis of biobutanol industry.

Keywords: Biobutanol, lignocellulosic biomass, sustainable, ABE fermentation, commercial, challenges, prospects

---

E-mail: mou.aru@gmail.com

11.1 Introduction

Energy is an indispensable component of modern life and is often considered vital for economic growth and development of a country. Studies carried out using World Bank Development Indicators have demonstrated that the use of energy is strongly associated with almost every conceivable facet of development. Energy has been shown to have a positive impact on various factors and needs of life, namely, health and nutrition, clean water, transport system, education, trade and commerce, industry and wealth, family life, and, even, life expectancy [1–4]. The global energy requirement is mostly fulfilled by nonrenewable resources. Fossil fuels, such as coal, natural gas, and oil, are crucial in contemporary society. They are predicted to continue exerting a dominant role in energy scenario and, in the process, may run out in not-so-distant future. In the face of declining supplies and mounting demands, oil prices are expected to go on escalating, much to the dismay of common public. Moreover, fossil fuels have significantly negative environmental impact and negative social consequences. Their extraction, refining, and burning release huge amount of greenhouse gases that contribute to the menace of climate change. Other adverse effects include oil spills, and considerable air and water pollution [3]. Oil has also been a major cause of modern age war and conflict. But every year global energy consumption is steadily going up, and trends show that energy demand is yet to reach its peak. Clearly, the world does not have enough known fossil fuel reserves to keep up with this growing need. We also must think of protecting the environment at the same time. Hence, current global policy is to gradually lower fossil fuel dependency as part of climate change mitigation and a concomitant, progressive transition toward safe, sustainable, universally accessible and affordable, ecofriendly energy, as per Sustainable Development Goal (SDG) 7 of United Nations [1, 3, 5–7].

For quite some time now, countries across the world have seriously been considering using alternative sources of energy in form of renewable biofuels, such as bioethanol, biobutanol, and biogas. Currently, these only provide 5% of world energy need, while nuclear energy provides 4% of energy. Nuclear power plants are also considered safe sources of energy. But storing of nuclear waste is a huge problem and a massive risk factor. So, the world is more inclined toward the "green way" to energy instead of the "nuclear way." Some nations have already specified a mandatory 5% to 10% blending of biofuels with petrol and diesel. This has triggered a fast expansion in the biofuel sector in the last decade.

However, the biofuel industry is still at a budding stage. For this sector to thrive, it needs government support in terms of lower taxes and other infant industry incentives. Since much of the raw materials for biofuels come from the agriculture and farming sector, it is important to ensure sufficient support system for farmers so that they may grow biofuel crops without compromising on food crops and food security. To make the entire process economically viable, countries need to devise appropriate strategies. Of late, increased research efforts are focusing on bioconversion of biomass to usable biofuels. In November 2022, Government of India granted approval for implementation of Biomass Programme under the Umbrella scheme of National Bioenergy Programme for duration of FY 2021–2022 to 2025–2026 [8, 9]. Back in 2007, United States mandated annual production of 16 billion gallons of cellulosic biofuels (especially from residues) out of total 36 billion gallons of renewable biofuels, including biobutanol, by 2022 [10]. Many other nations are also adopting new policies and schemes to make the production process of biofuels from biomass economically feasible and sustainable.

From the perspective of biotechnology, the term "biomass" generally refers to "all organic matter that grows by the photosynthetic conversion of solar energy." The solar power is converted to this usable organic form by green plants, algae, and photosynthetic bacteria. The annual amount of terrestrial and marine biomass is estimated to contain about 4,500 exajoules (4.5×10^{21} joules) of energy, which is almost 10 times the annual global energy consumption [11]. Huge amounts of organic wastes and surplus generated from forests, farming, milling, domestic household, gardens, and kitchens all form part of the biomass and generally add to the pollution level. Instead of spending huge sums of money on such organic waste management, these can be used to produce biofuels. Figure 11.1 summarizes a few different examples of biomass that may be used for biofuel production.

Among all the biofuels, biobutanol holds great promise as a substitute of fossil fuels, owing to better features such as its higher energy density, lower

Figure 11.1 Different examples of biomass for bio-fuel production.

cost, minimum chances of being hazardous, flexible blending, and also its mutual solubility with gasoline.

11.2 Biobutanol—Its Brief Background Story

Butanol is a four-carbon alcohol (C_4H_9OH). It has four structural isomers, the most important commercial isomer being the naturally occurring 1-butanol (n-butanol). It was primarily used as an industrial solvent or for surface coatings. It soon attracted attention as a highly potential biofuel candidate due to its superior performance and other advantages over other biofuels. Bio-butanol can be efficiently produced from industrial, agricultural, and domestic waste material through microbial fermentation, popularly known as the acetone–butanol–ethanol (ABE) pathway. Microbes that carry out the ABE fermentation are mostly obligate anaerobes, of which the most well-studied and widely used organism is *Clostridium acetobutylicum*. The ABE pathway can be divided into two major parts. At first, butyric and acetic acids are produced by *C. acetobutylicum* This is the acidogenesis step. In the subsequent phase, the solventogenesis step, butanol, acetone and ethanol are formed. Figure 11.2 gives a simplified schematic of ABE fermentation pathway.

The ABE fermentation produces solvents in a ratio of three parts acetone, six parts butanol to one part ethanol.

The process was initially reported by Louis Pasteur, as early as in 1861, but it was industrially developed by Russian-born biochemist, Chaim Azriel Weizmann, who later served as the first president of Israel. His pioneering work on industrial fermentation gave *Clostridium acetobutylicum*, its status

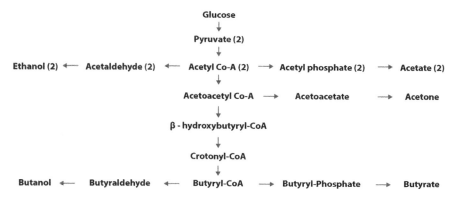

Figure 11.2 Simplified schematic of ABE fermentation pathway.

of being "commercially valuable" and the bacterium is often referred to as the "Weizmann Organism." This production method proved to be of great significance in the manufacture of cordite explosive propellants (which needed acetone), for the British war industry during World War I [12–14].

During the war, the requirement for acetone resulted in the accumulation of butanol as an unwanted by-product. The butanol was stored, and efforts were initiated to build a market for its use as a solvent, especially for the automobile industry. However, that did not yield fruitful result, at least from a commercial point of view. Thus, although the ABE method was one of the first large-scale industrial fermentation processes to be developed, it could not compete economically with the chemical synthesis processes used at that time to make solvents like butanol and the plants mostly shut down. Also, the petrochemical industry was prospering at an unprecedented rate during those years. However, as the need of biofuels gradually became urgent and the use of sustainable resources as feedstocks became popular, there was renewed interest in the ABE fermentation route. Moreover, the ever-increasing price of fossil resources, the uncertainty of the availability of fossil resources in near future, environmental concerns, related regulations and remarkable advances in biotechnology have led to intense research focus on cost-effective, ecofriendly biobutanol production by ABE fermentation, with improved yield and better commercial feasibility.

11.3 Commercial Aspect of Biobutanol Production from Biomass: Strength Analysis

Butanol has long been routinely used as an important solvent by many industries, for example, in cosmetics, hydraulic fluids, detergent formulations, many drugs, antibiotic preparations, hormones, vitamins, paints, lacquers, varnishes, natural and synthetic resins, gums, vegetable oils, dyes, camphor, alkaloids, pesticides, etc. It has also been used as a chemical intermediate in the production of butyl acrylate and methacrylate, and additionally as an extractant in the manufacture of pharmaceuticals [14–16]. However, it soon emerged that it may also be used as a very efficient and superior biofuel. Butanol, as compared to ethanol, is less volatile, less explosive, has higher flash point, lower vapour pressure. All these properties make it safer to handle, thus reducing the cost of precautionary measures and safety trainings. Detailed safety analysis has proved n-butanol to have minimum toxicity and associated hazard risk [16]. Also, it contains more energy, making it more efficient as a fuel due to the high calorific value. Besides, it also has low freezing point, high hydrophobicity, low flammability, less corrosiveness and is also less hygroscopic. Most importantly, it can easily mix with gasoline in any proportion. In addition, the air to fuel ratio and the energy content of butanol are close to gasoline. Butanol can be used either directly or can be blended with gasoline or diesel. There is almost no need for any vehicle modification, thus eliminating the need for retrofit expenses. Moreover, it can safely be supplied through the existing gasoline pipes [14–20]. This makes it an extremely promising, cost-efficient biofuel from industrial perspective.

A comparison with its commercial competitor bioethanol, further shows that biobutanol has better fuel economy (km/L) along with superior safety features as previously stated. It also has lower heat of vaporization, which enables the easier starting of a motor even in cold weather and decreases any ignition issue. In addition, it emits lower amounts of volatile organic compounds in internal combustion engines. Butanol is easily biodegradable under aerobic conditions which further adds to its financial viability [19, 21].

Thus, biobutanol has the potential to substitute gasoline and bioethanol on a commercial scale.

Furthermore, as mentioned above, biobutanol is an important solvent for paint, polymer, and plastic industries, whose global market was already over 3 million tons about a decade ago. Consequently, for the last couple of decades, production of butanol has been strongly supported by many governments around the globe.

The most commonly used butanol-producing Clostridial strains are *C. acetobutylicum, C. beijerinckii, C. saccharoperbutylacetonicum,* and *C. saccharobutylicum* [21]. Of these, the oldest and most popular organism, namely *C. acetobutylicum* utilizes ABE fermentation to completely ferment sugars, such as glucose, fructose, sucrose, lactose, mannose, starch, and dextrins, while galactose, arabinose, xylose, raffinose, insulin, glycerol, and mannitol are partially digested. Being an amylolytic bacterium, it mostly uses starch as the preferred substrate [21–23]. Most of the commercially available biofuels are produced from either starch- or sugar-rich crops or oilseeds. However, these bioenergy sources have serious drawbacks and scientists have turned their attention to the use of biomass like lignocellulosic feedstocks for fuel production [24].

The world is facing the challenge of feeding an ever-increasing population and simultaneously meeting its energy needs without exhausting the biological and physical resources of the earth. Achieving food and nutrition security for all, is central to the Sustainable Development

Goals (SDGs) of the United Nations and so is ensuring clean and renewable energy [7, 25]. First-generation biofuels are produced from biomass that includes edible energy crops, such as sugar-based crops (sugarcane, sugar beet, and sorghum), starch-based crops (corn, wheat, and barley), or oil-based crops (rapeseed, sunflower, and canola). Although they could minimize reliance on traditional fossil fuels and could effectively lower the emission of "Green House Gases," they inadvertently gave rise to serious concerns over food supply, food security, and land requirements. Thus, they are not compatible with the SDGs and inevitably led to the food versus fuel debate [26]. Second-generation biofuels look to be more promising owing to the practice of their bioconversion from organic waste material or residues. These include agricultural wastes, like wheat straw, corn cobs, oat hulls, and sugarcane bagasse; leftover residues from logging and rejects from timber milling, for instance, wood chips and sawdust; spoiled produce and food-processing wastes; urban solid waste materials such as paper, cardboard, kitchen garbage and garden litter. The use of these discarded waste products does not put food production at risk. Also, they do not need additional fertilizer, water, or land. The cellulosic butanol can be produced from agricultural residues, forestry wastes, grasses, wastepaper, and municipal wastes due to their easy availability, low cost and zero threat to food security. Butanol can also be produced from noncellulosic biomass, such as molasses, cheese whey. Lignocellulosic biomass is considered to be the source with the greatest potential for ABE-type fermentation. Lignocellulose is the major, nearly universal component of biomass, making up about half of all matter produced through photosynthesis [11, 22, 25–27].

Algal biomass is also a very promising sustainable and climate-friendly feedstock for producing third-generation biofuels that includes butanol [22]. Algal biomass includes both macroalgae and microalgae. Faster growth rate, high tolerance to poor environmental conditions, and their simple structure make algae a very promising feedstock for biobutanol production.

Waste is currently a major environmental problem and economic burden, both in developing and developed countries. Studies showed that the energy embedded in the food waste represented approximately 2% of annual energy consumption in the United States [28]. Many other nations also portray similar unfortunate scenario. Food waste includes unconsumed food that is discarded by food processing industries, retailers, restaurants, and domestic consumers. Disposal of these food waste in landfill or its incineration can cause severe adverse climatic effects. Their use in compost is not a pocket-friendly solution for average farmers, and additionally poses the risk of polluting both surface water and underground water. Thus, it is judicious to process food waste to produce high value-added products, which can be introduced into existing markets. With the growing interest in biobutanol as a superior and economically more advanced biofuel, researchers are increasingly focusing on converting food waste to produce butanol through the ABE fermentation pathway, using *Clostridium* species [28]. First, most food waste contains significant amounts of sugars and starch, which can be easily utilized by the butanol producing culture organism, without the need for extensive pre-treatment. Second, food waste comprises of significant quantities of other inherent organic molecules like proteins, fatty acids, minerals, which can provide nutrients to support the microbial growth. Therefore, it is commercially a most sensible and viable pathway of biobutanol manufacture.

Thus, industrial production of biobutanol involves several commercial advantages like superior properties, easily available cheap resources and raw materials, ecosafety and sustainability, adaptability to existing infrastructure and vehicle specification, etc.

The above strength analysis for biobutanol production from different types of biomass through microbial ABE fermentation process clearly favors the notion that this biofuel can indeed help shift world's energy future toward the use of sustainable and renewable resources in order to address the increasing demand for fuels at affordable price and to simultaneously preserve the environment for our future generations [19–22].

11.4 Commercial Aspect of Biobutanol Production from Biomass: Weakness Analysis

For decades, the biobutanol industry has been struggling to improve a few crucial issues in order to make a meaningful impact in the market as an alternative fuel. One of the major weak points is the cost of the feedstock and the associated pre-treatment steps involved. In fact, the ABE fermentation process itself, also has several constraints linked with it, like its sluggish nature, low uneconomical concentration of final product, impurities retained, energy inefficient product recovery, phage infection, contamination concerns, and, most importantly, degradation of product organisms due to butanol toxicity. All of these factors have impaired the development of the butanol fermentation industry [14].

Although there has been a significant increase in the number of metabolic engineering studies related to the production of butanol by fermentation, which can be traced back to the 1990s, yet major commercial success stays out of our reach. There is still substantial gap in our knowledge and also considerable lack of data on the mechanism of butanol tolerance in microbes at genetic and protein level. A detailed understanding of the best natural strains may help us construct improved strains and design innovative approaches in fermentation process.

Biobutanol production from different types of biomass involves a number of phases. The actual fermentation reaction is usually preceded by extensive pre-treatment steps. Initially, the feedstock consisting of starch-rich, sugar-rich, or lignocellulosic biomass has to undergo upstream processing to become suitable to be used as a substrate by the microorganism as shown in Figure 11.3. The pre-treatment strategy and expenses differ depending on the type of biomass used. Fermentation is followed by downstream processing that consists of desired product recovery and purification that would be commercially viable. Each stage has its own weak spot(s).

The first and foremost need is to have cheaper renewable nonedible feedstocks as the nature of the feedstock contributes most to the overall production cost (as much as 70–80%). Thus, profitability largely depends upon the price of feedstock and associated pre-treatment steps needed. Another pressing requirement is to accomplish an improved fermentation performance and to evolve progressively sustainable process operations related to solvent recovery and water recycling for effluent treatment. Water usage is another extremely sensitive issue related to UN Sustainable Development Goal and can influence the market [14, 17].

Among cheap, available, and viable feedstocks, lignocellulosic biomass (like corn stover) is becoming the pick of the day. It also helps exclude the risk of food insecurity connected to first-generation biofuel consumption at the expenses of food crops and pastureland [29]. However, bioconversion of this kind of raw material is complicated due to recalcitrance caused by the strong association of cellulose, hemicelluloses and lignin in the lignocellulosic biomass. Cellulose chains remain embedded in a cross-linked matrix of hemicellulose, in turn, wrapped by lignin on the outside,

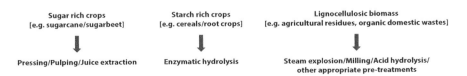

Figure 11.3 Transformation of different biomass to appropriate sugars that can be readily fermented.

making the sugars inaccessible to the microbial enzymes. Hence, extensive pre-treatment is required to reduce this resistant property by opening or partially breaking up their structure, while minimizing the risk of chemical degradation of fermentable sugars.

A key problem of the use of lignocellulosic biomass as a substrate for ABE fermentation is the lignin fraction, which cannot be used by the *Clostridium* species. Hence, pre-treatment and hydrolysis of lignocellulosic biomass before fermentation are essential to convert the complex structure of cellulose and hemicelluloses into simple sugars, that can be used by the bacteria to produce biobutanol. The pre-treatment processes destroy the lignin barrier and partially convert cellulose and hemicelluloses into fermentable sugars, such as glucose, xylose, arabinose galactose, and mannose. A subsequent enzymatic hydrolysis process transforms any remaining polymers into fermentable sugars. Thus, pre-treatment technologies are generally integrated with an added enzymatic hydrolysis process, which usually requires costly enzymes and also increases process time. The biotechnological processes used during pre-treatment often entail high fixed capital investment, which becomes a primary inhibitor to its economic feasibility. For instance, the pre-treatment and hydrolysis process using concentrated acid require expensive equipment capable of high tolerance to the corrosion caused by the acid. The exothermic reaction also needs a lot of cooling water to keep the equipment from overheating [20]. Such examples put a big question mark to the commercial success and sustainability of the entire operation (Figure 11.4).

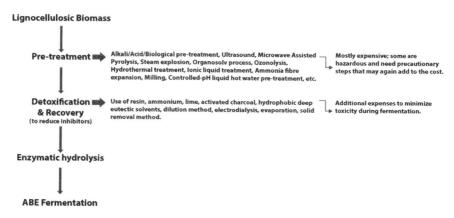

Figure 11.4 Upstream processing of lignocellulose includes pre-treatment and detoxification.

Moreover, pre-treatment process could also produce fermentation inhibitors, such as 5-hydroxymethylfurfural, furfural, acetic, ferulic, glucuronic, and phenolic compounds. These can inhibit microbial activity and may either halt the fermentation process or slow down reaction rates. Consequently, they reduce butanol production, thus additionally requiring a detoxification step to lessen or eliminate their toxic effects during fermentation (Figure 11.4) [30, 31]. This is yet another impediment on the way to commercial viability as it adds to the production cost.

Butanol producing cultures, such as *Clostridium acetobutylicum* and/or *Clostridium beijerinckii*, can work on both pentose and hexose sugars for fermentation. A major drawback of these species is that butanol itself is toxic to the *Clostridium* cells. Butanol is said to be the most hydrophobic and toxic fermentation end product. It is an amphipathic molecule that can partition in the hydrophobic regions of the cell membrane, increasing the polarity of its hydrophobic core. This may lead to severe membrane leakage accompanied by disruption of the membranes' phospholipids and protein content [32]. It impairs cellular metabolism and growth. Butanol concentration above 13 g/L in the fermentation broth is itself toxic to most *Clostridium* strain(s) [18, 33]. As a result, sporulation occurs, bacterial cells become dormant, and butanol production stops, thus creating a bottleneck.

Another vital limitation is low butanol titre. Conventional biobutanol recovery by distillation is an energy-intensive and expensive recovery process. Several butanol removal techniques that are currently in use, include pervaporation, liquid–liquid extraction, gas stripping, adsorption, and vacuum fermentation. It is essential to find an efficient way of integrating a competent butanol recovery method to the fermenter(s) to significantly increase productivity on a commercial scale.

11.5 Commercial Aspect of Biobutanol Production from Biomass: Opportunities and Challenges

Embracing challenges as opportunities and turning stumbling blocks to stepping stones are part of scientific research practices. Biobutanol production from biomass on a profitable commercial scale poses a number of difficulties. However, use of proper biomass substrate, opportunities in genetic engineering to design improved strains, integrated fermentation and modern product recovery techniques have the potential to beat the challenges.

Use of wastes (in form of agricultural, industrial, forestry, municipal and domestic surplus, leftover or reject) to obtain biofuel, is an excellent opportunity of its own accord, to mitigate pollution concerns. That itself can lessen economic burden of a society.

Utilization of lignocellulosic biomass involves the challenges of pre-treatment. However, current research offers abundant possibilities of improvement. The correct choice of substrate is very crucial from an economic point of view. Agricultural residues that have higher cellulose and hemicellulose content and lower lignin content are considered the most suitable substrate for ABE fermentation process. Barley straw has been reported to have the lowest lignin content and might be a good option. Then again, a single parameter cannot be used to define the most favourable substrate. This choice is equally influenced by other factors like yield rate per hectare, climatic condition, harvesting season, transport/delivery cost to the fermentation plant, the ease of hydrolysis, etc. [33]. For example, a recent study on economic analysis of butanol production using various lignocellulosic biomass (such as switchgrass, corn stover, wheat straw, barley straw) showed that corn stover had the lowest unit cost of $0.59/kg butanol yield, whereas barley straw had a cost of $0.75/kg butanol yield via ABE fermentation [34]. Thus, corn stover can be the most likely preference of many biorefineries as an efficient ABE fermentation substrate.

Size reduction of biomass is a necessary step of feedstock preparation prior to pre-treatment. This is another challenging area as it is a highly energy intensive and costly process [33, 35]. But this also opens a door to an excellent opportunity as determining the optimum particle size can yield more sugars, reduce inhibitors, and consequently lower production costs.

Severity factor (SF) is defined as the relationship between time and temperature of the pre-treatment process. Delignification of lignocellulosic biomass is specifically dependent on the SF, and it increases as the SF increases. Delignification is critical because it increases the biomass surface area, in turn, making it more accessible to hydrolytic enzymes so that most of the cellulose and hemicellulose fractions of the biomass can be converted into fermentable sugars. Every pre-treatment process has a critical SF and exceeding that value can adversely affect production. At the same time, finding the optimum SF can maximize yield, lower inhibitor concentration, and improve profitability [33, 36].

Currently, many laboratories are trying to optimize pre-treatment and hydrolysis parameters to reduce the risk of inhibitor formation and increase sugar production. Effort is on to find a better alternative to chemical pre-treatment processes. For example, biological pre-treatment of corn stover using ligninolytic enzymes have shown much reduced inhibitor formation thus decreasing the cost of detoxification [33].

Also, selection of proper nutrients and determining their most advantageous concentrations to favor optimum bacterial growth is crucial to augmenting butanol production and, with careful manipulation, may even give the opportunity to double the yield.

To deal with the issue of butanol toxicity, recent research efforts have focused on screening butanol-tolerant strain(s) from the parent *Clostridium* species. Emphasis is also being given in optimizing different fermentation processes, such as batch, fed batch, and continuous fermentation for the best possible outcome. One such approach involves integrated fermentation with simultaneous removal and recovery of butanol from the fermentation broth, so as to prevent butanol concentrations in the reactor to build up and cross the microbial culture's tolerance threshold.

Moreover, various recovery technologies have been developed, namely pervaporation, liquid–liquid extraction, adsorption, gas stripping, etc. These can be integrated with the conventional fermentation process. Among them, the adsorption technology is thought to offer the best opportunity to attain commercial viability because of its high energy efficiency and simplicity [20, 33].

Novel genetic engineering techniques are helping design newer and better inhibitor-tolerant and hyperbutanol producing microbial strains that can efficiently metabolize mixed sugars. Such approaches can accelerate the utilization of sustainable lignocellulosic biomass for industrial scale ABE fermentation and biofuel production. Recently, researchers have isolated a butanol-tolerant *Clostridium* strain 206 from *C. acetobutylicum* ATCC 824 using the N-methyl-N-nitro-N-nitroso guanidine (NTG) mutation method, which produced 39.3% higher butanol and 48.4% higher ABE than the parent strain [33].

Also, the practice of using a mixed culture (two strain coculture) is often found to be reproducibly more effective for ABE fermentation and can significantly boost up butanol production when compared to the pure culture [33].

With the advent of third-generation biofuel substrates, namely algae, a highly potential commercial prospect can be envisioned in near future [37, 38]. Currently, ongoing research targeting biobutanol production utilizing algal resources is in its nascent stages. But they are being extensively studied for production of other biofuels like biomethanol, bioethanol, biohydrogen, and biosyngas as they have huge advantages. For example, algae have no competition with agricultural crops for resource and land allotment. They can be grown easily without much capital investment, can thrive in different regions and climate and have faster harvest cycles. They possess simple structural characteristics that make them less recalcitrant as a biomass feedstock that does not necessitate extensive preprocessing and conversion steps. High content of starch and convertible sugars and absence of lignin and hemicellulose make microalgae an ideal substrate for fermentation. *Tetraselmis subcordiformis*, different species of *Chlorella* can provide great commercial opportunity, bypassing the usual challenges of butanol production. Unfortunately, only a limited number of studies have used microalgal biomass for butanol production. Moreover, most of the studies are only confined to the laboratory stage and no commercial scale production have been seriously attempted or considered so far. Merging algal cultivation technology with biobutanol production process may help rejuvenate the commercial prospect of biobutanol industry and should be explored in greater details.

11.6 Discussion: Evaluating the Future Prospects of Biobutanol

The current world owes this to the future generation that they inherit a planet where they do not have to worry about equal access to natural resources, or be threatened by environmental degradation and menace of unsustainable development [7]. Ever-increasing population, extensive urbanization, modernization, industrialization, booming transportation sector, technological innovations, and gadgets that run most metropolitan households, continue to strain and drain global energy supplies. With the steady depletion of petroleum and fossil fuels and consequent increase in their prices, accompanied by escalating concerns of global warming and environmental pollution, the scientific community and industrial sector are opting for renewable ecofriendly alternative energy sources like biofuel.

Biobutanol may play a pivotal role in the overall success of the biofuel industry because it offers more benefits compared to ethanol as discussed above. Butanol can be produced through petrochemical and biochemical routes. The petrochemical route is less expensive but is predicted to soon become unsustainable in near future. Microorganisms, such as *Escherichia coli*, *Clostridium acetobutylicum*, *Bacillus subtilis*, *Clostridium beijerinkckii*, *Pseudomonas putida*, and *Saccharomyces cerevisiae*, etc., can use carbohydrate-rich waste material to produce biobutanol through aerobic and anaerobic fermentation [39]. *Clostridium*-mediated ABE fermentation pathway is the most preferred process for biobutanol production and the substrate of choice is biomass. About 60% of EU renewable energy is reported to originate from biomass. It remains to be seen whether the world can produce enough biomass to sustain all the food, animal feed, and bioenergy needed for the future generations.

Recent improved technologies to give higher butanol yield and concentration, use of process by-products in other commercial applications, modified bacterial strains and culture strategies, better separation and purification techniques together show great promise for the biobutanol industry. Ongoing research efforts, support and funding by governments, ecofriendly energy policies, all offer opportunities for future breakthrough that may give additional boost to its successful commercial scale production. According to a recent report, biobutanol market size is estimated to reach US$1.8 billion by 2027 after growing at a compound annual growth rate of around 7.3% from 2022 to 2027.

Along with biofuel driven automobiles, there is also a growing demand for electric vehicles that may hinder the growth of the biobutanol market to some extent. However, electric vehicles have their own challenges like much shorter driving range, long, multiple charging time, very few charging stations, risk of break down on charge exhaustion, etc. Thus, biofuel offers a better sustainable alternative of which biobutanol shows the most encouraging future prospect. The main advantages of biobutanol that promises commercial success are summarized in Figure 11.5.

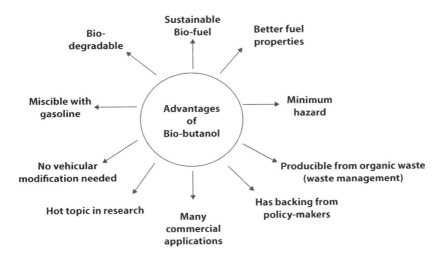

Figure 11.5 Advantages of biobutanol that adds to its future prospect.

Acknowledgment

The author would like to thank Dr. Sasabindu Jana, Principal, Raidighi College, West Bengal, India and the Department of Microbiology, Raidighi College for their support. The author would also like to thank the editors of the book.

References

1. Lloyd, P.J., The role of energy in development. *J. Energy South. Afr.*, 28, 1, 54–62, 2017. https://dx.doi.org/10.17159/2413-3051/2017/v28i1a1498.
2. https://databank.worldbank.org/source/world-development-indicators
3. Basavaraj, G., Rao, P.P., Reddy, R., Kumar, A.A., Srinivasa Rao, P., Reddy, B.V.S., *A review of national biofuel policy in India: A critique of the need to promote alternative feedstocks*, Journal of Biofuels, 3, 2, 65–78. July-December 2012.
4. https://www.worldbank.org/en/topic/energy/overview
5. https://sdgs.un.org/goals/goal7
6. Bosettia, V., Carrarob, C., Tavoni, M., Climate change mitigation strategies in fast-growing countries: The benefits of early action. *Energy Econ.*, 31, Supplement 2, S144–S151, 2009. http://dx.doi.org/10.1016/j.eneco.2009.06.011.
7. *World* summit on sustainable development, WSSD, Johannesburg, 2002, https://documents-dds-ny.un.org/doc/UNDOC/GEN/N02/636/93/PDF/N0263693.pdf?OpenElement.
8. Administrative approval for implementation of Biomass Programme under the Umbrella scheme of National Bioenergy Programme for duration of FY 2021-22 to 2025-26 (Phase-I)-Reg, Ministry of New and Renewable Energy, Govt. of India, 23rd Nov. 2022. https://mnre.gov.in/img/documents/uploads/file_f-1669179617561.pdf.
9. https://mnre.gov.in/Bio%20Energy/policy-and-guidelines
10. Energy Independence and Security Act (EISA), 2007, Retrieved from http://www.gpo.gov/fdsys/pkg/BILLS-110hr6enr/pdf/BILLS-110hr6enr.pdf.
11. Glazer, *Microbial Biotechnology*, 2nd Edition, Cambridge University Press, New York, 2007. ISBN-13 978-0-521-84210-5.
12. Jones, D.T. and Woods, D.R., Acetone-butanol fermentation revisited. *Microbiol. Rev.*, 50, 4, 484–524, December 1986, https://doi.org/10.1128/mr.50.4.484-524.1986.
13. Rose, N., *A Biography*, C. Weizmann (Ed.), pp. 55–56, Viking Penguin, Inc., New York, New York, 1986.
14. García, V., Päkkilä, J., Ojamo, H., Muurinen, E., Keiski, R.L., Challenges in biobutanol production: How to improve the efficiency? *Renewable Sustain. Energy Rev.*, 15, 2, 964–980, 2011. https://doi.org/10.1016/j.rser.2010.11.008.

15. Dürre, P., Biobutanol: An attractive biofuel. *Biotechnol. J.*, 2, 12, 1525–34, 2007 Dec.
16. https://www.epa.gov/sites/default/files/2015-04/documents/butanol.pdf
17. Green, E.M., Fermentative production of butanol–The industrial perspective. *Curr. Opin. Biotechnol.*, 22, 3, 337–43, 2011 Jun, Epub 2011 Mar 1. PMID: 21367598.
18. Baral, N.R. and Shah, A., Techno-economic analysis of cellulosic butanol production from corn stover through acetone-butanol-ethanol (ABE) fermentation. *Energy Fuels*, 30, 7, 5779–5790, 2016. https://doi.org/10.1021/acs.energyfuels.6b00819.
19. Rathour, R.K., Ahuja, V., Bhatia, R.K., Bhatt, A.K., Biobutanol: New era of biofuels. *Int. J. Energy Res.*, 42, 4532–4545, 2018. https://doi.org/10.1002/er.4180.
20. Jang, M.-O. and Choi, G., Techno-economic analysis of butanol production from lignocellulosic biomass by concentrated acid pre-treatment and hydrolysis plus continuous fermentation. *Biochem. Eng. J.*, 134, 30–43, 2018. https://doi.org/10.1016/j.bej.2018.03.002.
21. Antil, S., Biobutanol: Production, scope and challenges. *Int. J. Curr. Microbiol. App. Sci.*, 8, 11, 580–584, 2019, https://doi.org/10.20546/ijcmas.2019.811.070.
22. Abo, B.O., Gao, M., Wang, Y., Wu, C., Wang, Q., Ma, H., Production of butanol from biomass: Recent advances and future prospects. *Environ. Sci. Pollut. Res. Int.*, 26, 20, 20164–20182, 2019 Jul.
23. Lee, S.Y., Park, J.H., Jang, S.H., Nielsen, L.K., Kim, J., Jung, K.S., Fermentative butanol production by Clostridia. *Biotechnol. Bioeng.*, 101, 2, 209–28, 2008 Oct 1. PMID: 18727018.
24. Popp, J., Lakner, Z., Harangi-Rákos, M., Fári, M., The effect of bioenergy expansion: Food, energy, and environment. *Renewable Sustain. Energy Rev.*, 32, 559–578, 2014. https://doi.org/10.1016/j.rser.2014.01.056.
25. Muscat, A., de Olde, E.M., de Boer, I.J.M., Ripoll-Bosch, R., The battle for biomass: A systematic review of food-feed-fuel competition. *Global Food Secur.*, 25, 100330, 2020. https://doi.org/10.1016/j.gfs.2019.100330.
26. Moodley, P. and Ray, R.C., *1 - Sustainable biofuels: Opportunities and challenges*, Applied Biotechnology Reviews, Sustainable Biofuels, pp. 1–20, Academic Press (Elsevier), Headquarters Cambridge, Massachusetts, 2021, https://doi.org/10.1016/B978-0-12-820297-5.00003-7.
27. Dahman, Y., Syed, K., Begum, S., Roy, P., Mohtasebi, B., 14 - Biofuels: Their characteristics and analysis, in: *Biomass, Biopolymer-Based Materials, and Bioenergy*, D. Verma, E. Fortunati, S. Jain, X. Zhang (Eds.), pp. 277–325, Woodhead Publishing (Elsevier). Headquarters: Sawston, Cambridge, https://doi.org/10.1016/B978-0-08-102426-3.00014-X.
28. Huang, H., Singh, V., Qureshi, N., Butanol production from food waste: A novel process for producing sustainable energy and reducing environmental

pollution. *Biotechnol. Biofuels*, 8, 147, 2015. https://doi.org/10.1186/s13068-015-0332-x.
29. Rulli, M., Bellomi, D., Cazzoli, A. et al., The water-land-food nexus of first-generation biofuels. *Sci. Rep.*, 6, 22521, 2016. https://doi.org/10.1038/srep22521.
30. Galbe, M. and Wallberg, O., Pre-treatment for biorefineries: A review of common methods for efficient utilisation of lignocellulosic materials. *Biotechnol. Biofuels*, 12, 294, 2019. https://doi.org/10.1186/s13068-019-1634-1.
31. Birgen, C., Dürre, P., Preisig, H.A. et al., Butanol production from lignocellulosic biomass: Revisiting fermentation performance indicators with exploratory data analysis. *Biotechnol. Biofuels*, 12, 167, 2019. https://doi.org/10.1186/s13068-019-1508-6.
32. Amiri, H., Recent innovations for reviving the ABE fermentation for production of butanol as a drop-in liquid biofuel. *Biofuel Res. J.*, 7, 4, 1256–1266, 2020.
33. Baral, N.R. and Shah, A., Microbial inhibitors: Formation and effects on acetone-butanol-ethanol fermentation of lignocellulosic biomass. *Appl. Microbiol. Biotechnol.*, 98, 22, 9151–72, 2014 Nov.
34. Kumar, M., Goyal, Y., Sarkar, A., Gayen, K., Comparative economic assessment of ABE fermentation based on cellulosic and noncellulosic feedstocks. *Appl. Energy*, 93, 193–204, 2012.
35. Zhu, Z., Sathitsuksanoh, N., Vinzant, T., Schell, D.J., McMillan, J.D., Zhang, Y.H.P., Comparative study of corn stover pretreated by dilute acid and cellulose solvent-based lignocellulose fractionation: Enzymatic hydrolysis, supramolecular structure, and substrate accessibility. *Biotechnol. Bioeng.*, 103, 715–724, 2009.
36. Baral, N.R., Li, J., Jha, A.K., Perspective and prospective of pre-treatment of corn straw for butanol production. *Appl. Biochem. Biotechnol.*, 172, 840–853, 2014.
37. Yeong, T.K., Jiao, K., Zeng, X., Lin, L., Pan, S., Danquah, M.K., Microalgae for biobutanol production–Technology evaluation and value proposition. *Algal Res.*, 31, 367–376, 2018. https://doi.org/10.1016/j.algal.2018.02.029.
38. Kongjan, P., Usmanbaha, N., Khaonuan, S., Jariyaboon, R., Reungsang, A., Chapter 18 - Butanol production from algal biomass by acetone-butanol-ethanol fermentation process, in: *Clean Energy and Resources Recovery*, V. Tyagi and K. Aboudi (Eds.) pp. 421–446, Elsevier Headquarter, Amsterdam, Netherlands, 2021, https://doi.org/10.1016/B978-0-323-85223-4.00014-2.
39. Rathour, R.K., Ahuja, V., Bhatia, R.K., Bhatt, A.K., Biobutanol: New era of biofuels. *Int. J. Energy Res.*, 42, 4532–4545, 2018. https://doi.org/10.1002/er.4180.

12

Current Status and Challenges of Biobutanol Production from Biomass

Ram Bhajan Sahu and Priyanka Singh*

Institute of Allied Medical Science and Technology, NIMS University, Rajasthan, India

---

Abstract

Butanol has been used as advance fuel due to its high energy content, corrosiveness, and volatility properties. The physiochemical pre-treatment methodologies for extraction of butanol from lignocellulose biomaterials are very costly and could be preferred at greater extent. Therefore, the microbial fermentation process has attracted scientist to explore some efficient microbes for the production of biobutanol. Acetone–butanol–ethanol (ABE) fermentation process has commonly used for their production but this methodology has major limitation in terms of lower productivity of butanol, high cost of pre-treatment methodology for hydrolysis of cellulosic feedstocks, and accumulation of toxic inhibitors components in feedstock hydro lysates. These challenges have been extensively addressed using different downstream processing strategies. In this book chapter, we have described the physiochemical properties and industrial application of biobutanol. The comparative analysis of different pre-treatment methodologies based on chemical, physical and biological process has been subsequently studied. The development and challenges for scarification, enzyme hydrolysis, and microbial fermentation process have been also extensively highlighted.

Keyword: Butanol, pre-treatment methodology, fermentation process, cellulosic feedstock

12.1 Introduction

Petroleum derived fuels has not been currently preferred due to rising global warming and environmental issues, and high price of oil [1].

Corresponding author: priyay20@gmail.com

Arindam Kuila and Mainak Mukhopadhyay (eds.) Production of Biobutanol from Biomass, (301–322) © 2024 Scrivener Publishing LLC

Renewable substrates are being currently preferred for production of biofuel by fermentative approach than petroleum derived substrates. Butanol has been considered as promising biofuel in compare to bioethanol. These biobutanol fuels caused lesser ignition problems due to their longer carbon chain length, higher heating value, lower volatility, and heat of vaporization. In recent years, diesel engines are consuming pure butanol or blended biobutanol without any apparent modification [2]. Diesel blended with gasoline has been commonly used in transportation instead of bioethanol [3, 4]. Butanol (C_4H_9OH) has four structural isomers, like n-butanol, secondary-butanol, iso-butanol, and tertiary-butanol [5]. Butanol can be used as solvent, intermediate compound for chemical synthesis, and as a transport fuel [6, 7]. *Clostridium species* are generally used for fermentative production of biobutanol through acetone, butanol, and ethanol (ABE) fermentation process [8–10]. The economy of bioseparation process will depend on the cost of removal of acetone and ethanol as by-products from butanol main product during distillation process. The first commercial plants have been established in 1910 for production of biobutanol using molasses and corn as substrate. This plant has commercially produced about two-thirds of butanol and one-tenth of acetone required all around the world. However, the higher cost of molasses and corns has decreased the popularity of this method for production of butanol since 1960.

Different types of agricultural biomasses are being commercially used for the production of biofuels at a very low cost [11]. The production of bioethanol has been improved with fermentation process with agricultural waste product such as corn stover, rice straw, grasses and wood chips. Their availability with renewable nature has been considered as cheap raw material for the fermentative production of biofuel [10, 12].

This lignocellulosic biomass has resulted the production of 50% to 80% bioethanol with low emission of Green House Gases (GHG). The hydrolysis of these residues has also released recalcitrant molecules as by-products due to impregnation of lignin content. Enzymatic treatment and scarification process has further removed the lignin content from other hydrolyzed components. Various pre-treatment methodologies have been employed for the delignification of hydrolyzed cellulosic biomaterials for reducing their degree for recalcitrant nature. Cellulose hydrolysis process has been subsequently improved by targeting hemicellulose conversion using pentose sugar based specific enzymes. The combinatorial approaches for pre-treatment methodologies including physical, chemical, biological has effectively reduced the level of recalcitrant at a very low extent. Physical treatment will improve overall crystallinity of biomaterials, chemical treatment for hemicellulose bioconversion and biological methodologies for delignification and polymerization of cellulosic polymer [13]. Microbial fermentation approaches have been used for the production of bioethanol from hydrolyzed sugar components [6]. In the current scenario, the production of cellulosic ethanol from cellulose hydrolysis is not considered as a cost-effective method due to obtaining lower yield of ethanol [14]. Has reported the production of bioethanol by simultaneous scarification and fermentation process. The higher percentage of bioethanol (0.96%) has been produced by mixed culture fermentation approach using *Saccharomyces cerevisiae* and *Zymomonas mobilis* from groundnut shell. The global demand of bioethanol production has caused the scarcity of food crops in whole world. Therefore, alternative resources have been explored to avoid a situation of competition between food and bioenergy production [10]. The biofuels of first and second generations showed higher moisture content. These biofuels need high transportation costs and high management skill for avoiding methodologies and hydrolytic process for bioconversion of fermentable sugars into Butanol. Its production has major challenge for bioconversion of lignocellulose to monomeric glucose unit as compared to bioethanol. These bioconversion process required higher capital investment in terms of equipment, raw materials, pre-treatment process and strain development, cost for sales and marketing of Butanol. The production cost can be minimized using low-cost raw materials as substrate, selecting an efficient pre-treatment and hydrolysis method, use of recombinant strains with tolerance against butanol.

This book chapter has focussed on the sources, limitations for the production of different biofuels. The pre-treatment methodologies including physical, chemical and biological processes has been extensively described for efficient production of butanol. Authors would get information for all stages of production of butanol viz. raw materials, pre-treatment methodologies, downstream proceeding unit, and upstream process for their fermentation, and recovery. We have also discussed the factors contributing cost for its production and analyzed the strategies of research unit.

12.2 Overview of Biofuel

Over the last few decades, bioethanol has been considered as effective alternative source for fossil-based petrochemical-derived fuel in the transport sector. The use of ethanol-blended gasoline as transport fuel has continuously reduced emissions of greenhouse gases. The use of renewable energy resources has decreased the values of crude petroleum for increasing energy security and diversifying energy supplies. Brazil and United

State of America are first countries for starting production of ethanol from biological sources. Due to having high value of octane rating, ethanol has been commercially preferred in spark-ignition engines. The quality of ignition has been reported as poor due to low cetane number, which might cause low vapour pressure and high latent heat of vaporization. Gasoline has been added as the most cost-effective volatile fuel in small proportion to the ethanolic mixture. The most common blends have been categorized on the basis of different percentage of ethanol and gasoline as E5G to E26G, E85G, E15D, and E95D. The light duty flexible fuel vehicles (FFV) have commonly used bioethanol blend of E85G. Ethanolic blend with 10% to 15% gasoline has enhanced the octane rating of fuel and reduced emission of greenhouse gases. E22G blends can be used effectively in spark ignition engines without any operating problems. E22D Blends of diesel with 15% ethanol has not caused any technical engine problem and not required to improve ignition efficiency. The leading countries for production of bioethanol are mainly Brazil and USA. All Asian countries account for 14% of world's production of bioethanol.

12.2.1 History for Biofuel

From last few decades, many significant strategies have been applied for the production of biofuels. Louis Pasteur has firstly explored the production of ethanol in 1861 during fermentation process of glucose under anaerobic condition [17]. The production of butanol has been reported by Fernbach (1911) due to fermentative process for potatoes substrate. The commercial production of butanol has been further carried out through mixed ABE fermentation process in United Kingdom (1912) [18].

Weizmann has also reported the production of butanol by ABE fermentation process using *Clostridium acetobutylicum*. They have commercially used the fermentative product containing acetone, butanol and ethanol as gunpowder. The specific plant for production of ethanol has firstly established by Weizmann in the year 1920 at the Terre Haute plant on the basis of ABE fermentation process [18]. Thereafter, other ABE plants have been established in different places around the world like India, Australia, Philadelphia, America, Japan, and Africa during 1936 to 1941 periods [19]. The industrial fermentation process based on consumption of food crops was not popular since World War II. Chemical methods are become more popular for the production of butanol as transport fuel from 1960 onward [2].

Genetically improved strains are being used for the improved production of butanol and efficient downstream processing techniques has been designed for higher rate of recovery of butanol from 1960 to 1990 [20]. In 20th century, researchers have used starch feedstock as raw materials for carrying ABE fermentation process [21, 22]. The use of agrochemical waste residues for butanol production has improve the circular economy and sustainable development. In current scenario, municipal wastes, industrial wastes, and algae-based biomasses are used as effective source for the production of butanol. Fernbach and Strange has filed patent for the production of butanol in the United Kingdom (UK). Weizmann had successfully isolated *Clostridium acetobutylicum* as new fungal strain for ABE fermentation process to produce large amounts of acetone as product [23].

During World War I, mixture of acetone, butanol, and ethanol is used as effective gunpowder, and therefore, its production was in huge demand in Britain. The first commercial ABE plant had been established in Canada and USA to fulfil large demand of butanol. After 1936, other industries for production of ABE were established in Soviet Union, South Africa, China, and Egypt. During the start of the Second World War (1945), Japan had started the production of butanol from food energy crop of sugar [24]. In the 1950s, petrochemical derived materials are also being used for the production of *n*-butanol using aldol condensation process for acetaldehydes, followed by dehydration and hydrogenation of croton aldehyde [25]. The chemical methodology for the production of butanol has been preferred instead of ABE fermentation process after 1960 due to low-cost oil prices. The chemical route had not lasted for a long time due to rising price of crude oil. The industrial production of butanol by ABE fermentation had been also started in China and Brazil. Thereafter, many factories had been established globally in different locations like USA, France, Slovakia, and UK for production of biobutanol as transport fuel. In 2005, n-butanol has been started to use as biofuel in different countries. The consumption of butanol as fuel increased 9% higher than conventional petrochemical derivatives [26]. The emissions of greenhouse gases include hydrocarbons, carbon monoxide (CO), and nitrogen oxides (NOx) were substantially increased in environment. The first new commercial plant for production of n-butanol through ABE fermentation process has been established by BP in Salted (UK) to achieve its production up to 420 million liters [27].

The fermentation process for bioconversion of glucose sugar into ethanol has been firstly suggested for the production of simple biofuel agent. Ethanol has been commonly used as spirit and intoxicating ingredient for alcoholic beverages. In the 12th century, distillation of wine with salt had been used for the production of aqua Arden in a number of Latin works. After the 13th century, ethanol had been used as common agent by Western European. Zimbabwean Triangle Ethanol Plant in 1980 has produced large amount of bioethanol (approximately 120,000 litres daily and 40 million liters annually) for 12 years. Sugar cane molasses had been used as main feedstock raw materials for commercial production of bioethanol [13]. The historical events in ABE fermentation process have been listed in Table 12.1. Butanol can be effectively used as crucial fuel in an internal combustion engine. Its properties were found to be similar with gasoline in compare to ethanol. The fuel agent of n-butanol and isobutanol could be effectively produced from fossil fuels and lignocellulosic biomasses.

12.3 Classification of Bioethanol

Bioethanol has been classified into the following two categories according to availability of different sources of raw materials and production process:

(a) The chemical route results the synthesis of ethanol by converting nonrenewable sources of ethylene into ethanol after treatment with steam under optimum condition

$$\text{Ethylene } (C_2H_4) + \text{steam } (H_2O) \rightarrow C_2H_5OH \text{ (Ethanol)}$$

(b) Fermentation process for production of Ethanol using microbial cellular metabolism process for bioconversion of sugar sources into bioethanol through biochemical/microbial/enzymatic/biotechnological route. The production of bioethanol has been carried out by anaerobic fermentation process of sugar in the presence of yeast culture by following equation:

$$C_6H_{12}O_6 + \text{Yeast} \rightarrow 2C_2H_5OH + 2CO_2$$

Biofuel has been classified into different types of generation fuels on the basis of feedstocks and biological conversion processes.

12.3.1 First-Generation of Ethanol

The first-generation ethanol has been produced from fermentation process of food crop based feedstock including starch, sugar molecules. These are produced by enzymatic hydrolysis along with microbial fermentation process of raw material of starches and sugar. Food crop

Table 12.1 Historical events for production of acetone through ABE fermentation process.

Years	Scientists contributions	References
1861	Microbial production of vibron butyrique as butanol	[28]
1910	Cellular metabolism for biosynthesis of isoprene or butadiene from isoamyl alcohol or butanol	[18, 29]
1911	Fermentation process for production of butanol from potatoes	[18]
1912–1914	ABE fermentation process for production of Butanol	[18, 30]
1913	Filed patent on production of butanol from microbial fermentation of potato substrate	[18]
1916	ABE fermentation process using maize as raw materials	[31]
1918	Production of acetone for cordite and airplane In Canada, chemical synthesis of butanol and their storage in vats Bioconversion of butanol to methyl ethyl ketone	[18]
1919–1920	Weizmann had produced butanol at the Terre Haute plant in U.S.A. They started operation according to US license and later got worldwide process rights of this patent.	[18]
1935	Distillation unit had been used for separation of butanol from other alcoholic components	[30]
1936–1941	Exploration of new microbial strains for ABE fermentation process	[19, 32]
1936–1960	Construction of bioreactor for production of solvents had been started in different part of world, such as South Africa, Australia, Japan, and India.	[31, 33]

(Continued)

Table 12.1 Historical events for production of acetone through ABE fermentation process. (*Continued*)

Years	Scientists contributions	References
1950–1960	ABE fermentation process has been less preferred due to increased cost of substrate and increased competition for availability of raw substrate material.	[30]
1970–1982	Chemical Factories of Egyptian Sugar and Distillation Company had become partial functional. Plant unit for production of solvent had been fully functional in South African until 1982.	[20, 34]
1980–1990	Genetic improved strains had been used for enhancement of production of biobutanol.	[22]

like sugar cane, sugar beet and sweet sorghum has been mainly composed of glucose, fructose, and sucrose [15]. These sugars crops have undergone pre-treatment process followed by enzymatic hydrolysis and microbial fermentation approach for the commercial production of ethanol. Distillation unit has been further used for separation of ethanol from steam followed by dehydration. Corn and wheat grains are mainly composed of starch as polysaccharide with glucose units linked by α (1–4) and α (1–6) glycosides bonds. These starches are being hydrolyzed into glucose monomers using specific enzymes containing α-amylase and glucoamylase. Glucose has been finally fermented into ethanol by *Saccharomyces cerevisiae* under anaerobic condition. The high-cost production of ethanol by using food crop-based feedstock and high use of cultivated land has been criticized. These drawbacks of first-generation bioethanol has given rise to explore alternative non–food-based feedstocks for production of bioethanol.

12.3.2 Second-Generation Bioethanol

The fermentation process for feedstock of nonfood crops are mainly used for production of second-generation (2G) bioethanol. These feedstocks are composed of cellulosic biomass derived from energy crops (switch grass, miscanthus), agricultural residues (woodchips, corns over, sugarcane bagasse, and sawdust). These cellulosic biomasses are mainly composed of cellulose, hemicellulose, and lignin polymers interlinked in a heterogeneous matrix. Cellulose as polysaccharide has been polymerized with many β (1–4) linked D-glucose units. Hemicellulose has been considered as heteropolymer of xylose, mannose, galactose, rhamnose and arabinose. Lignin is a complex polymer with cross-linked aromatic compounds. These lignin molecules as protective barrier, has hindered the depolymerization of polymeric chain of cellulose and hemicellulose to fermentable sugars.

12.3.3 Third-Generation Bioethanol

Algal biomasses are being mainly used for production of third-generation bioethanol through fermentation approach. These microalgae can utilize sugars units for conversion into butanol along with hydrogen, diesel, and isobutene. These autotrophic algae have ability for conversion of sunlight and atmospheric CO_2 into starch and cellulose through photosynthetic process [35]. On the other hand, these heterotrophic algae species have efficiency for conversion of small organic carbon compounds into lipids, protein, and oils. Some algae like red, brown, and green algae could be used for the production of hydrocolloids constituting 10% to 40% of their biomass. These macro algae have low concentration of lipids, 35% to 74% carbohydrates and 5% to 35% proteins. The oil content has been reported 20% to 50% in microalgae, like *Botryococcus braunii*, *Chlorella* sp., *Nannochloris* sp., *Nietzsche* sp., *Schizochytrium* sp. The higher yield for bioethanol has been reported from both microalgal and macroalgal biomasses.

12.3.4 Fourth-Generation Bioethanol

Advanced technologies like electrochemical synthesis, oxide electrolysis, and petroleum hydro processing are being currently used for conversion of captured carbon dioxide into fourth-generation bioethanol. These fourth-generation bioethanol's are being considered carbon negative due to equal production yield of carbon in compare to consumption of captured carbon molecules. The comparative analysis of all three generation of bioethanol has been compared in Figure 12.1.

12.4 Production of Biobutanol

The metabolic pathway for production of butanol was mainly based on the biorefinery approaches for conversion of cassava waste residues into butanol at analytical concentration level. The fermentation process results the

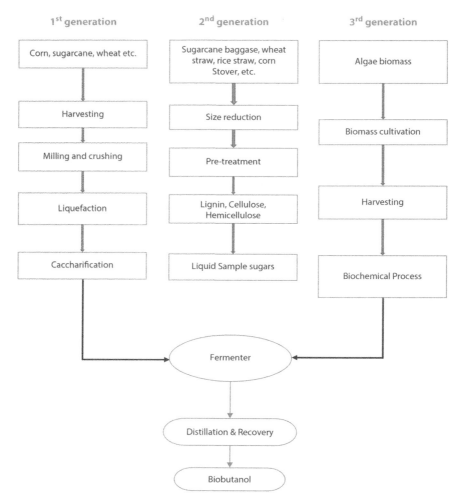

Figure 12.1 Production path of biobutanol from various generations of feedstock.

secretion of butanol as main product and acetone-ethanol as by-products. ABE fermentation unit has been comprised of several interconnected processing units to accomplish the production of butanol with other by products as shown in Figure 12.2.

12.4.1 Pre-Treatment Stages

This fermentation process has been started with pre-treatment stage, where, raw materials are being crushed into fine particle size by mechanical

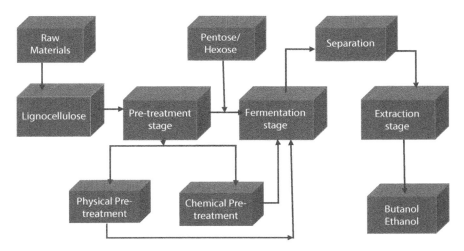

Figure 12.2 Different steps for processing of biobutanol in industrial sectors.

methodology. The sizes of particles are being reduced to 4 mm after crushing process [36]. These fine particles are mixed with 1.05% (w/w) sulphuric acid in tank bioreactor.

The process has been continued to the acid pre-treatment reactor for hydrolysis of xylene into xylose (approximately 90% yield), glycan into glucose (approximately 7%), and partial solubilization of lignin (approximately 5%) (Lou et al., 2010). This process has also released other by products like acetic acid, furfural derivatives. The steam would be released in reactor after performing acid pre-treatment at 190°C temperature and 13 atm pressure. The steam has been collected in flash cooler for reducing pressure up to atmospheric level. The product collected after biorefinery treatment process includes solid-phase chemicals like hydrolyzed five-carbon sugars and acids. These residues have been passed through filtration unit for separating solid components from liquid phases. The liquid stream containing water and xylose are being finally passed to the fermentation unit. Solid stream like lignin, cellulose, and undesirable residues has been passed to an ion-exchange unit for separating acetic and sulphuric acids from the solid residues [28]. The reacidification of system has been conducted with sulphuric acid for achieving acidic condition (pH 2). The alkaline condition has been maintained by adding small concentration of Lime in to mixture and pH value has been increased up to 10. Gypsum could be produced as by product during over liming process, which could be effectively separated from mixture using press filter and hydrocyclone unit.

The hydro cyclone unit would effectively remove 99.5% of gypsum and solid stream containing 20% liquid mixture [37]. The cellulose stream has been further carried for delignification and enzymatic hydrolysis stage to enhance production of glucose in bioreactor. The main flow leaving this stage are mainly composed of 25% of glucose, 74% of water, and >1% of impurities.

12.4.2 Enzymatic Hydrolysis Stage

The main flow of cellulose stream has been directed to enzymatic hydrolysis stage after neutralization and over liming stages. Lignin has been initially separated from crude cellulosic mixture and suitable biorefinery process has been applied for conversion of cellulose into glucose monomer units [38]. Using 2% NaOH (w/w). The delignified components has been further processed for enzymatic hydrolysis in specific bioreactor for the bioconversion into glucose molecules. This cellulosic enzymatic hydrolysis process has been carried out at optimum temperature (45°C) and optimum pressure (1.68 atm) to achieve 90% glucose yield [17]. In bioreactor. Mixture of enzyme containing endoglucanase, exoglucanases, and β-glucosidase has been generally used in enzymatic process unit [17]. The outlet stream with high glucose content has been filtered for removing undesirable impurities, and they are subsequently used for next fermentation stage.

12.4.3 Fermentation Stage

The fermentation stage has been started with supplementation of xylose and cellulose streams for production of biobutanol in bioreactor [8]. Acetone–butanol–ethanol (ABE) fermentation process has been carried out after inoculation with *Clostridium saccharobutylicum*. This fermentation process has resulted the production of butanol (63%), acetone (32%), and ethanol (1.4%) [37]. They are being further supplied to distillation unit for separation of acetone and ethanol from butanol.

12.4.4 Separation Stage

The fermentation broth stream has been further processed to gas stripping unit for removal of maximum water content and by-products (butyric and acetic acid). Acetone, butanol and ethanol as volatile components are being entrapped in the vapour phase in distillation units and subsequently entrapped in gas stripping unit. The vapour stream has been further passed through condenser unit for separation of fermentation gases and solvents at very low temperature. Condensate stream has been finally processed through Beer column extraction unit attached with two sections for collecting rich-solvent streams and rich-water stream separately. The water stream has been passed for wastewater treatment and solvent stream has been passed through double distillation unit for separation of solvents on the basis of their boiling points. The mainstream has been connected with separate column unit for collecting 98% acetone-rich distillate stream.

12.4.5 Production of Butanol from Genetically Improved Strains

Genetic engineering approaches are being applied for obtaining higher yields of butanol [1] the fermentation technology has not been applied effectively for the production of butanol efficiently. Their yields are very low and separation techniques opted for their purification are being very costly. Microbial production of butanol has been considered as a low-cost approach relative to their production from fossil fuel based petroleum-derived substrates [39, 40], *the combinatorial approach of electrochemical and microbial fermentation process has effectively improved the production of butanol after fermentation process with renewable substrates* [41].

Escherichia coli as Gram-negative bacterium has been reported for commercial production of isobutanol. *Genetically improved strains for E. coli cell have ability for production of higher yield of isobutanol.* The metabolic efficiency of *E. coli* cells could be effectively improved by elementary mode analysis to achieve higher production of isobutanol [42]. *These genetically improved E.* coli cells have tendency for consumption of Lignocellulosic polymeric constituents from agricultural plant for synthesis of isobutanol products [43, 42]. Genetic manipulation technology has been applied for developing recombinant strains of *E. coli* for ability to consume lignocellulose at higher rate for production of higher yield of isobutanol [42]. Random mutagenesis has been applied for developing mutant strains with higher degree of tolerance against secreted isobutanol as main product [43, 45].

The fermentation approaches of ABE have been applied for production of n-butanol in bioreactor inoculated with *Clostridium acetobutylicum, Clostridium beijerinckii* after hydrolysis of starch substrate molecules [8, 46, 47]. Energy food crops (sugar cane, sugar beets, wheat, corn grain, and cassava) have been also used as feedstocks for production of n-butanol from fermentative approaches. The limited use of energy food crops for production of butanol may be due to high-cost values of these food crops and scarcity of global food. In current scenario, these feedstocks have been replaced with no food-energy crops like switch grass, bagasse, straw and corn stalks. The production of n-butanol from these agricultural by-products have been reported as low cost and more efficient technique [48]. *Clostridium cellulolyticum* has efficiently produced isobutanol by fermentation process with cellulosic polymer [49].

Metabolic pathway has been manipulated in *Clostridium kluyveri* for conversion of succinate into butyrate product under specific aerobic condition. These butyrate molecules will be further act as precursor agent for production of n-butanol. *Clostridium acetobutylicum* and *Clostridium saccharobutylicum* under anaerobic conditions has converted succinate into 4-hydroxybutyrate by two-step biochemical reaction and further metabolized to crotonyl-coenzyme A (CoA) for conversion into butyrate as main product [29].

Cyanobacteria as photosynthetic bacteria has also showed ability for production of isobutanol and its corresponding aldehydes after undergoing genetically improvement [40, 50]. Cyanobacteria can grow faster in compare to plant growth and efficiently absorb sunlight for their photosynthetic process [50]. These Cyanobacterial species could be replenished at a faster rate than Lignocellulosic plant matter used for production of biofuel. These Cyanobacterial species has been easily grown in nonarable land and therefore, recently preferred for production of bioethanol [50]. This prevents competition between food energy crop sources and fuel sources [51]. The production of butanol form these Cyanobacterial strains has two advantages as follows:

- Due to uptake of CO_2 directly from atmosphere, Cyanobacteria do not need plant matter for biosynthesis of isobutanol [13].
- Since plant matter has not been used for the production of isobutanol, the necessity of cellulosic raw materials from food energy crop sources and designing food-fuel price relationship should be avoided [50].
- The possibility of bioremediation can exist due to absorption of CO_2 from the atmosphere by Cyanobacteria [13].

The primary drawbacks for using cyanobacteria for production of biofuels are as follows:

- Cyanobacteria are sensitive to strong intensity of sunlight, inadequate concentration of CO_2, inappropriate amount of salinity. These factors have affected the production of isobutanol in cyanobacteria [52].
- Cyanobacteria bioreactors require constant mixing, and consumption of high energy for harvesting of biosynthetic products. These parameters have reduced the efficiency of production of isobutanol via cyanobacteria [52].
- Genetically improved strains of cyanobacteria are being developed with the significant application of ATP and cofactors as driving forces for higher yield of isobutanol. Many organisms are using acetyl-CoA dependent pathway for production of butanol, but major challenge has been observed for first reaction for synthesis of acetoacetyl-CoA due to condensation of two acetyl-CoA molecules. This reaction requires positive value of Gibbs free energy ($\Delta G=6.8$ kcal/mol) and therefore, called as thermodynamically nonspontaneous reaction [53]. *Bacillus subtilis* has been reported to produce very low yield of isobutanol as compare to *E. coli* cells [42] after utilizing Lignocellulosic biomaterials. These strains could be genetically improved for improving the yield of isobutanol from lignocellulose *by applying* elementary mode analysis [42]. *Saccharomyces cerevisiae* has specific valine biosynthetic pathway for production of isobutanol in small quantities during cellular metabolism process [54]. They can be efficiently grown in acidic condition at low pH value and may not be affected by bacteriophages [42]. The yield of isobutanol can be enhanced by genetic improvement for overexpression of enzymes in the valine biosynthetic pathway in cells of *S. cerevisiae* [54]. Due to having eukaryotic characteristics, it is difficult to manipulate *S. cerevisiae* genetically than prokaryotic cells of *E. coli* or *B. subtilis* [42]. *S. cerevisiae* has not metabolic pathway for consumption of pentose sugars and therefore, they are restricted for consumption of lignocellulose for production of higher yield of isobutanol. This will cause an unfavorable relationship of food/fuel

price during production of isobutanol from *S. cerevisiae* [42, 55]. *Ralstonia eutropha* (gram-negative soil bacterium) of β-proteobacteria class has tendency to produce isobutanol by electrochemical methodology. In this step, anodes have been kept in a mixture of H_2O and CO_2. An electric current has been passed through anodes and formic acid has been synthesized due to combination of H_2O and CO_2 through an electrochemical process. The mixture has been further inoculated with culture of *Ralstonia eutropha* (strain tolerant to electricity) for conversion of formic acid into isobutanol biofuel. The high cost of raw material has been considered as major obstacles for the commercial production of butanol. The economic viability of process has been improved using low-cost agricultural plant residues [51]. Metabolic engineering can be applied in cellular metabolism process allow microbes for utilizing cheaper substrate such as glycerol instead of glucose efficiently [56]. The use of food energy crop for production of butanol through fermentative approach has limit the availability of food crop across whole world. Therefore, the consumption of glycerol by metabolically engineered microbes has been proposed as good alternative source for production of butanol. Glycerol has a very low market price due to its secretion as a waste product during production of biodiesel. The production of Butanol from glycerol has been economically viable using metabolically engineered *Clostridium pasteurianum* [57].

Biobutanol has great global demand as a potential alcohol fuel in compare to other lower grade alcoholic fuels like ethanol and methanol. Many solid wastes residues have been integrated for the production of different generation of biofuels [49]. The yield of biobutanol has been improved using different strategies for the bioconversion, production methods and their appropriateness in automotive engines. For commercialization in market, the production of these biobutanol have faced many challenges as addressed in Figure 12.3.

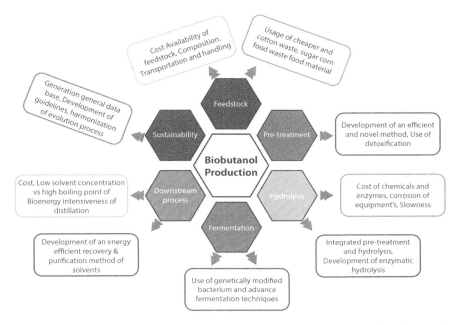

Figure 12.3 Limitations and suggestions on the various stages involved in the biobutanol production.

12.5 Conclusion

Fermentative production has been explored as new strategy for production of butanol from Lignocellulosic biomass. It is crucial parameters for collecting information of relevant process aspects and physiological characteristics of specific microbes. This strategy has main challenge of efficient bioconversion of costly cellulosic feedstock to fermentable sugars. This cost could be efficiently minimized by opting cost effective pre-treatment methodology and downstream process for achieving higher yield of glucose or xylose monomers. The higher rate of bioconversion of these monomers into biofuel component can make this process more economically feasible. However, specific downstream processing technique has been used for minimizing the concentration of undesirable solvent of acetone and ethanol as by products. Genetic improved non–acetone-forming strains have been recently used for the higher production of butanol. This book chapter has subsequently provided the detail information about pre-treatment process for hydrolysis of solid plant residues into fermentable sugars. Fermentation process and downstream processing techniques has been also discussed for conversion of these monomeric units into biobutanol at a greater extent for biofuel applications in the future.

References

1. Ranjan, A. and Moholkar, V.S., Biobutanol: Science engineering and economics review essay. *Int. J. Energy Res.*, 38, 3, 277–323, 2012. https://doi.org/10.1002/er.1948.
2. Campos-Fernandez, J., Arnal, J.M., Gómez, J., Dorado, M.P., A comparison of performance of higher alcohols/diesel fuel blends in a diesel engine. *Appl. Energy*, 95, 267–275, 2012. https://doi.org/10.1016/j.apenergy.2012.02.051.
3. Bharathiraja, B., Jayamuthunagai, J., Sudharsanaa T., Bharghavi A., Praveenkumar, R., Chakravarthy, M., Yuvaraj, D., Biobutanol - An impending biofuel for future: A review on upstream and downstream processing tecniques. *Renewable Sustain. Energy Rev.*, 68, 788–807, 2017. https://doi.org/10.1016/j.rser.2016.10.017.
4. Da Silva Trindade, W.R. and dos Santos, R.G., Review on the characteristics of butanol, its production and use as fuel in internal combustion engines. *Renewable Sustain. Energy Rev.*, 69, 642–51, 2017.
5. Jiang, Y., Liu, J., Jiang, W., Yang, Y., Yang, S., Current status and prospects of industrial bio-production of n-butanol in China. *Biotechnol. Adv.*, 33, 1493–1501, 2015. https://doi.org/10.1016/j.biotechadv.2014.10.007.
6. Wang, S., Zhang, Y., Dong, H., Mao, S., Zhu, Y., Wang, R. et al., Formic acid triggers the "acid crash" of acetone-butanol-ethanol fermentation by Clostridium acetobutylicum. *Appl. Environ. Microbiol.*, 77, 1674–80, 2011, http://dx.doi.org/10.1128/AEM.01835-10.
7. Singhal, V. and Rai, J.P.N., Biogas production from water hyacinth and channel grass used for phytoremediation of industrial effluents. *Bioresour. Technol.*, 86, 221–225, 2003. https://doi.org/10.1016/S0960-8524(02)00178-5.
8. Ibrahim, M.F., Ramli, N., Kamal Bahrin, E., Abd-Aziz, S., Cellulosic biobutanol by clostridia: Challenges and improvements. *Renewable Sustain. Energy Rev.*, 79, 1241–1254, 2017. https://doi.org/10.1016/j.rser.2017.05.184.
9. Roffler, S.R., Blanch, H.W., Wilke, C.R., *In situ* extractive fermentation of acetone and butanol. *Biotechnol. Bioeng*, 31, 2, 135–143, 1988. https://doi.org/10.1002/bit.260310207.
10. Talebnia, F., Karakashev, D., Angelidaki, I., Production of bioethanol from wheat straw: An overview on pre-treatment hydrolysis and fermentation. *Bioresour. Technol.*, 101, 13, 4744–4753, 2010. https://doi.org/10.1016/j.biortech.2009.11.080.

11. Liu, G., Zhang, J., Bao, J., Cost evaluation of cellulase enzyme for industrial-scale cellulosic ethanol production based on rigorous aspen plus modeling. *Bioprocess. Biosyst. Eng.*, 39, 133–40, 2016. https://doi.org/10.1007/s00449-015-1497-1.
12. Sarangi, P.K., Nanda, S., Mohanty, P., *Recent advancements in biofuels and bioenergy utilization*, Springer, Singapore, 2018, https://doi.org/10.1007/978-981-13-1307-3.
13. Varman, A.M., Xiao, Y., Pakrasi, H.B., Tang, Y.J., Metabolic engineering of synechocystis sp. strain pcc 6803 for isobutanol production. *Appl. Environ. Microbiol.*, 79, 3, 908–14, 2013. https://doi.org/10.1128/AEM.02827-12.
14. Lay, C.-H., Sen, B., Chen, C.-C., Wu, J.-H., Lee, S.-C., Lin, C.-Y., Co-fermentation of water hyacinth and beverage wastewater in powder and pellet form for hydrogen production. *Bioresour. Technol.*, 135, 610–615, 2013. https://doi.org/10.1016/j.biortech.2012.06.094.
15. Ferrari, M.D., Guigou, M., Lareo, C., Energy consumption evaluation of fuel bioethanol production from sweet potato. *Bioresour. Technol.*, 136, 377–384, 2013. https://doi.org/10.1016/j.biortech.2013.03.045.
16. Schubert, C., Can biofuels finally take center stage? *Nat. Biotechnol.*, 24, 777–84, 2006.
17. Daramola, M.O. and Ayeni, A.O., *Valorization of biomass to value-added commodities: Current trends challenges and future prospects*, Springer, Cham, 2020, Retrieved September 16, 2022 from https://public.ebookcentral.proquest.com/choice/publicfullrecord.aspx?p=6181584.
18. Gabriel, C.L. and Crawford, F.M., Development of the butyl-acetonic fermentation industry. *Ind. Eng. Chem. Chem.*, 22, 11, 1163–1165, 1930. https://doi.org/10.1021/ie50251a014.
19. Beesch, S.C., Acetone-butanol fermentation of sugars. *Ind. Eng. Chem.*, 44, 7, 1677–1682, 1952. https://doi.org/10.1021/ie50511a054.
20. Chiao, J.-S. and Sun, Z.-H., History of the acetone-butanol-ethanol fermentation industry in china: Development of continuous production technology. *J. Mol. Microbiol. Biotechnol.*, 13, 12–14, 2007. https://doi.org/10.1159/000103592.
21. Bai, F.W., Anderson, W.A., Moo-Young, M., Ethanol fermentation technologies from sugar and starch feedstocks. *Biotechnol. Adv.*, 26, 1, 89–105, 2008.
22. Nimcevic, D. and Gapes, J.R., The acetone-butanol fermentation in pilot plant and pre-industrial scale. *J. Mol. Microbiol. Biotechnol.*, 2, 1, 15–20, 2000.
23. Wyman, C.E., Dale, B.E., Elander, R.T., Holtzapple, M., Ladisch, M.R., Lee, Y.Y., Comparative sugar recovery data from laboratory scale application of leading pre-treatment technologies to corn stover. *Bioresour. Technol.*, 96, 18, 2026–2032, 2005. https://doi.org/10.1016/j.biortech.2005.01.018.
24. Ezeji, T., Milne, C., Price, N.D., Blaschek, H.P., Achievements and perspectives to overcome the poor solvent resistance in acetone and butanol-producing microorganisms. *Appl. Microbiol. Biotechnol.*, 85, 1697–712, n.d. https://doi.org/10.1007/s00253-009-2390-0.

25. Uyttebroek, M., Van Hecke, W., Vanbroekhoven, K., Sustainability metrics of 1-butanol. *Catal. Today*, 239, 7–10, 2015. https://doi.org/10.1016/j.cattod.2013.10.094.
26. Durre, P., Fermentative butanol production: Bulk chemical and biofuel. *Ann. N.Y. Acad. Sci.*, 1125, 1, 353–62, 2008. https://doi.org/10.1196/annals.1419.009.
27. Wackett, L.P., Biomass to fuels via microbial transformations. *Curr. Opin. Chem. Biol.*, 12, 2, 187–93, 2008. https://doi.org/10.1016/j.cbpa.2008.01.025.
28. Jones, D.T. and Woods, D.R., Acetone-butanol fermentation revisited. *Microbiol. Rev.*, 50, 4, 484–524, 1986.
29. Killeffer, D.H., Butanol and acetone from corn 1: A description of the fermentation process. *Ind. Eng. Chem.*, 19, 1, 46–50, 1927. https://doi.org/10.1021/ie50205a013.
30. Hastings, J.J., Development of the fermentation industries in great britain. *Adv. Appl. Microbiol.*, 14, 1–45, 1971.
31. Rose, A.H., *Economic microbiology: Primary products of metabolism*, Elsevier Science, Saint Louis, USA, 2014.
32. McCutchan, W.N. and Hickey, R.J., The butanol acetone fermentations. *Ind. Ferment.*, 1, 347–388, 1954.
33. Spivey, M.J., The acetone/butanol/ethanol fermentation. *Proc. Biochem.*, 13, 2–5, 1978.
34. Abou-Zeid, A.A., Fouad, M., Yassein, M., Microbial production of acetone-butanol by Clostridium acetobutylicum. *Zbl. Bacterial. Hyg. II. Abt.*, 133, 125–134, 1978.
35. Singh, N.K. and Dhar, D.W., Microalgae as second generation biofuel. A review. *Agron. Sustain. Dev.*, 605–629, 2011. https://doi.org/10.1007/s13593-011-0018-0.
36. Panpatte, D.G., Jhala, Y.K., Agricultural waste: *A Suitable Source For Biofuel Production*. In: *Prospects of Renewable Bioprocessing in Future Energy Systems. Biofuel and Biorefinery Technologies,* Rastegari, A., Yadav, A., Gupta, A. (eds.), vol. 10, Springer, Cham, 2019. https://doi.org/10.1007/978-3-030-14463-0_13
37. Higashide, W., Li, Y., Yang, Y., Liao, J.C., Metabolic engineering of clostridium cellulolyticum for production of isobutanol from cellulose. *Applied and Environmental Microbiology*, 77, 8, 2727–33, 2011. https://doi.org/10.1128/AEM.02454-10
38. Pfeffer, M., Wukovits, W., Beckmann, G., Friedl, A., Analysis and decrease of the energy demand of bioethanol-production by process integration. *Appl. Therm. Eng.*, 27, 16, 2657–2664, 2007. https://doi.org/10.1016/j.applthermaleng.2007.04.018.
39. Berezina, O.V., Zakharova, N.V., Yarotsky, C.V., Zverlov, V.V., Microbial producers of butanol. *Appl. Biochem. Microbiol.*, 48, 625–638, 2012. https://doi.org/10.1134/S0003683812070022.

40. Lan, E., I and Liao, J.C., ATP drives direct photosynthetic production of 1-butanol in cyanobacteria. *Proc. Natl. Acad. Sci. U. S. A.*, 109, 16, 6018–6023, 2012.
41. Lee, W.-H., Seo, S.-O., Bae, Y.-H., Nan, H., Jin, Y.-S., Seo, J.-H., Isobutanol production in engineered saccharomyces cerevisiae by overexpression of 2-ketoisovalerate decarboxylase and valine biosynthetic enzymes. *Bioprocess. Biosyst. Eng.*, 1467–1475, 2012. https://doi.org/10.1007/s00449-012-0736-y.
42. Peralta-Yahya, P.P., Zhang, F., del Cardayre, S.B., Keasling, J.D., Microbial engineering for the production of advanced biofuels. *Nature*, 488, 320–328, 2012. https://doi.org/10.1038/nature11478.
43. Blaschek, H.P., Ezeji, T.C., Scheffran, J., *Biofuels from agricultural wastes and byproducts*, Wiley-Blackwell, 2010, https://doi.org/10.1002/9780813822716.
44. Li, S., Huang, D., Li, Y., Wen, J., Jia, X., Rational improvement of the engineered isobutanol-producing bacillus subtilis by elementary mode analysis. *Microb Cell Fact*, 11, 101, 2012. https://doi.org/10.1186/1475-2859-11-101.
45. Chong, H., Geng, H., Zhang, H., Song, H., Huang, L., Jiang, R., Enhancing E. coli isobutanol tolerance through engineering its global transcription factor camp receptor protein (crp). *Biotechnol. Bioeng.*, 111, 4, 700–708, 2014. https://doi.org/10.1002/bit.25134.
46. Monot, F., Martin, J.R., Petitdemange, H., Gay, R., Acetone and butanol production by clostridium acetobutylicum in a synthetic medium. *Appl. Environ. Microbiol.*, 44, 6, 1318–1324, 1982.
47. Higashide, W., Li, Y., Yang, Y., Liao, J.C., Metabolic engineering of clostridium cellulolyticum for production of isobutanol from cellulose. *Appl. Environ. Microbiol.*, 77, 8, 2727–2733, 2011. https://doi.org/10.1128/AEM.02454-10.
48. Jiang, Y., Liu, J., Jiang, W., Yang, Y., Yang, S., Current status and prospects of industrial bio-production of n-butanol in china. *Biotechnol. Adv.*, 33, 7, 1493–1501, 2015. https://doi.org/10.1016/j.biotechadv.2014.10.007.
49. Birgen, C., Duurre, P., Preisig, H.A., Wentzel, A., Butanol production from lignocellulosic biomass: Revisiting fermentation performance indicators with exploratory data analysis. *Biotechnol. Biofuels*, 12, 167, 1–15, 2019. https://doi.org/10.1186/s13068-019-1508-6.
50. Machado, I.M.P. and Atsumi, S., Cyanobacterial biofuel production. *J. Biotechnol.*, 162, 1, 50–56, 2012. 10.1016/j.jbiotec.2012.03.005. PMID 22446641.
51. Yu, T., Zhou, Y.J., Huang, M., Liu, Q., Pereira, R., David, F., Nielsen, J., Reprogramming yeast metabolism from alcoholic fermentation to lipogenesis. *Cell*, 174, 6, 1549–1558.e14, 2018. https://doi.org/10.1016/j.cell.2018.07.013.
52. Pandiyan, K., Singh, A., Singh, S., Saxena, A.K., Nain, L., Technological interventions for utilization of crop residues and weedy biomass for second generation bio-ethanol production. *Renewable Energy*, 132, 723–741, 2019. https://doi.org/10.1016/j.renene.2018.08.049.

53. Stern, J.R., Coon, M.J., Campillo, A.D., Acetoacetyl coenzyme a as intermediate in the enzymatic breakdown and synthesis of acetoacetate 1. *J. Am. Chem. Soc.*, 75, 6, 1517–1518, 1953. https://doi.org/10.1021/ja01102a540.
54. Kondo, T., Tezuka, H., Ishii, J., Matsuda, F., Ogino, C., Kondo, A., Genetic engineering to enhance the ehrlich pathway and alter carbon flux for increased isobutanol production from glucose by saccharomyces cerevisiae. *J. Biotechnol.*, 159, 32–37, n.d. https://doi.org/10.1016/j.jbiotec.2012.01.022.
55. Matsuda, F., Kondo, T., Ida, K., Tezuka, H., Ishii, J., Kondo, A., Construction of an artificial pathway for isobutanol biosynthesis in the cytosol of saccharomyces cerevisiae. *Biosci. Biotechnol. Biochem.*, 76, 11, 2139–41, 2012.
56. Yoshikawa, K., Toya, Y., Shimizu, H., Metabolic engineering of synechocystis sp. pcc 6803 for enhanced ethanol production based on flux balance analysis. *Bioprocess Biosyst. Eng.*, 40, 791–796, 2017. https://doi.org/10.1007/s00449-017-1744-8.
57. Lee, S.Y., Kim, H.U., Chae, T.U., Cho, J.S., Kim, J.W., Shin, J.H., Kim, D., I, Ko, Y.-S., Jang, W.D., Jang, Y.-S., A comprehensive metabolic map for production of bio-based chemicals. *Nat. Catal.*, 2, 18–33, 2019. https://doi.org/10.1038/s41929-018-0212-4.

13

Biobutanol: A Promising Liquid Biofuel

Aakansha Raj[1], Tasnim Arfi[2] and Satyajit Saurabh[3]*

[1]Department of Botany, Patna University, Patna, Bihar, India
[2]Department of Bioengineering and Biotechnology, Birla Institute of Technology, Mesra, Ranchi, Jharkhand, India
[3]DNA Fingerprinting Laboratory, Bihar State Seed and Organic Certification Agency, Mithapur, Patna, Bihar, India

Abstract

Large-scale production of liquid biofuels, for example, biobutanol, can be considered as a sustainable measure to assuage concerns about rising crude oil costs, climate emergencies, such as global warming and declining petroleum reserves. Microbiological conversion of agriculture residues produced as wastes in the course of processing crops to fruitful products, especially cost-effective biofuels is presently a pressing problem of biotechnology. Second-generation biobutanol is obtained from lignocellulosic feedstock using saccharification, which is followed by fermentation using microbes, and thereafter recovery of product is done. The paper enlightens the process involved in the second-generation biobutanol production, and the latest advancements in the pre-treatment of substrates and *Clostridia* sp. involved in the industrial production of biobutanol.

Keywords: Biofuels, biobutanol, biorefinery, acetone, butanol, ethanol, fermentation, *Clostridium*

13.1 Introduction

The demand for food and energy is rising along with the population, which presents significant difficulties for the 21st century. The primary energy sources that run our daily lives are fossil fuels including coal, oil, and natural gas. As the dependency on energy is largely based on nonrenewable fossil

---

Corresponding author: satyajitsaurabh@gmail.com

Arindam Kuila and Mainak Mukhopadhyay (eds.) Production of Biobutanol from Biomass, (323–354) © 2024 Scrivener Publishing LLC

fuels, there are growing concerns globally such as depleting fossil resources, oil price hikes, energy security, policies to limit the use of fossil fuels, and climate change [1]. Overuse of fossil fuels increases the amount of CO_2 in the atmosphere, causing global warming and eventually climate change. Ultimately, this causes increased extreme weather conditions, disasters in communities and countries, and substantial economic losses [3]. So, there is increasing interest to explore the production of alternative, more carbon-neutral, and renewable sources of energy. To counterbalance the CO_2 emission (a greenhouse gas), renewable biomass could be utilized. As, CO_2 released can eventually be recaptured by plants, cyanobacteria, and eukaryotic microalgae.

Biofuels are the most needed alternative energy source to address the energy crisis and changing climate; by replacing fossil fuels with clean, sustainable, and renewable forms of energy. It can be liquid (like biobutanol, bioethanol, biodiesel, etc.) or gas (like syngas, hydrogen gas, etc.). It can be described as the fuels that are developed from biological materials – often known as biomass [4]. Biomass (substrates or raw materials) are the organic substance, which could be derived from agricultural produce or its wastes, e.g., sugarcane, corn, bagasse, hay, straw, seaweed, etc. Several crops are being used for the synthesis of biofuels, viz., sugarcane, corn, soybeans, rapeseed, wheat, sugar beet, sorghum, cassava, switch grass, silver grass, willows, etc. In addition to these crops, algae and municipal wastes are also being used for biofuel production.

Biofuels can be classified to different generations depending upon the availability of the raw materials required for the production of different biofuels. The tremendous market growth indicates higher demand for the production of technologically economical, reliable, and practical technique for the utilization of carbon biomass. Biofuels can be both liquid biofuels and gas biofuels. Biobutanol, bioethanol, and biodiesel are liquid biofuels whereas syngas, hydrogen and methane are gaseous biofuels [5]. The development in biofuel research has led us to different generations of biofuels, on the basis of feedstock, and its potential advantages and disadvantages (Table 13.1).

13.1.1 First-Generation Biofuels

The first-generation biofuels were obtained from starch-based foods and grains such as sugarcane, cassava, corn, potato, etc [2, 5, 6]. Utilization of these crops in biofuel production started influencing the food supply globally, as these crops were not only high-demanding food materials but also consumed large specific areas of cultivation lands [2, 5]. Nearly, 94% to 96% of the world's total biofuel production in 2020 was generated from

Table 13.1 Comparison between different generations of biofuels.

Generation	Biofuels	Feedstock used	Processing method	Advantage	Disadvantage
First	Biobutanol, biodiesel, bioethanol	Edible oil seeds, grains, or starch-based crops,	Fermentation, distillation, and transesterification	Commercially economical	Interferes with global food supply, less yield
Second	Bioethanol, biobutanol, biodiesel	Lignocellulosic wastes: agricultural wastes, forest residues, nonedible oil seeds	Physical, chemical, and biological pret-reatment of feedstock followed by fermentation	No clash with food materials	Requires extensive processing for an efficient enzymatic breakdown of lignohemicellulose
Third	Biodiesel, biobutanol, bioethanol, biohydrogen, syngas, methane	Algal biomass	Cultivation and harvesting of algae, oil extraction, esterification, or fermentation	No interference with food, easy and rapid cultivation	Risk of contamination, high variation in lipid content, the high energy input for harvesting, and extensive downstream processing.
Fourth	Biodiesel, bioethanol, syngas, methane, biobutanol, biohydrogen	Algae and microbes	Genetic engineering of algae	High yield of algal biomass as well as high yield of lipid content	The high initial investment and regulatory clearance for GMOs are required

first generation. However, the raw materials utilized and the massive arable land occupied in the production of first-generation biofuels made them less potential than second-generation biofuels [7].

13.1.2 Second-Generation Biofuels

The early 21st century marked the dawn of second-generation biofuels in the market [8]. The conflict between the policymakers and the public over food and fuels made researchers conduct experiments to find potential sources of biofuels by utilizing lignocellulosic biomass and organic wastes (such as sugarcane bagasse, corn cobs, wheat straw, rice straw, etc). These second-generation sources of biofuels are the best way out as they are in abundance, affordable, renewable, and do not get in the way with the food supply [2, 5]. Some other substrates which can be used for the production of second-generation biofuels include cash crops, honge, maize, cotton stalk, and mixed paper waste or any other cellulosic biomass [2]. Although the second generation is preferred over the first generation owing to the advantages discussed above, the basic challenge with this generation is the utilization of resources in the context of the quality of the substrate and to produce cheap and sustainable biofuel through different pre-treatment and fermentation processes [2, 5].

13.1.3 Third-Generation Biofuels

Biofuels obtained from algal biomass (macroalgae or microalgae) by making use of either traditional transesterification or hydro treatment of algal oil are considered a third-generation biofuel. Third-generation biofuels offer many advantages over first and second gen biofuels as they do not interfere the human or animal food chain, can be grown on nonarable or uncultivated land, and freshwater is generally not needed as algal biomass can be cultivated using brackish, saline, or wastewater [9]. In addition, microalgae show a very rapid growth which making them suitable feedstock for the production of biofuel [9]. However, the large-scale production of biofuels from microalgae still possesses several technical constraints, such as variation in lipid content according to species, as well as different physical parameters, such as pH, stress, etc. [2, 9]. Moreover, it requires a lot of energy and is currently unprofitable.

13.1.4 Fourth-Generation Biofuels

Fourth-generation biofuels are generally an extended version of third-generation biofuels where the microalgae are modified using genetic engineering. The fourth-generation biofuel production combines synthetic biology and technology where algae are genetically engineered to improve their yield and lipid content. This technology holds a basic purpose of producing sustainable energy as well as sequestering carbon dioxide [2, 5, 9].

13.2 Biobutanol

Butanol production has been in practice since 1916, but then it was generally put in use as a solvent feedstock. However, after decades, now, butanol is widely used as a fuel enhancer apart from being a basic feedstock in chemical industries [15]. The market for biobutanol fuel has shown a tremendous increase and is continuously at a compound annual growth rate [CAGR] of 8.34% from 2019 to 2024.

Biobutanol or butyl alcohol (N-butanol) is a four-carbon chain, renewable, climate-friendly, and cost-effective chemical compound which could be obtained through fermentation of biomass. It could be utilized as an organic solvent or as an alternative fuel material in internal combustion [IC] engines just like other biofuels namely bioethanol, methanol, etc. [10, 11]. In comparison with conventional fuels (coal, gasoline, or diesel fuels), biobutanol holds many positive aspects which makes it a potential future fuel in terms of reducing vehicle emissions, limiting conventional energy demand, and thereby improving global atmosphere quality [11]. The comparative specifications of different isomers of butanol with gasoline and ethanol have been tabulated in Table 13.2.

Bioethanol when compared to biobutanol is more corrosive in nature and thus ends up causing corrosion of fuel injectors and electric fuel pumps (owing to its hygroscopic nature). Moreover, bioethanol has a high latent heat of vaporization and is difficult to evaporate, which makes it difficult to start engines in cold weather. In addition, the characteristics and functionality of engine oil are considerably diminished and interfered with when bioethanol is used as a lubricant. Nonetheless, bioethanol is soluble in water but insoluble in oil. Thus, there is a high probability of bioethanol forming emulsion (bioethanol–water–oil) which could ultimately end up causing severe engine failures [12]. Attributing to all the above reasons, biobutanol is being considered a superior automobile fuel and is gaining demand as a potentially clean and green substitute for conventional fuels.

Now, biobutanol is being produced commercially as a solvent from propylene utilizing the Oxo method. This process involves two major steps: hydroformylation and hydrogenation. Gasoline, however, is obtained from the fractionation of crude petroleum. Therefore, replacing gasoline

Table 13.2 Specifications of isomers of butanol with other fuels.

Specification	Gasoline	ethanol	n-Butanol	s-Butanol	i-Butanol	t-Butanol
Chemical formula	C_nH_{2n+2} (n=4–12)	C_2H_5OH	$C_4H_{10}O$	$C_4H_{10}O$	$C_4H_{10}O$	$C_4H_{10}O$
Molecular weight (g/mol)	111.19	46.07	74.11	74.11	74.11	74.11
Octane number	81–89	89–103	78	91	94	89
Density (g/ml)	0.72–0.78	0.789	0.808	0.808	0.805	0.800
Boiling point (°C)	25–215	78.5	117.7	99.51	107.89	82.4
Freezing point (°C)	–22.2	–96.1	–89.8	–115	–89.8	
Flash point (°C)	–44 to –38	14	35	34	24	28
Autoignition temperature (°C)	275	368–375	314–320	314–322	314–320	380–387
Latent vaporization heat (kJ/kg)	289	846–854	43.29	43.29	43.29	43.4
Solubility in water (at 25° C)	0.046–0.077	∞	73	185.1	89.4	∞
Viscosity (mm²/s) at 25°C	0.4–0.8	1.07	2.63	3.1	4	4.31
Air–fuel ratio	14.6	9.0	11.2	NA	NA	NA

by deriving butanol through petrol is definitely not a feasible option, neither economically nor from an environmental perspective [27]. In contrast, butanol obtained from the fermentation process is both a benign and cost-effective process. At present, China is leading the global market of biobutanol, whereas BASF and Dow Chemicals are the top producers of n-butanol (using the Oxo method) [26, 27]. Some other companies involved in biobutanol production are BP, GEVO, etc. (Table 13.3).

13.3 Biorefinery and Biobutanol Production

Butanol production and consumption through ABE fermentation (also known as 'solvent fermentation') by deploying microbial route and aseptic techniques was the first commercial fermentation and was generated during World War I. In general, biobutanol is obtained through ABE fermentation utilizing *Clostridia* species [28].

The commercial production of biofuels and chemicals from biomass, through biorefining technology, has shown great advancement in recent years. Biorefinery development objectifies attaining the maximum sustainable criterion for biofuel production [13, 93]. Making the best use of the raw material components, biorefineries are integrative and multifunctional concepts that provide facilities for the separation and conversion of biomass for the sustainable production of a wide range of intermediates and finished products (chemicals, heat, power, biofuels, etc.) [93, 94]. Generally, the biorefinery process comprises primary refining which involves pretreating and purifying the raw materials as well as separating the biomass components into intermediates (cellulose, starch, sugar, oils, etc.), and secondary refining which involves the further conversion of the intermediate products and other processing and product purification steps (Figure 13.1).

13.3.1 Substrates and Their Pre-Treatment for Biobutanol Production

13.3.1.1 Substrate

Starch and sugar
Biobutanol production in the early 20th century was done utilizing substrates consisting of sugar and starch such as sugarcane, cereal grains, and wastes from food industries. However, owing to the increasing global population and rising demand for food, this production process was no longer

Table 13.3 Some companies involved in butanol production [13–16].

Company	Country	Production details	References
Agrosys Products India Pvt. Ltd	Tamil Nadu, India	Biobutanol production enhancement research	https://venturecenter.co.in/campaigns/bioenergy/bioenergy-companies.php
GEVO	CO, USA	Isobutanol is produced from glucose by utilizing genetically modified yeast	
Butyl Fuel, LLC	Columbus, USA	n-butanol produced from genetically modified and patented *Clostridium* strain [high butanol titer 1.3–1.9 times]	
Green Biologics	Abingdon, UK	n-butanol produced from genetically modified microorganisms	
Tetraviate Bioscience	Chicago, USA	n-butanol produced from mutated *Clostridium beijerinckii*	
Metabolic Explorer	Clermont-Ferrand, France	Especially designed microorganisms are utilized to produce biobutanol from lignocellulosic feedstock	
Butala Co.	Zug, Switzerland	Butanol is produced from genetically modified yeast [yeast is modified to produce higher butanol titer and efficient enough to utilize C5/C6 sugars]	

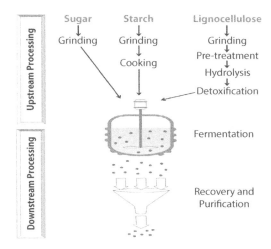

Figure 13.1 Schematic representation of upstream and downstream processes involved in the production of biobutanol.

ethical and economical [17]. Wastes emerging from food industries were reported to be an attractive substrate for biobutanol production serving as an additional advantage of waste management [18]. Sugarcane molasses, as well as cheese whey (over lactose substrate), are some other substrates that were studied by various researchers to be potential feedstock for biobutanol production [19–22]. Cane molasses were reported to produce a total solvent of 19.8 g/l with 13.4 g/l of butanol production within a period of 72 hours in batch culture mode. Semicontinuous mode showed better solvent productivity (1.05 g/l/hr), enhanced sugar utilization, and required shorter fermentation time, thereby showing the process feasibility at an industrial scale [23].

Cellulose
One amongst the other strategies to reduce the production cost for making the product economically viable is to use cheap feedstock. A significant feedstock for the production of second-generation biobutanol is lignocellulosic biomass. Being the most abundant renewable resource on the planet, it is widely present in crop residues and forest resources [24]. A steep rise in biobutanol production has been reported in nations possessing large cultivating areas [25]. Data published by UNEP in 2015, reported approximately 140 billion tons of agrowastes being produced annually which is equivalent to around 50 million tons of oil [26]. Moreover, Costa Rica, Cambodia as well as India were nations listed by UNEP that generated a

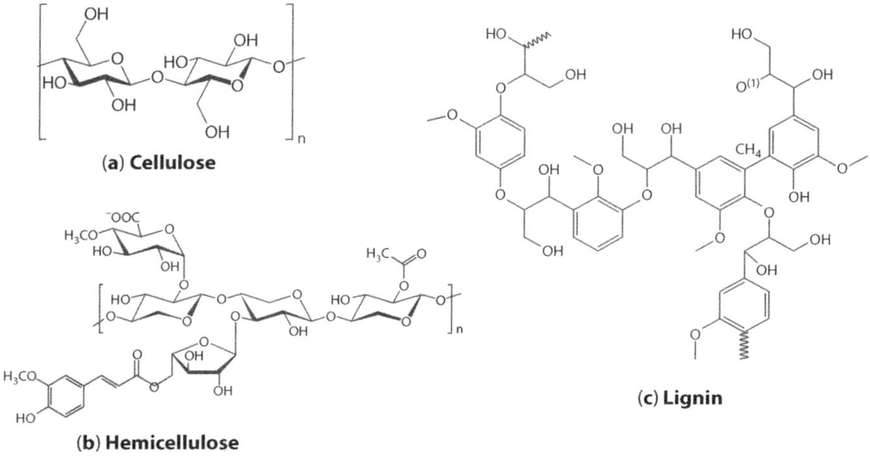

(a) Cellulose

(b) Hemicellulose

(c) Lignin

Figure 13.2 Three types of polymers and their structures constituting lignocellulosic biomass. **a.** Cellulose is a straight chain polymer of glucose, comprising 50% wood mass. **b.** Hemicellulose is a short, branching heteropolymer that mostly consists of the sugars xylose, glucose, mannose, galactose, and arabinose. Uronic acid can occasionally be found in hemicellulose. **c.** Lignin is basically constituted of phenyl propanoid units which are derivatives of p-hydroxycinnapyl alcohols.

large amount of stubble (agricultural wastes). India alone generated 415.5 million tonnes among these countries, which is equivalent to almost 103.9 million tonnes of oil [27]. Amongst these, rice straw and wheat straw are the major stubble producers with global yearly availability of 731.3 million tons and 354.34 million tons, respectively [49]. In developing nations like India, an ample amount of stubble is left over after harvesting various crops like cotton, wheat, rice, maize, sorghum, oilseeds, pulses, etc. these agricultural wastes are either left in the fields or the stubble is put on fire, leading to global warming and increasing soil salinity [50]. This stubble could be used well as a substitute resource for producing biobutanol and other biofuels, in an ecofriendly manner.

However, there are a number of issues associated with the effective use of lignocellulosic biomass. In terms of structure, these raw materials primarily consist of three different types of polymers: cellulose, hemicelluloses, and lignin (Figure 13.2). The recalcitrant property of these lignocellulosic biomasses prevents their breakdown into favorable byproducts. In addition, the hydrophobic nature of lignin and its resistance to chemicals adds to the hindrance in its degradation [26]. Researchers are constantly exploring into *Clostridium beijerinckii* as a possible strain for the industrial synthesis of biobutanol from lignocellulosic biomass [48].

Algal biomass

Algae are considered one of the pioneer living microorganisms on this planet [51]. They possess an exceptional growth rate, around 100 folds more than the terrestrial plants. Additionally, they can double their biomass in less than 24 hours [52]. Nevertheless, a number of microalgal strains have the ability to accumulate huge quantities of lipids into their cells, wherein these lipids could be converted to biobutanol [53].

Owing to the limitations of lignocellulosic feedstock researchers have diverted their attention towards microalgal feedstock that is available worldwide in large quantities with no or minimal arable land requirement [54]. The potential and sustainability of algal feedstock in the generation of biofuels, particularly biobutanol, are increased by its higher lipid and lower lignin content [55, 56]. Microalgal biomass in comparison to macroalgae (brown macroalgae, red macroalgae, green macroalgae, mixed macroalgae) is more feasibly accessible and convertible to biofuels due to the presence of alginate in the latter [57]. Although many studies have been carried out to increase the carbohydrate as well as the lipid content in algal biomass, not much information on biobutanol production from these substrates is available [58–60]. However, the search for desirable algal species for the industrial production of biobutanol is still on. Table 13.4 shows the major feedstock, globally used for the production of biobutanol.

13.3.1.2 Pre-treatment of Substrates

Preprocessing of biomass is the most essential and major challenge in butanol production. This upstream process is required to increase the yield of sugar which could consequently increase the fermentation efficiency [27]. Several methods for the pre-treatment (physical, chemical, physiochemical, biological, or combinations of these) of biomass have been suggested by researchers depending on the different types and compositions of feedstock [2]. First-generation feedstock due to the large availability of sugars requires minimal or no complex preprocessing technique [62]. Simple pre-treatment processes like heating, dilute acid pre-treatment, deproteinization and tyndallization have been reported to be used for the first-generation feedstock [63]. Among all the abovementioned techniques, deproteinization of cheese whey was reported in the maximum yield concentration of butanol (8.9 g/l). It has been established that dry tyndallization and heat sterilization are efficient pre-treatment methods for food crops. However, in the case of cheese whey, these methods resulted in clot formation [63, 64]. A more comprehensive pre-treatment technique is required for the second-generation feedstock due to the presence of complex molecules

Table 13.4 Major feedstock, their composition and hydrolyzing bacteria.

Feedstock	Composition	Strain	Hydrolysis required	Reference
Wheat straw	38% Cellulose, 29% hemicellulose, 24% lignin, 6% ash	C. beijerincki	Yes	[29–31]
Corn fiber	20% Starch, 50–60% nonstarch polysaccharides	C. beijerincki	Yes	[32]
Barley straw	42% Cellulose, 28% hemicellulose, 7% lignin, 11% ash	C. beijerincki	yes	
Corn strover	38% Cellulose, 26% hemicellulose, 23% lignin, 6% Ash	C. beijerincki	Yes	[33]
Switchgrass	37% Cellulose, 29% hemicellulose, 19% lignin	C. beijerincki	Yes	[33]

(Continued)

Table 13.4 Major feedstock, their composition and hydrolyzing bacteria. (Continued)

Feedstock	Composition	Strain	Hydrolysis required	Reference
Sugarcane bagasse	47% Cellulose, 16% hemicellulose, 27% lignin, 19.7% reducing sugar	C. acetobutylicum	Yes	[43, 46]
Sugarcane straw	43% Cellulose, 15% hemicellulose, 23% lignin, 25.1% reducing sugar	C. acetobutylicum	Yes	[43, 46]
Corn cob	40–44% Cellulose, 31–33% hemicellulose, 16–18% lignin, 2.3% ash, 12.77% moisture	C. beijerinckii	Yes	[39, 40]
Rice straw	32% Cellulose, 24% hemicellulose, 13% lignin, 15% ash, 11.695 moisture	C. acetobutylicum	Yes	[40, 44]

(Continued)

Table 13.4 Major feedstock, their composition and hydrolyzing bacteria. (*Continued*)

Feedstock	Composition	Strain	Hydrolysis required	Reference
Rice bran	34–52% carbohydrates, 7–24% total fiber, 8.13% minerals, 15–20% lipid, 8.41% moisture	*C. saccharoperbutylacetonicum*	Yes	[42]
Soy molasses	60% carbohydrates, 10% each protein & mineral, 20% fat & lipid	*C. beijerincki*	Yes	[40]
Degermed corn	73% Starch, 3% ash, 13% proteins	*C. beijerincki*	Yes	[42]
Rye straw	37–38% Cellulose, 37–40% hemicellulose, 18–22% lignin, ash, carbohydrates	*C. acetobutylicum*	Yes	[45]
Domestic organic waste [34]	59% Sugars, 13% lignin, 17% ash	*C. acetobutylocum*	Yes	[34]
Sago	86% Starch, small amounts of mineral and nitrogenous matters	*C. saccharobutylicum*	Yes	[35]

(*Continued*)

Table 13.4 Major feedstock, their composition and hydrolyzing bacteria. (*Continued*)

Feedstock	Composition	Strain	Hydrolysis required	Reference
Defidered sweet-potato-slurry (DSPS)	Starch	*C. acetobutylicum*	No	[92]
Extruded corn	61% Starch, 3.8% corn oil, 8.0% Protein, 11.2% fiber	*C. acetobutylocum*	Yes	[38]
Cassava [37]	70% Starch, 2.7% protein, 2.4% fiber, 0.2% ash	*C. acetobutylicum*	No	[37]
Whey permeate [36]	5% Lactose, 0.36% fat, 0.86% protein	*C. acetobutylocum*	No	[36–38]
Liquefied corn starch	39% Starch, 45% moisture	*C. beijerincki*	Yes	[41]
Algal biomass [47]	7.1% Cellulose, 16.3% hemicellulose, 1.52% lignin, upto60% carbohydrates	*C. saccharoperbutylacetonicum*	Yes	

such as cellulose, hemicellulose, and lignin. In the course of this pre-treatment process, the cellulose–lignin matrix bounded by hemicellulose is broken thereby reducing the crystallinity of cellulose and increasing the amorphous fraction of cellulose for better digestion activity [27]. The third-generation feedstock offers minimum complexity thus making it a feasible contender for butanol production. The schematic representation for different methods used in the pre-treatment of lignocellulosic biomass is illustrated in Figure 13.3.

Physical pre-treatment
Almost every feedstock undergoes physical pre-treatment methods before being exposed to any other treatment strategies. Basically, this method is a dry process that serves in reducing the degree of polymerization and increases the surface area of substrates by reducing the particle size, facilitating the access of cellulase to the substrate surface thus increasing cellulose conversion [65]. Physical treatment includes comminution (ball or hammer milling, colloid or vibro-energy milling, and two-roll milling) to produce particles that could easily pass through a 3- to 5-mm diameter sieve. Many other physical pre-treatment methods include pyrolysis, irradiation (gamma rays, electron beam, microwave, infrared, or sonication),

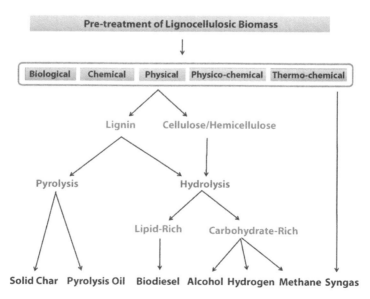

Figure 13.3 Schematic representation for processes involved in the production of different end-products from lignocellulosic biomass. *Alcohol denotes ethanol/butanol *Syngas is also known as synthetic gas.

hydrothermal, expansion, extrusion, etc [13]. For the conversion of soybean hulls to fermentable sugars, extrusion methods were employed and compared with two conventional pre-treatment methods, dilute acid (1% w/v H_2SO_4) and alkali (1% w/v NaOH). Extrusion demonstrated the highest cellulose-to-glucose conversion (95% by weight) under ideal conditions [66].

Physiochemical pre-treatment
Physiochemical methods are considered to be more efficient than physical techniques of preprocessing feedstock. This technique is solely applied as well as initiated by chemical reactions with the only motive to disrupt the harsh structure of biomass. This method includes AFEX, steam explosion (autohydrolysis), SO_2 steam explosion, liquid hot water (LHW), supercritical CO_2 explosion, and wet air oxidation [27, 65].

Steam explosion is considered to be the most viable technique under physiochemical treatment. It involves autohydrolysis (involving the formation of acetic acid) and depressurization (involves rupturing the bonds of complex feedstock [67]. Under steam explosion, the substrate is decompressed explosively as a result of the abrupt release of pressure after treating the ground biomass with highly pressurized steam. Temperature is maintained at 160°C to 220°C at a pressure of 0.69 to 4.83 MPa for a short time interval before exposing the biomass to atmospheric pressure [68]. In the course of this pre-treatment degradation of hemicellulose and lignin takes place. Although being a cost-effective process, this process at times destroys a part of xylan-fraction and also results in the partial breakdown of the lignin–carbohydrate matrix, which causes the release of a significant number of inhibitory chemicals [69, 70].

Ammonia fiber explosion (AFEX) technique makes use of high temperature and high pressure (physical) along with ammonia (chemical) to attain an effective biomass hydrolysis. AFEX's benefits include increasing the surface area for hydrolysis, boosting cellulose decrystallization, hemicellulose depolymerization (partially), and lowering lignin recalcitrance in the processed biomass. Nevertheless, this technique is inappropriate for feedstock with a high lignin content [69].

Supercritical fluids are substances that are gaseous in nature and are compressed beyond their critical point temperature to liquid-like density [65]. In addition to carbon dioxide, water and ammonia in their supercritical state are frequently employed. Due to its nontoxic, cost-friendly, easily available, and noninflammable, with no report of the formation of inhibitory compounds, carbon dioxide in its supercritical form is preferred over water and ammonia [65].

Microwaves also serve as one of the physiochemical pre-treatment of feedstock to a great extent. Several studies have shown the localized heating effect of microwaves on substrates that disrupted the lignocellulosic structure thereby making cellulose and hemicellulose more feasible for enzymatic hydrolysis [65, 71].

For woody biomass (both hardwood and softwood), sulfite pre-treatment was reported to be effective enough to overcome the recalcitrance of lignocellulose [65, 72–75]. The efficient working temperature range and time interval reported for SPORL are 160°C to 190°C for 10 to 30 minutes, respectively. Adding sulfite to the medium raises the pH which thereby reduces the release of toxic substances [65, 72, 73]. In addition, partially sulfonating lignin results in wood softening which ultimately cuts down the energy demand for pre-treatment [65].

Chemical pre-treatment
Chemical treatment is reported to be the most effective technique for the lignin degradation of biomass [13]. Chemical pre-treatment includes ionic liquids/green solvents, strong and dilute acid treatments, organosolv treatment (methanol, ethanol, ethylene glycol, etc.) at times gas treatments (chlorine, nitrogen, or sulfur dioxide) in addition to oxidizing agents (O_2, O_3, or H_2O_2) and ionic liquids are also included for chemically pretreating biomass. The main objective of this pre-treatment is to carry out the reduction of hemicellulose to xylose as well as to make cellulose more accessible for hydrolysis [13].

The use of concentrated acids (sulfuric, hydrochloric, phosphoric, or nitric acid) lay a major drawback as these acids in combination with high temperature and high pressure release large quantities of inhibitor compounds for example furfuryl (from pentoses), HMF (from hexoses) as well as acetic acid during downstream processing [91]. In addition pH adjustment prior to hydrolysis is also tough during acid pre-treatment. To minimize this, dilute acid treatment is preferred since they are less toxic and corrosive. Dilute acids might also be employed at low temperatures with high solid loading (10–40%) as well as high temperatures with low solid loading (5–10%) [27, 76]. High sugar yield (~83%) was confirmed when feedstock was treated with 1% H_2SO_4 (temperature, 160–180°C) followed by immediate enzymatic (β-galactosidase) hydrolysis (in less than 5 minutes) [27, 77]. The efficiency of acid hydrolysis where 95% xylose was recovered from Kraft black liquor and was reported to be 9.4g/l of ABE production [27, 78]. This method is known to be most efficient for noncellulosic feedstock or low lignin-containing biomass (algal biomass) [27].

Alkali pre-treatment
Alkaline pre-treatments are carried out under milder conditions, utilizing less corrosive products (potassium hydroxide, sodium hydroxide, calcium hydroxide, sodium carbonate, etc). Alkaline pre-treatment increases cellulose digestibility and facilitates delignification but generally requires a long time interval for the release of a sufficient quantity of sugars [79]. Acid pre-treatment is preferred for low lignin biomasses (hardwood) whereas alkali treatment is preferably done for softwood [30, 80]. Calcium hydroxide (at 95°C) and sodium hydroxide (at 55°C) treatment on rice straw was reported to significantly improve delignification in the biomass. Feedstock such as rice straw, pinus as well as barley straw were comparatively investigated using pre-treatment of various alkaline (sodium hydroxide, sodium carbonate, potassium hydroxide, and aqueous ammonia) agents [30, 81]. Barley straw treated with 15% aqueous ammonia showed a higher amount of enzymatic digestibility (95%). Sodium hydroxide, on the other hand, was reported to effectively increase the surface area of cellulose, by rupturing the lignin and causing a reduction in the polymerization of biomass structure [30, 81].

Ozonolysis brings about a reduction in lignin content and is reported to produce negligible or no amount of inhibitor compounds. However, extensive use of ozone makes this method uneconomical [13].

Nowadays, to make use of the advantages offered by alkalis as well as the physical pre-treatment methods, mechanochemical pre-treatment techniques have attracted researchers to a great extent [81–84]. Sodium solution and ammonia (5:1 w/v ratio) mixed with wheat straw (5% w/w) at 25°C, followed by ball milling after drying the mixture at 105°C in an oven was reported to effectively enhance the enzymatic hydrolysis [81, 85]. However, this process involves a long duration of alkali treatment followed by drying of the mixture, making the procedure nonviable for industrial processes. Thus, to lessen the treatment duration, reduce the dosage of high-cost alkalis, and elude energy consumption while drying the mixture, a combination of *wet al*kaline mechanical pre-treatment at room temperature was proposed by Yang *et al.* [81].

Organosolv pre-treatment
It has also been extensively researched to replace chemical compound hydrolysis methods with the use of ionic liquids (ILs) and organic or aqueous solvents for the pre-treatment and processing of biomass. ILs are composed of ordered ions (large organic cations and small inorganic anions) and are liquid below 100°C. They can be reused because of their low flammability, wide liquid temperature range, ability to dissolve both polar and

nonpolar species, thermal and chemical stability, and low vapor pressure [13, 86, 87]. These liquids are termed "green" solvents owing to the fact that they do not form any type of toxic inhibitors during their use [13, 88]. A combination of ILs and ammonia for rice straw pre-treatment showed 82% of cellulose recovery with around 97% of glucose conversion [13]. Pre-treatment of switchgrass with ILs alone showed a 64% increase in xylose yield and 96% of glucose in 1 day [13, 89].

The organosolv pre-treatment is based on utilizing an organic or aqueous organic phase to recalcitrate lignin for efficient accessibility of cellulose. Ethanol being less toxic and possessing low combustion potential makes it the most commonly used solvent for this process. In addition to ethanol, methanol, acetone, butanol, and ethylglycol are also used as solvents for organosolv pre-treatment procedures [13]. Organosolv pre-treatment involves combining water with the specific solvents in different proportions, mixing them with biomass followed by heating, at 100°C to 250°C [13, 90].

Pre-treatment of biomass with nanoadditives
Nanotechnology is based on designing devices or substances in a nanoscale (10^{-9} m). The addition of nanoadditives, for example, nanocrystals, nanomagnets, nanofibers, nanodroplets, etc to accelerate the yield of biofuels is being incorporated into use by researchers, recently [13]. These nanoadditives being used as catalysts are reusable and are capable of altering the chemistry at the molecular level, thereby reducing environmental pollution as well as increasing hydrolysis of biomass. Many metal nanoadditives are capable of easily penetrating the cell wall of the feedstock where they easy interaction with the biomolecules efficiently releases carbohydrates which be utilized for butanol production [38].

13.3.2 Microorganisms

Solventogenic bacteria (native butanologens) of the genus *Clostridium* via ABE fermentation, using suitable substrates and under anaerobic conditions are known to produce biobutanol [61, 62]. Some *Clostridia* sp. used are *C. acetobutylicum, C. sporogenes, C. beijerinckii, C. pasteurianum, C. saccharoperbutylacetonicum,* and *C. saccharobutylicum*. Among all the Clostridia species, *C. acetobutylocum* is the most popularly used strain for biobutanol production [61]. *C. acetobutylocum* is an amylase-secreting bacterium that satisfactorily produces butanol utilizing starch as feedstock. Microalgal species such as *Chorella* possess around 40% sugar in their dry matter, which makes them even more potential organisms for biobutanol production. Thus, it becomes essential to select the most appropriate fermentation substrate for commercial and economical production of biobutanol [61].

13.3.3 Acetone–Butanol–Ethanol Fermentation

Biobutanol production utilizing *Clostridium* species through acetone–butanol–ethanol (ABE) production is being used since World War II and was considered the most dominant production process for biobutanol. However, this route of production was phased out when more cost-effective petrochemical routes emerged. Currently, approximately all the butanol produced worldwide comes from petrochemical feedstock. In the past two decades, a number of research and development efforts have been carried out that focus on several aspects of the ABE fermentation process [96]. The major issue that comes along with the ABE fermentation process by bacteria is that the process gets self-inhibited due to toxicity caused by butanol to the culture [95]. Molecular biology researchers have been successful in achieving prominent breakthroughs in developing strains/mutants that are seen to microbe tolerance to butanol toxicity thereby resulting in a remarkable rise in the yield of ABE solvent production [95].

Since fermentation processes are often exothermic, the final products have relative energy levels below those of the substrates. Theoretically, during the ABE fermentation process, the yields for mass and energy are 37% and 94%, respectively. These calculations were done taking into consideration the combustion energy and the ratio of products that were obtained during the fermentation process. It cannot be denied that substrate cost has a potential hand in the economics of fermentation. Owing to the recent needs of the market neither starch- or sugar-based crops could make fermentation economically feasible. Thus, agricultural wastes (lignocellulosic substrates) and several industrial wastes can turn out to be possible alternatives for substrates used in ABE fermentation [97]. The characteristics of products obtained from ABE fermentation are tabulated in Table 13.5.

13.4 Commercial Importance of Biobutanol

Biobutanol is four-carbon alcohol (butyl alcohol, butanol), the term itself defines that it has been produced by microbes using biomass feedstock through the process of microbial fermentation. The most popular bacterial species used for fermentation is *Clostridium acetobutylicum*. Acetone,

Table 13.5 Characteristics of products obtained from ABE fermentation [95].

Products	Density	LHV(Btu/gal)	Uses
Acetone	2,964	83,127	Paint or varnish remover
Butanol	3,065	99,837	Used in surface coatings
Ethanol	2,988	76,330	Used as a potential biofuel
Gasoline	2,819	116,090	Transportation fuel
DDGS	-	8,703 (Btu/lb)	

butanol, and ethanol are the principal byproducts of this process, which is sometimes referred to as ABE fermentation. Biobutanol has several uses that range from fuel for different industries to the solvent used in different industrial sectors (Figure 13.4). In addition to being used as fuel, it has a wide range of other uses, including as a chemical intermediate in the production of butyl acrylate and methacrylate, as a solvent in cosmetics, hydraulic fluids, detergent formulations, drugs, antibiotics, hormones, and vitamins, and as an extracting agent in the processing of various

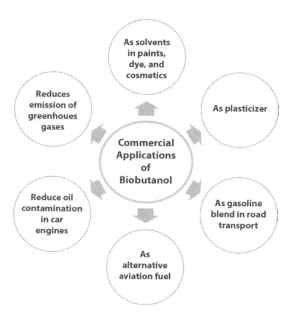

Figure 13.4 Applications of biobutanol.

pharmaceutical compounds after they have been removed from the production process.

Biobutanol in road transport
By combining biobutanol with regular petrol, spark ignition in engines can be produced. Without modifying the engine, its concentration in the gasoline can reach up to 30% v/v. It can be directly incorporated into the current petrol supply infrastructure, perhaps saving money. The risk of air pollution is decreased because it is nontoxic, noncorrosive, and readily biodegradable. Moreover, full combustion of fuel results in the least amount of CO and NO emissions.

1. Biobutanol can use the existing pipeline distribution infrastructure
   Butanol can be transported via pipelines without endangering the reliability, efficiency, or integrity of the pipeline system in the transportation fuels sector. It may be delivered in the current pipeline network, both incoming and outgoing, potentially resulting in cost savings, flexibility, and effective market access.
2. Biobutanol helps in reducing oil contamination
   The use of biobutanol can resolve the oil additive contamination in car engines as it is volatile, it dissipates when the engine warms up after oil contamination caused by frequent cold starts. This phenomenon occurs during winter conditions when there is leakage of fuel into the engine oil through piston rings.
3. Biobutanol as an alternative aviation fuels
   Biobutanol is an ideal molecule to produce renewable iso-paraffinic kerosene (IPK), a blend stock for jet fuel, and it can be easily transformed to a mixture of predominantly C12/C16 hydrocarbons used as fuels. Due to the excessive emission of greenhouse gases (GHG), the overall quality of air is decreasing. As a result, the airline industry is approaching towards viable and efficient alternative fuels to reduce operating costs and environmental impacts, thus improving local air quality.
4. Biobutanol in reducing ozone level from the surface
   Due to its low blend volatility, biobutanol offers more flexibility in meeting requirements for both lower volatility and renewable fuels. Breathing ground-level ozone can be

dangerous, because it harms plants like trees and crops. The addition of Biobutanol with Gasoline can control the release of gases responsible for the formation of the ozone hole by controlling ozone precursors.

13.5 Conclusion

Biobutanol is alcohol, obtained by fermentation of organic substances, comparatively better than ethanol. It is a major end-product of the acetone–butanol–ethanol (ABE) fermentation of biomass by certain microbes. For the production of biobutanol through anaerobic fermentation, mostly used biomass is lignocellulose and microbes are *Clostridium* species. Biobutanol is considered an efficient new generation fuel. Butanol is similar to gasoline; it has high calorific value and low hydrophilicity. It can be easily mixed with other fuels. However, the low efficiency of biobutanol production still limits large-scale production. Due to its chemical and physical nature, it has the potential to be utilized as an alternative and most promising renewable source of energy to address the concerns associated with the energy crisis and climate change being economical and sustainable. However, the major challenge that needs to be addressed is production cost and time. Several research studies are being carried out to reduce the problems and challenges. Comprehensive studies are being carried out to improve biobutanol production through rerouting the methodology with statistical tools, designing bioreactors, and utilizing the approaches involved in genetic manipulations with computational studies. Still, the fuel characteristics and engine compatibility, and other difficulties are also being addressed for its commercialization.

Abbreviations

GHGs: Green house gases
CAGR: Compound Annual Growth Rate
IC: Internal Combustion
ABE: Acetone–butanol–ethanol
UNEP: United Nations Environment Program
SPORL: Sulfite treatment to overcome recalcitrance of lignocellulose
LHW: Liquid hot water
AFEX: Ammonia fiber explosion
O_3: Ozone

H_2O_2: Hydrogen Peroxide
HMF: 5-hydroxy methyl furfuryl
ILs: Ionic Liquids
DDGS: distiller's dried grains with solubles

References

1. Girard, P. and Fallot, A., Review of existing and emerging technologies for the production of biofuels in developing countries. *Energy Sustain Dev.*, 10, 2, 92–108, 20062006. https://doi.org/10.1016/S0973-0826(08)60535-9.
2. Prasad, R.K., Chatterjee, S., Mazumder, P.B., Gupta, S.K., Sharma, S., Vairale, M.G., Datta, S., Dwivedi, S.K., Gupta, D.K., Bioethanol production from waste lignocelluloses: A review on microbial degradation potential. *Chemosphere*, 231, 588–606, 2019.
3. Field, C.B. ed., *Managing the risks of extreme events and disasters to advance climate change adaptation: special report of the intergovernmental panel on climate change*. Cambridge University Press, 2012.
4. da Maia, J.L., Cardoso, J.S., da Silveira Mastrantonio, D.J., Bierhals, C.K., Moreira, J.B., Costa, J.A.V., de Morais, M.G., Microalgae starch: A promising raw material for the bioethanol production. *Int. J. Biol. Macromol.*, 165, 2739–2749, 2020.
5. Sharma, B., Larroche, C., Dussap, C.G., Comprehensive assessment of 2G bioethanol production. *Bioresour. Technol.*, 313, 123630, 2020.
6. Robak, K. and Balcerek, M., Review of second generation bioethanol production from residual biomass. *Food Technol. Biotechnol.*, 56, 2, 174, 2018.
7. Robak, K. and Balcerek, M., Current state-of-the-art in ethanol production from lignocellulosic feedstocks. *Microbiol. Res.*, 240, 126534, 2020.
8. Pandey, R. and Kumar, G., A comprehensive review on generations of biofuels: Current trends, development and scope. *Int. J. Emerg. Technol.*, 8, 1, 561–565, 2017. https://www.researchtrend.net/ijet/pdf/121-S-857a.pdf.
9. Jeswani, H.K., Chilvers, A., Azapagic, A., Environmental sustainability of biofuels: A review. *Proc. R. Soc A*, 476, 2243, 20200351, 2020.
10. *Biobutanol market: global industry trends, share, size, growth, opportunity and forecast 2021-2026*. https://www.researchandmarkets.com/reports/5642280/biobutanol-market-global-industry-trends. Accessed on 21 August 2022.
11. Zhen, X., Wang, Y., Liu, D., Bio-butanol as a new generation of clean alternative fuel for SI (spark ignition) and CI (compression ignition) engines. *Renewable Energy*, 147, 2494–2521, 2020.
12. Bušić, A., Marđetko, N., Kundas, S., Morzak, G., Belskaya, H., IvančićŠantek, M., Komes, D., Novak, S., Šantek, B., Bioethanol production from renewable raw materials and its separation and purification: A review. *Food Technol. Biotechnol.*, 56, 3, 289–311, 2018.

13. Obergruber, M., Hönig, V., Procházka, P., Kučerová, V., Kotek, M., Bouček, J., Mařík, J., Physicochemical properties of biobutanol as an advanced biofuel. *Materials*, 14, 4, 914, 2021.
14. Antoni, D., Zverlov, V.V., Schwarz, W.H., Biofuels from microbes. *Appl. Microbiol. Biotechnol.*, 77, 1, 23–35, 2007.
15. Bharathiraja, B., Jayamuthunagai, J., Sudharsanaa, T., Bharghavi, A., Praveenkumar, R., Chakravarthy, M., Yuvaraj, D., Biobutanol–An impending biofuel for future: A review on upstream and downstream processing tecniques. *Renewable Sustain. Energy Rev.*, 68, 788–807, 2017.
16. Hahn, H.D., Dämbkes, G., Rupprich, N., Bahl, H., Frey, G.D., Butanols, in: *Ullmann's Encyclopedia of Industrial Chemistry*, Wiley, Germany, 2013, Wiley-VCH, Verlag GmbH & Co. KGaA.
17. Zhang, Z., Lohr, L., Escalante, C., Wetzstein, M., Food versus fuel: What do prices tell us? *Energy Policy*, 38, 1, 445–451, 2010.
18. Stoeberl, M., Werkmeister, R., Faulstich, M., Russ, W., Biobutanol from food wastes–Fermentative production, use as biofuel an the influence on the emissions. *Proc. Food Sci.*, 1, 1867–1874, 2011.
19. Foda, M.I., Dong, H., Li, Y., Study the suitability of cheese whey for biobutanol production by Clostridia. *J. Am. Sci.*, 6, 8, pp.39–46, 2010.
20. Becerra, M., Cerdán, M.E., González-Siso, M.I., Biobutanol from cheese whey. *Microb. Cell Fact.*, 14, 1, 1–15, 2015.
21. Jiang, L., Wang, J., Liang, S., Wang, X., Cen, P., Xu, Z., Butyric acid fermentation in a fibrous bed bioreactor with immobilized Clostridium tyrobutyricum from cane molasses. *Bioresour. Technol.*, 100, 13, 3403–3409, 2009.
22. Van der Merwe, A.B., Cheng, H., Görgens, J.F., Knoetze, J.H., Comparison of energy efficiency and economics of process designs for biobutanol production from sugarcane molasses. *Fuel*, 105, 451–458, 2013.
23. Ni, Y., Wang, Y., Sun, Z., Butanol production from cane molasses by Clostridium saccharobutylicum DSM 13864: Batch and semicontinuous fermentation. *Appl. Biochem. Biotechnol.*, 166, 8, 1896–1907, 2012.
24. Gyalai-Korpos, M., Mangel, R., Alvira, P., Dienes, D., Ballesteros, M., Réczey, K., Cellulase production using different streams of wheat grain-and wheat straw-based ethanol processes. *J. Ind. Microbiol. Biotechnol.*, 38, 7, 791–802, 2011.
25. Kumar, M., Goyal, Y., Sarkar, A., Gayen, K., Comparative economic assessment of ABE fermentation based on cellulosic and non-cellulosic feedstocks. *Appl. Energy*, 93, 193–204, 2012.
26. Visioli, L.J., Enzweiler, H., Kuhn, R.C., Schwaab, M., Mazutti, M.A., Recent advances on biobutanol production. *Sustain. Chem. Processes*, 2, 1, 1–9, 2014.
27. Kushwaha, D., Srivastava, N., Mishra, I., Upadhyay, S.N., Mishra, P.K., Recent trends in biobutanol production. *Rev. Chem. Eng.*, 35, 4, 475–504, 2019.
28. Dürre, P., Fermentative production of butanol—The academic perspective. *Curr. Opin. Biotechnol.*, 22, 3, 331–336, 2011.

29. Qureshi, N., Saha, B.C., Hector, R.E., Hughes, S.R., Cotta, M.A., Butanol production from wheat straw by simultaneous saccharification and fermentation using Clostridium beijerinckii: Part I—Batch fermentation. *Biomass Bioenergy*, 32, 2, 168–175, 2008.
30. Qureshi, N., Saha, B.C., Cotta, M.A., Butanol production from wheat straw hydrolysate using Clostridium beijerinckii. *Bioprocess Biosyst. Eng.*, 30, 6, 419–427, 2007.
31. Qureshi, N., Saha, B.C., Hector, R.E., Cotta, M.A., Removal of fermentation inhibitors from alkaline peroxide pretreated and enzymatically hydrolyzed wheat straw: Production of butanol from hydrolysate using Clostridium beijerinckii in batch reactors. *Biomass Bioenerg.*, 32, 12, pp.1353–1358, 2008.
32. Qureshi, N., Ezeji, T.C., Ebener, J., Dien, B.S., Cotta, M.A., Blaschek, H.P., Butanol production by Clostridium beijerinckii. Part I: Use of acid and enzyme hydrolyzed corn fiber. *Bioresour. Technol.*, 99, 13, 5915–5922, 2008.
33. Qureshi, N., Saha, B.C., Hector, R.E., Dien, B., Hughes, S., Liu, S., Iten, L., Bowman, M.J., Sarath, G., Cotta, M.A., Production of butanol (a biofuel) from agricultural residues: Part II–Use of corn stover and switchgrass hydrolysates. *Biomass Bioenerg.*, 34, 4, pp.566–571, 2010.
34. Claassen, P.A., Budde, M.A., López-Contreras, A.M., Acetone, butanol and ethanol production from domestic organic waste by solventogenic clostridia. *J. Mol. Microbiol. Biotechnol.*, 2, 1, pp.39–44, 2000.
35. Liew, S.T., Arbakariya, A., Rosfarizan, M., Raha, A.R., Production of solvent (acetone-butanol-ethanol) in continuous fermentation by Clostridium saccharobutylicum DSM 13864 using gelatinised sago starch as a carbon source. *Malaysian J. Microbiol.*, 2, 2, pp.42–50, 2006.
36. Qureshi, N. and Maddox, I.S., Continuous production of acetone-butanol-ethanol using immobilized cells of Clostridium acetobutylicum and integration with product removal by liquid-liquid extraction. *J. Ferment. Bioeng.*, 80, 2, 185–189, 1995.
37. Qureshi, N. and Maddox, I.S., Reduction in butanol inhibition by perstraction: Utilization of concentrated lactose/whey permeate by Clostridium acetobutylicum to enhance butanol fermentation economics. *Food Bioprod. Process.*, 83, 1, 43–52, 2005.
38. Hossain, N., Mahlia, T.M.I., Saidur, R., Latest development in microalgae-biofuel production with nano-additives. *Biotechnol. Biofuels*, 12, 1, 1–16, 2019. https://doi.org/10.1186/s13068-019-1465-0.
39. Wang, L. and Chen, H., Increased fermentability of enzymatically hydrolyzed steam-exploded corn stover for butanol production by removal of fermentation inhibitors. *Process Biochem.*, 46, 2, 604–607, 2011.
40. Biswas, B., Pandey, N., Bisht, Y., Singh, R., Kumar, J., Bhaskar, T., Pyrolysis of agricultural biomass residues: Comparative study of corn cob, wheat straw, rice straw and rice husk. *Bioresour. Technol.*, 237, 57–63, 2017.
41. Ezeji, T.C., Qureshi, N., Blaschek, H.P., Production of acetone butanol (AB) from liquefied corn starch, a commercial substrate, using Clostridium

beijerinckii coupled with product recovery by gas stripping. *J. Ind. Microbiol. Biotechnol.*, *34*, 12, 771–777, 2007.

42. Ezeji, T., Qureshi, N., Blaschek, H.P., Production of acetone–butanol–ethanol (ABE) in a continuous flow bioreactor using degermed corn and Clostridium beijerinckii. *Process Biochem.*, *42*, 1, 34–39, 2007.

43. de Souza Moretti, M.M., Perrone, O.M., Nunes, C.D.C.C., Taboga, S., Boscolo, M., da Silva, R., Gomes, E., Effect of pre-treatment and enzymatic hydrolysis on the physical-chemical composition and morphologic structure of sugarcane bagasse and sugarcane straw. *Bioresour. Technol.*, *219*, 773–777, 2016.

44. Karthikeyan, O.P. and Visvanathan, C., Bio-energy recovery from high-solid organic substrates by dry anaerobic bio-conversion processes: A review. *Rev. Environ. Sci. Biotechnol.*, *12*, 3, 257–284, 2013.

45. Monlau, F., Barakat, A., Trably, E., Dumas, C., Steyer, J.P., Carrère, H., Lignocellulosic materials into biohydrogen and biomethane: Impact of structural features and pre-treatment. *Crit. Rev. Environ. Sci. Technol.*, *43*, 3, 260–322, 2013.

46. Pratto, B., Chandgude, V., de Sousa Junior, R., Cruz, A.J.G., Bankar, S., Biobutanol production from sugarcane straw: Defining optimal biomass loading for improved ABE fermentation. *Ind. Crops Prod.*, *148*, 112265, 2020.

47. Kumar, M. and Gayen, K., Developments in biobutanol production: New insights. *Appl. Energy*, *88*, 6, 1999–2012, 2011.

48. Sarkar, N., Ghosh, S.K., Bannerjee, S., Aikat, K., Bioethanol production from agricultural wastes: An overview. *Renewable Energy*, *37*, 1, 19–27, 2012.

49. Abdurrahman, M.I., Chaki, S., Saini, G., Stubble burning: Effects on health & environment, regulations and management practices. *Environ. Adv.*, *2*, 100011, 2020.

50. Song, D., Fu, J., Shi, D., Exploitation of oil-bearing microalgae for biodiesel. *Chin. J. Biotechnol.*, *24*, 3, 341–348, 2008.

51. Tredici, M.R., Photobiology of microalgae mass cultures: Understanding the tools for the next green revolution. *Biofuels*, *1*, 1, 143–162, 2010.

52. Lam, M.K. and Lee, K.T., Microalgae biofuels: A critical review of issues, problems and the way forward. *Biotechnol. Adv.*, *30*, 3, 673–690, 2012.

53. Ullah, K., Ahmad, M., Sharma, V.K., Lu, P., Harvey, A., Zafar, M., Sultana, S., Assessing the potential of algal biomass opportunities for bioenergy industry: A review. *Fuel*, *143*, 414–423, 2015.

54. Chen, W.H., Chen, Y.C., Lin, J.G., Evaluation of biobutanol production from non-pretreated rice straw hydrolysate under non-sterile environmental conditions. *Bioresour. Technol.*, *135*, 262–268, 2013.

55. Suutari, M., Leskinen, E., Fagerstedt, K., Kuparinen, J., Kuuppo, P., Blomster, J., Macroalgae in biofuel production. *Phycol. Res.*, *63*, 1, 1–18, 2015.

56. Kushwaha, D., Upadhyay, S.N., Mishra, P.K., Growth of cyanobacteria: Optimization for increased carbohydrate content. *Appl. Biochem. Biotechnol.*, *184*, 4, 1247–1262, 2017b, 2018.

57. Ho, S.H., Chen, C.Y., Chang, J.S., Effect of light intensity and nitrogen starvation on CO2 fixation and lipid/carbohydrate production of an indigenous microalga Scenedesmus obliquus CNW-N. *Bioresour. Technol.*, *113*, 244–252, 2012.
58. Sun, X., Cao, Y., Xu, H., Liu, Y., Sun, J., Qiao, D., Cao, Y., Effect of nitrogen-starvation, light intensity and iron on triacylglyceride/carbohydrate production and fatty acid profile of Neochlorisoleoabundans HK-129 by a two-stage process. *Bioresour. Technol.*, *155*, 204–212, 2014.
59. Depraetere, O., Deschoenmaeker, F., Badri, H., Monsieurs, P., Foubert, I., Leys, N., Wattiez, R., Muylaert, K., Trade-off between growth and carbohydrate accumulation in nutrient-limited Arthrospira sp. PCC 8005 studied by integrating transcriptomic and proteomic approaches. *PLoS One*, *10*, 7, e0132461, 2015.
60. Zhang, W.L., Liu, Z.Y., Liu, Z., Li, F.L., Butanol production from corncob residue using Clostridium beijerinckii NCIMB 8052. *Lett. Appl. Microbiol.*, *55*, 3, 240–246, 2012.
61. Pugazhendhi, A., Mathimani, T., Varjani, S., Rene, E.R., Kumar, G., Kim, S.H., Ponnusamy, V.K., Yoon, J.J., Biobutanol as a promising liquid fuel for the future-recent updates and perspectives. *Fuel*, *253*, 637–646, 2019.
62. Mohanram, S., Amat, D., Choudhary, J., Arora, A., Nain, L., Novel perspectives for evolving enzyme cocktails for lignocellulose hydrolysis in biorefineries. *Sustain. Chem. Processes*, *1*, 1, 1–12, 2013.
63. Li, L., Ai, H., Zhang, S., Li, S., Liang, Z., Wu, Z.Q., Yang, S.T., Wang, J.F., Enhanced butanol production by coculture of Clostridium beijerinckii and Clostridium tyrobutyricum. *Bioresour. Technol.*, *143*, 397–404, 2013.
64. Raganati, F., Olivieri, G., Procentese, A., Russo, M.E., Salatino, P., Marzocchella, A., Butanol production by bioconversion of cheese whey in a continuous packed bed reactor. *Bioresour. Technol.*, *138*, 259–265, 2013.
65. Saini, J.K., Saini, R., Tewari, L., Lignocellulosic agriculture wastes as biomass feedstocks for second-generation bioethanol production: Concepts and recent developments. *3 Biotech.*, *5*, 4, 337–353, 2015.
66. Yoo, C.G., Meng, X., Pu, Y., Ragauskas, A.J., The critical role of lignin in lignocellulosic biomass conversion and recent pre-treatment strategies: A comprehensive review. *Bioresour. Technol.*, *301*, 122784, 2020.
67. Liu, C.G., Xue, C., Lin, Y.H., Bai, F.W., Redox potential control and applications in microaerobic and anaerobic fermentations. *Biotechnol. Adv.*, *31*, 2, 257–265, 2013.
68. Öhgren, K., Vehmaanperä, J., Siika-Aho, M., Galbe, M., Viikari, L., Zacchi, G., High temperature enzymatic prehydrolysis prior to simultaneous saccharification and fermentation of steam pretreated corn stover for ethanol production. *Enzyme Microb. Technol.*, *40*, 4, 607–613, 2007.
69. Kumar, P., Barrett, D.M., Delwiche, M.J., Stroeve, P., Methods for pretreatment of lignocellulosic biomass for efficient hydrolysis and biofuel production. *Ind. Eng. Chem. Res.*, *48*, 8, pp.3713–3729, 2009.

70. Sun, Y. and Cheng, J., Hydrolysis of lignocellulosic materials for ethanol production: A review. *Bioresour. Technol.*, 83, 1, 1–11, 2002.
71. Sarkar, N., Ghosh, S.K., Bannerjee, S., Aikat, K., Bioethanol production from agricultural wastes: An overview. *Renewable Energy*, 37, 1, 19–27, 2012.
72. Wang, G.S., Pan, X.J., Zhu, J.Y., Gleisner, R., Rockwood, D.J.B.P., Sulfite pretreatment to overcome recalcitrance of lignocellulose (SPORL) for robust enzymatic saccharification of hardwoods. *Biotechnol. Progr.*, 25, 4, 1086–1093, 2009.
73. Gao, J., Anderson, D., Levie, B., Saccharification of recalcitrant biomass and integration options for lignocellulosic sugars from Catchlight Energy's sugar process (CLE Sugar). *Biotechnol. Biofuels*, 6, 1, 1–7, 2013.
74. Zhu, J.Y., Zhu, W., OBryan, P., Dien, B.S., Tian, S., Gleisner, R., Pan, X.J., Ethanol production from SPORL-pretreated lodgepole pine: Preliminary evaluation of mass balance and process energy efficiency. *Appl. Microbiol. Biotechnol.*, 86, 5, pp.1355–1365, 2010.
75. Zhu, W., Zhu, J.Y., Gleisner, R., Pan, X.J., On energy consumption for size-reduction and yields from subsequent enzymatic saccharification of pretreated lodgepole pine. *Bioresour. Technol.*, 101, 8, 2782–2792, 2010.
76. Esteghlalian, A., Hashimoto, A.G., Fenske, J.J., Penner, M.H., Modeling and optimization of the dilute-sulfuric-acid pre-treatment of corn stover, poplar and switchgrass. *Bioresour. Technol.*, 59, 2-3, pp.129–136, 1997.
77. Hsu, T.C., Guo, G.L., Chen, W.H., Hwang, W.S., Effect of dilute acid pretreatment of rice straw on structural properties and enzymatic hydrolysis. *Bioresour. Technol.*, 101, 13, 4907–4913, 2010.
78. Kudahettige-Nilsson, R.L., Helmerius, J., Nilsson, R.T., Sjöblom, M., Hodge, D.B., Rova, U., Biobutanol production by Clostridium acetobutylicum using xylose recovered from birch Kraft black liquor. *Bioresour. Technol.*, 176, 71–79, 2015.
79. Zhao, Y., Wang, Y., Zhu, J.Y., Ragauskas, A., Deng, Y., Enhanced enzymatic hydrolysis of spruce by alkaline pre-treatment at low temperature. *Biotechnol. Bioeng.*, 99, 6, 1320–1328, 2008.
80. Zhu, J.Y. and Pan, X.J., Woody biomass pre-treatment for cellulosic ethanol production: Technology and energy consumption evaluation. *Bioresour. Technol.*, 101, 13, 4992–5002, 2010.
81. Park, Y.C. and Kim, J.S., Comparison of various alkaline pre-treatment methods of lignocellulosic biomass. *Energy*, 47, 1, 31–35, 2012.
82. Yang, J., Gao, C., Yang, X., Su, Y., Shi, S., Han, L., Effect of combined wet alkaline mechanical pre-treatment on enzymatic hydrolysis of corn stover and its mechanism. *Biotechnol. Biofuels Bioprod.*, 15, 1, pp.1–11, 2022.
83. Perona, A., Hoyos, P., Farrán, Á., Hernáiz, M.J., Current challenges and future perspectives in sustainable mechanochemical transformations of carbohydrates. *Green Chem.*, 22, 17, 5559–5583, 2020.
84. Shen, F., Xiong, X., Fu, J., Yang, J., Qiu, M., Qi, X., Tsang, D.C., Recent advances in mechanochemical production of chemicals and carbon materials

from sustainable biomass resources. *Renewable Sustain. Energy Rev.*, *130*, 109944, 2020.
85. Baig, K.S., Wu, J., Turcotte, G., Future prospects of delignification pretreatments for the lignocellulosic materials to produce second generation bioethanol. *Int. J. Energy Res.*, *43*, 4, 1411–1427, 2019.
86. Barakat, A., Chuetor, S., Monlau, F., Solhy, A., Rouau, X., Eco-friendly dry chemo-mechanical pre-treatments of lignocellulosic biomass: Impact on energy and yield of the enzymatic hydrolysis. *Appl. Energy*, *113*, 97–105, 2014.
87. Hayes, D.J., An examination of biorefining processes, catalysts and challenges. *Catal. Today*, *145*, 1-2, 138–151, 2009.
88. Banerjee, C., Mandal, S., Ghosh, S., Kuchlyan, J., Kundu, N., Sarkar, N., Unique characteristics of ionic liquids comprised of long-chain cations and anions: A new physical insight. *J. Phys. Chem. B*, *117*, 14, 3927–3934.2013.
89. Nguyen, T.A.D., Kim, K.R., Han, S.J., Cho, H.Y., Kim, J.W., Park, S.M., Park, J.C., Sim, S.J., Pre-treatment of rice straw with ammonia and ionic liquid for lignocellulose conversion to fermentable sugars. *Bioresour. Technol.*, *101*, 19, 7432–7438, 2010.
90. Zhao, H., Baker, G.A., Cowins, J.V., Fast enzymatic saccharification of switchgrass after pre-treatment with ionic liquids. *Biotechnol. Progr.*, *26*, 1, 127–133, 2010.
91. Park, N., Kim, H.Y., Koo, B.W., Yeo, H., Choi, I.G., Organosolv pre-treatment with various catalysts for enhancing enzymatic hydrolysis of pitch pine (Pinus rigida). *Bioresour. Technol.*, *101*, 18, 7046–7053, 2010.
92. Jin, Y., Zhang, L., Yi, Z., Fang, Y. and Zhao, H., Waste-to-energy: biobutanol production from cellulosic residue of sweet potato by Clostridia acetobutylicum. *Environmental Engineering Research*, 27(5), 2022.
93. Badr, H.R., Toledo, R., Hamdy, M.K., Continuous acetone–Ethanol–Butanol fermentation by immobilized cells of Clostridium acetobutylicum. *Biomass Bioenergy*, *20*, 2, 119–132, 2001.
94. Bušić, A., Marđetko, N., Kundas, S., Morzak, G., Belskaya, H., Ivančić Šantek, M., Komes, D., Novak, S., Šantek, B., Bioethanol production from renewable raw materials and its separation and purification: A review. *Food Technol. Biotechnol..*, 56, 3, 289–311, 2018.
95. Wu, M., Wang, M., Liu, J., Huo, H., *Life-cycle assessment of corn-based butanol as a potential transportation fuel (No. ANL/ESD/07-10)*, Argonne National Lab.(ANL, Argonne, IL (United States), 2007.
96. Kujawska, A., Kujawski, J., Bryjak, M., Kujawski, W., ABE fermentation products recovery methods—A review. *Renewable Sustain. Energy Rev.*, *48*, 648–661, 2015.
97. Kumar, M. and Gayen, K., Developments in biobutanol production: New insights. *Appl. Energy*, *88*, 6, 1999–2012, 2011.

Index

"Carbon-negative" biofuel, 117

ABE fermentation, 112, 133, 137, 138, 141, 155–157, 210, 213, 219, 226, 228–230
ABE producing Clostridia, 8–9
ABE production, 4–8
Acetone butanol ethanol (ABE), 107
ABE-producing clostridia, 111
Acetone-butanol-ethanol fermentation, 340, 342, 343
Acid method, 57
Acid treatments, 210
Acidic strength, 262
Acidogenesis, 2, 4, 8, 218, 220–221, 109
Acidogenic phase, 193
Adsorption, 113, 114, 177, 178, 182, 183, 185
Advance technology,
batch fermentation, 12–16
continuous fermentation, 17–27
fed batch fermentation, 16–17
Algae, 285, 289, 295–296
Algal biomass, 109, 333, 337
Alkali method, 56
Alkaline pre-treatments, 341
Ammonia fiber explosion (AFEX) technique, 48, 54, 62, 339
Anaerobic consortia, 256
Analysis of variance (ANOVA), 118
Autohydrolysis, 339
Autolithotrophic microorganisms, 220

Barley straw, 334
BASF, 329
Biobased products, 209
Biobutanol, 167–185
commercial importance of, 343–346
commercial applications of, 109
first-generation biofuels, 324, 326
fourth-generation biofuels, 326–327
introduction, 323
production and biorefinery, 327
acetone-butanol-ethanol fermentation, 343
microorganisms, 342–343
substrates and their pre-treatment for, 329, 331–333, 338–342
recovery from fermentation broth, 112–114
second-generation biofuels, 326
third-generation biofuels, 326
Biobutanol production, 107, 109–110
ABE fermentation, 112
challenges in, 115
microbes and biobutanol production, 110–111
recovery of biobutanol from fermentation broth, 112–114
substrate for biobutanol production, 111–112
Biobytenol production,
C. beijerinckii, 242
Clostridium acetobutylicum, 241
Clostridium tyrobutyricum, 242
Klebsiella pneumoniae, 242

355

Biochemical, 257
Biocompatible, 259
Biofuel, 167–169, 172, 184, 284–291, 294–297
Biofuel, fuel characteristics of different, 108
Biofuels, 323
  first-generation, 324, 326
  fourth-generation, 326–327
  second-generation, 326
  third-generation, 326
Biogas fuel, 258
Biogenesis, 226
Biological pre-treatment, 54, 64, 66
Biologices, 214
Biomass, 167, 168, 171, 172, 176, 178, 180–182, 323
Biomass feedstock, 209
Biomass product, 209, 269–270, 273
Bioreactor, 171–173, 176, 177, 179, 214
  batch, 197
  continuous, 198
  two-stage continuous, 199
Biorefinery methods, 214
Butamax, 214–215
Butanol, 107, 111, 112
  companies involved in production of, 330
  isomers of, 328
Butanol advantages, 157
Butanol fermentation,
  biobutanol production, 107, 109–110
    ABE fermentation, 112
    challenges in, 115
    microbes and biobutanol production, 110–111
    recovery of biobutanol from fermentation broth, 112–114
  ubstrate for biobutanol production, 111–112
  perspectives, 115

alleviate carbon catabolite repression, 117
  butanol recovery, 122
  fermentation improvement, 118
  strain development, 118–122
  substrate, 116–117
Butanol producing micro organisms, genetically modified micro organisms, 3–7
  wild type micro organisms, 3
Butanol stress, 227

Cane molasses, 109
Carbon, 169, 170, 177, 178, 179, 184
Carbon catabolite repression (CCR), 29–30, 117
Carbonization, 263
Carboxylic acids, 109
Cassava, 109, 337
Catalytic activity, 258, 267, 271, 273, 278
Cation-exchange resin, 262
Cell viability, 214
Cellular regulatory, 212, 214
Cellulolytic enzyme, 227
Cellulose, 48–50, 54, 64–65, 331–332
Cheese whey, 109
Chemical treatment, 340
Chorella, 342–343
Classification of biofuel, 306–309
  first-generation of ethanol, 306–308
  fourth-generation of ethanol, 309
  second-generation of ethanol, 308–309
  third-generation of ethanol, 309
Clostridia species, 329
Clostridium, 119–120, 286, 290, 292–293, 295–296
Clostridium acetobutylicum, 110–111, 112, 117, 119–120, 121
Clostridium acetobutylicum, 342, 343–344
Clostridium beijerinckii, 110, 112, 332, 342

Clostridium butylicum, 136, 159
Clostridium pasteurianum, 342
Clostridium saccharobutylicum, 110, 342
Clostridium saccharoperbutylacetonicum, 110, 342
Clostridium sporogenes, 342
Clostridium tyrobutyricum, 120
Corn cob, 335
Corn fiber, 334
Corn stover, 107, 334
Cosolvent enhanced lignocellulosic fractionation, 48, 54, 61
Crotonaldehyde hydrogenation, 153
Cyanobacteria, 120, 151–152

Deep eutectic solvents, 54, 63
Defidered sweet-potato-slurry (DSPS), 337
Degermed corn, 336
Delignification, 51, 59, 61–62, 64
Depressurization, 339
Designed biomass, 10–11
Detrimental effect, 258, 274
Dispersion barrier, 259
Distillation, 172, 175, 182, 183, 185
DNA constrain, 222
Domestic organic waste, 336
Dow chemicals, 329

Ecofriendly, 284, 287, 296–297
Economics of biobutanol, 31–33
Elastomers, 156
Emden-Meyerhof-Pranas (EMP) pathway, 109
Energy, 167–169, 171, 172, 174, 175, 176, 178, 179, 182–185, 284–285, 288–290, 293–297
Energy crops, 116
Engineered strain, 222, 228
Enzyme, 170, 180, 181, 184, 185
Escherichia coli, 117, 120

Extraction, 171, 172, 174–177, 182, 183, 185
Extruded corn, 337
Extrusion, 54

Fed-batch, 135, 140,
Feedstock, 194, 333, 334
Feedstock for ABE fermentation, 9
Fermentation, 167, 168, 170–185
Food waste, 109
Fossil fuel, 284–285, 289, 296
Fractional distillation, 153

Gas stripping, 27–28, 113, 140, 158, 161
Gasoline, 327, 329
Genetic modification, 212, 221, 229
Genome editing, 244–245
Genomic tools, 222, 229
Genotoxicity, 141
Geobacillus thermoglucosidasius, 121
Green house gases, 284, 289
Greenhouse gases, regulated emissions and energy use in transportation (GREET) model, 116–117

Hemicellulose, 48–50, 53–54
Heterofermentative, 220
Heterologous hosts, 222
Hydrogenation, 257, 272–273
Hydrolysis, 291–292, 294
Hydroxyapatite nanoparticle, 268, 269
Hygroscopic activity, 211

Immobilized enzyme, 278
Immobilized protein, 257
Integrated biomass supply analysis and logistics (IBSAL), 116
Integration, 167–169, 172, 178, 185
Inulin, 107
Ionic liquids (ILs), 48, 54, 58, 341–342

Klebsiella pneumoniae, 121

Lactobacillus, 148, 152
Lignin, 48–51, 53–54, 58–59, 62, 65
Lignocellulose, 107, 289, 292
Lignocellulosic, 170, 181, 182
Lignocellulosic biomass, 48–49, 54
Lignocellulosic material, 209
Limitation of ABE fermentation, 9–10
Liquefied corn starch, 337
Liquid hot water (LHW), 339
Liquid-liquid extraction, 27–28, 113–114
Liquid membrane extraction, 27–28
Low temperature steep delignification, 48, 54, 62

Membrane, 172–175, 176, 178, 182–185
Metabolic modification, 227
Metabolic rate, 230
Metal nanoadditives, 342
Methyl viologen (MV), 118
Methylene blue (MB), 118
Microalgae, 261, 277–278
Microalgal biomass, 112
Microalgal feedstock, 333
Microbes and biobutanol production, 110–111
Microbial growth, 227
Microbial species, 229
Microbial strain, 196
Microemulsion, 263
Micronutrient, 258
Microorganism, 170, 172, 174, 176, 177, 180,
Microwave method, 56
Microwaves, 340
Milling, 54–55
Modern biofuels, 216

Nanoadditives, pre-treatment of biomass with, 342
Nanofiber, 256, 274

Nondiagnostic, 274
Nonrenewable, 284

Off targets, 248–249
Oleyl alcohol, 114
Organosolv method, 58
Organosolv pre-treatment, 341–342
Overview on biofuels and its classification, 79–87
   first-generation biofuels, 79–82
   fourth-generation biofuels, 87
   second-generation lignocellulosic biofuel, 82–85
   third-generation biofuels, 85–86
Oxo method, 327
Ozonolysis, 270, 341

Palm oil mill effluent, 109
Paper chromatography, 136
Perstraction/membrane extraction, 114
Pervaporation, 27–28, 113, 172–174, 182
Physical pre-treatment methods, 338–339
Physiochemical methods, 339
Plasmids, 222
Plasticizers, 134, 156
Pre-treatment, 168, 170, 171, 176, 178, 180, 181, 185, 290–294
Pre-treatment methodologies for hydrolysis of lignocellulose biomass, 87–97
   chemical process for pre-treatment of lignocellulose, 91–93
   ionic liquid as pre-treatment agent, 93–94
   overview, 87–90
   pre-treatment process with alkali agents, 94–96
   pre-treatment process with ultrasonic waves, 96–97
   structural analysis for cellulosic hydrolysis, 90–91

Problem of ABE fermentation, 9–10
Process intensification, 168, 169, 172–178
Production of biobutanol, 309–317
  enzymatic-hydrolysis stages, 310–312
  fermentation stage, 312
  pre-treatment stages, 310
  production of butanol from genetically improved strains, 313–317
  separation stage, 312–313
Pyrococcusfuriosus, 121–122
Pyrolysis, 261, 273

Recombinant DNA technology, 148–149
Renewable, 284–285, 289–291, 296–297
Renewable biomass, 209, 229
Renewable material, 209, 260
Resonance surface methodology (RSM), 118
Reverse osmosis (RO), 113
Rice bran, 336
Rice straw, 335
Rye straw, 336

Saccharification, 29, 135, 161, 168, 171, 178–181
Saccharomyces cerevisiae, 122
Sago, 336
SDG, 284, 289
Separation, 168, 171, 172–179, 182–185
Separation techniques,
  adsorption, 202
  gas stripping, 201
  hybrid separation process, 203
  liquid-liquid extraction, 200
  perstration, 202
  pervaporation, 200
SO_2 steam explosion, 339
Solvent toxicity, 196

Solventogenesis, 2, 4, 8, 218, 220–221
Solventogenic phase, 193
Soy molasses, 336
Specific surface area, 262
Spontaneous mutations, 142–143
Starch, 109, 288–291, 296
Starch and sugar, 329, 331
Steam explosion (autohydrolysis), 339
Sugar, 285, 288–296
Sugarcane bagasse, 107, 335
Sugarcase straw, 335
Sulfite pre-treatment, 340
Supercritical CO_2 explosion, 339
Supercritical fluids, 48, 60, 339
Sustainable organic material, 260
Switchgrass, 334
Synechococcus elongates, 120

Thermal durability, 264
Thermoanaero bacterium, 121
Thermoanaerobacter tengeongensis, 122
Toxic, 287, 290, 292–293, 295
Transesterification procedure, 260
Transesterification process, 261, 265, 267

Ultrasound, 54–55

Valorization, 51, 60, 66
Valorization process, 154
Van der Waals force, 261
Vibrion butyriqu, 210

Waste, 284–286, 289–291, 294, 296–297
Water solubility, 258
Weizmann, 286–287
Weizmann organism, 109
Wet air oxidation, 339
Wheat straw, 334
Whey permeate, 337
Woody biomass, 340

Yeast, 122

Printed and bound by CPI Group (UK) Ltd, Croydon, CR0 4YY

29/11/2023

08198636-0001